R

SPATIAL ANALYSIS AND
LOCATION-ALLOCATION MODELS

SPATIAL ANALYSIS AND LOCATION-ALLOCATION MODELS

Edited by

Avijit Ghosh

New York University

Gerard Rushton

University of Iowa

VNR VAN NOSTRAND REINHOLD COMPANY

New York

Manufactured in the United States of America.

Published by Van Nostrand Reinhold Company Inc.
115 Fifth Avenue
New York, New York 10003

Van Nostrand Reinhold Company Limited
Molly Millars Lane
Wokingham, Berkshire RG11 2PY, England

Van Nostrand Reinhold
480 La Trobe Street
Melbourne, Victoria 3000, Australia

Macmillan of Canada
Division of Canada Publishing Corporation
164 Commander Boulevard
Agincourt, Ontario MIS 3C7, Canada

15   14   13   12   11   10   9   8   7   6   5   4   3   2   1

Library of Congress Cataloging-in-Publication Data
Spatial analysis and location-allocation models.

Includes bibliographies and index.
1. Industry—Location—Mathematical models.   2. Public buildings—
Location—Mathematical models.   3. Space in economics—Mathematical
models.   I. Ghosh, Avijit.   II. Rushton, Gerard.
HD58.S6738   1987   338.6′042   86-15929
ISBN 0-442-22803-1

CONTENTS

PREFACE

Location-allocation models, developed in the 1960s, offered the first practical methods for finding optimal locations for many public and private facilities. The growing interest in these methods reflects technical progress in the field and the increasing variety of practical problems that can be solved with their use. The methods involve simultaneously selecting a set of locations for facilities and assigning demand to these facilities to optimize some specified measurable criteria. Such models have been used, for example, in siting public facilities like libraries and day-care centers, locating warehouses and corporate receivables systems, selecting retail locations, designing emergency medical and fire service systems, and siting waste disposal systems, to name only a few applications.

The use of location-allocation models is rapidly becoming expected in modern planning practice. Yet a number of important improvements are needed in these methods to make them more suitable for their present uses and for new areas of application. Although the early literature was devoted almost entirely to evaluating alternative methods of modeling, the past decade has seen an increasing interest in the systematic study of practical problems of application. It is now recognized that the context of any applied problem must be correctly represented if the problem is to be solved. The variety of decision-making procedures used, the kinds of people and institutions who are involved, the uncertainty in knowing the pertinent characteristics of the environment, both currently and in the future, the way in which consumers will choose between alternative locations—these are only a few of the context-related problems with which the location-allocation modeler must deal in any practical setting.

In the context of the past decade's applications orientation, new applications have appeared as the various professions, as well as business, have recognized location-allocation modeling's relevance for improving the performance of their systems. There is an increasing interest in using location-allocation models to evaluate the efficiency of decision-making processes for location decisions or to determine the characteristics of the spatial structures that will emerge if certain processes of location and allocation occur. These applications exemplify new problems that emerge when location-allocation is recognized as a process rather than, as originally defined in classic literature, as an outcome in which certain cross-sectional relationships were optimized. The link between process and outcome is now regarded as an important subject for research, since experience shows that in practice most outcomes are brought about by changes in the processes of locational choice and of individuals' spatial choices between alternative locations. Not surprisingly, therefore, location-allocation modeling is attracting a broader audience and its application is growing in sophistication. This diversity of developments, which adds to the viability of the methods, also raises the need for a forum to bring together recent advances on the subject.

In developing this book we tried to identify the problems whose solution could increase the range of legitimate application areas for these models. We identified individuals whose work has been a significant contribution to the advancement of knowledge in this area. Here, these authors have related their own contributions to those of other researchers, so that a larger portrait of the research field appears than the particulars of chapter titles might indicate. We hope that this portrait will provide a source with which informed scholars of location-allocation modeling can compare their knowledge and interest with that of other scholars in the field. We feel safe in predicting that such readers will, as they have in the past, see beyond the particulars of any application area and recognize the general principles of location and allocation that are being discussed. This level of abstraction, as much as any particulars of solution methods, has made the field of location-allocation modeling an especially stimulating research area in the past ten years.

We hope that this book will appeal to a broad audience in the fields of geography and regional science, operations research, urban and regional planning, transportation planning, management of health care systems, environmental science, marketing, management science, and archaeology, among others. The essays here represent the state of the art in applications and developments and should greatly facilitate the dissemination of information and the growth of the field of location-allocation modeling.

Many colleagues assisted us in preparing this book. We are especially grateful to the authors, who met our deadlines, and to the reviewers, who assisted us with their evaluations and suggestions.

AVIJIT GHOSH
GERARD RUSHTON

CONTRIBUTORS

John R. Beaumont, University of London, 100 Stamford Avenue, Springfield Milton Keynes MK6 3LQ, England

Thomas L. Bell, Department of Geography, University of Tennessee, Knoxville, Tennessee 37916

Paul Casillas, Department of Mathematics, Oregon State University, 933 South Farmer Avenue #4, Tempe, Arizona 85281

Richard L. Church, Department of Geography, University of California, Santa Barbara, California 93106

Mark S. Daskin, Department of Civil Engineering, Northwestern University, Evanston, Illinois 60201

David Eaton, University of Texas at Austin, Lyndon B. Johnson School of Public Affairs, Drawer Y, University Station, Austin, Texas 78713

Avijit Ghosh, Journal of Retailing, 202 Tisch 40 West 4th Street, New York University, New York, New York 10003

Michael F. Goodchild, Department of Geography, University of Western Ontario, London, Ontario, Canada N6A 5C2

Victor Ginsburgh, Université Libre de Bruxelles, Belgium

Sidheswar Jena, Centre for Human Settlements and Environmental Studies, Indian Institute of Management, 1/G5 Assaye Road, Bangalore, India 56042

Michael Kuby, Department of Geography, Boston University, 648 Beacon Street, Boston, Massachusetts 02215

Keumsoik Lee, Department of Geography, Boston University, 648 Beacon Street, Boston, Massachusetts 02215

Sara L. McLafferty, Department of Geography, Columbia University, New York, New York, 10003

Pitu Mirchandani, Department of Electrical, Computer, and Systems Engineering, Rensselaer Polytechnic Institute, Troy, New York, 12181

Valerian T. Noronha, Department of Geography, University of Western, London, Ontario, Canada N6A 5C2

Morton E. O'Kelly, Department of Geography, Ohio State University, 103 Administration Building, 190 North Oval Mall, Columbus, Ohio 43210-1361

Jeffrey P. Osleeb, Center for Energy and Environmental Studies, Boston University, 648 Beacon Street, Boston, Massachusetts 02215

Pierre Pestieau, Université de Liege, Belgium

Samuel Ratick, Department of Geography, Boston University, 648 Beacon Street, Boston, Massachusetts 02215

John M. Reilly, Capital District Transportation Authority, 110 Watervliet Avenue, Albany, New York 12206

Gerard Rushton, Department of Geography, University of Iowa, Iowa City, Iowa 52242

Vinod K. Tewari, Centre for Human Settlement and Environmental Studies, Indian Institute of Management, 1/G5 Assaye Road, Bangalore, India 56042

Jacques Thisse, Université Catholique de Louvain, Louvain-la-Neuve (CORE), 34 Voie due Roman Pays, 1348 Louvain-la-Neuve Belgium

INTRODUCTION

PROGRESS IN LOCATION-
ALLOCATION MODELING

Avijit Ghosh

New York University

Gerard Rushton

University of Iowa

Space has a profound impact on the organization of economic and social activities. Location theory developed as a discipline that addressed questions related to the spatial organization of activities. For example, one of the earliest practitioners of this discipline, Alfred Weber, attempted to find the most efficient point of production between raw material sources and market locations (Friedrich, 1929) in order to build a normative model of industrial location. He developed a system of *isodapanes* to determine the location that would minimize the total cost of transporting raw materials and finished goods. He recognized, however, that the geometrical procedures used in the construction of isodapanes and the mechanical principles used in the physical analog constructions, such as the *Varignon Frame* (Launhardt, 1882), were limited to dealing with simple transport-cost functions. Neither more complex single-location problems nor the multifacility case could be solved by these methods. Despite numerous attempts to remedy these shortcomings in the ensuing 50 years, no substantial progress was made. Consequently, location theorists turned their backs on the development of models to solve facility location problems for empirical conditions, believing that such complexity could not be unraveled analytically. Instead, using primarily graphical approaches, theorists developed idealized patterns of spatial organization that, they argued, were optimal in the simplified environments for which the problems could be solved (Christaller, 1933; Losch, 1954; Isard, 1956). If only, wrote Losch, "We had a method that combined the generality of equations with the clarity of geometrical figures! Such a combination would have weaknesses, of course, since in a strict sense it is impossible" (1954, p. 100).

1

Losch's wish was to come true when in the early 1960s several researchers, working independently, formulated solutions to the general facility location problem (Bindschedler and Moore, 1961; Kuhn and Kuenne, 1962; Cooper, 1963; Kuehn and Hamburger, 1963; Maranzana, 1964). These approaches not only provided a methodology for solving Weber's problem for complex environments but also extended the problem to deal with the location of multiple facilities. With multiple facilities the objective goes beyond finding the optimal locations to also determining the allocation of demand to those locations. Since the optimal locations depend on the allocation rule, both features must be determined simultaneously. This process of recognition was the genesis of the location-allocation problem.

Since the 1960s the literature on location-allocation models has rapidly increased and a number of surveys of this literature have now been published (see, for example, Scott, 1970; ReVelle, Marks, and Liebman, 1970; Lea, 1973; Rushton, Goodchild, and Ostresch, 1973; Francis and White, 1974; ReVelle et al., 1977; Hodgart, 1978; Handler and Mirchandani, 1979; Leonardi, 1981; Hansen, Peeters, and Thisse, 1983). These surveys provide a taxonomy of the different types of problems that have been proposed and the methods that are used to solve them. This introductory chapter is organized differently. Our purpose is to identify significant developments that have occurred in the types of problems solved with location-allocation models. At the same time, we wish to identify important issues that need to be addressed to broaden the range of problems that these methods can be used to solve. Finally, this introduction relates the contributions in this volume to these issues.

LOCATION-ALLOCATION MODELING:
THE CLASSIC PHASE

The Weber problem's extension to multiple supply points (Cooper, 1963) resulted in the well-known *p-median* location-allocation model. The *p*-median problem seeks the locations of *p* supply centers that minimize the aggregate distance separating them from a set of demand points (see Beaumont, chap. 1 herein). The formulation of this problem was a critical event in the development of the location-allocation literature. The following years witnessed applications of the *p*-median model to a wide variety of problem contexts. These applications were furthered by a number of critical developments. The first was the problem's transference from continuous space to discrete networks where facility locations were restricted to the vertices of a transportation network (Maranzana, 1964; Hakimi, 1965; ReVelle and Swain, 1970). (The mathematical formulation of the network *p*-median problem is shown herein in Beaumont, chap. 1; Mirchandani and Reilly, chap. 8; Daskin, chap. 9.) The development of the network formulation of the *p*-median problem greatly extended the range of situations in which location-allocation models could be applied.

The design of efficient algorithms for solving facility location problems also furthered the application of location-allocation models. Because of the combinatorial nature of location-allocation problems it is difficult to obtain globally optimal solutions using exact algorithms. For example, the network problem, although similar in structure to an integer program, is NP-hard (Kariv and Hakimi, 1979), and no polynomial time exact algorithm is known to exist. As a result, much effort has been devoted to designing enumerative and heuristic procedures. Cooper's (1964, 1967) alternating approach, for example, provided a general procedure for solving the continuous-space location problem (for a description of the algorithm, see Beaumont, chap. 1 herein). For the network problem, most heuristics involve the addition, deletion, or interchange of facilities one at a time. The addition, or GREEDY algorithm (Kuehn and Hamburger, 1963) locates facilities incrementally until p facility locations are chosen. The objective at each stage is to find the location that results in the least incremental cost. The deletion, or DROP algorithms, on the other hand, start with facilities located at all possible sites (see, for example, Feldman, Lehrer, and Ray 1966). The algorithms proceed iteratively, dropping at each stage the facility that has the least impact on the objective function. The INTERCHANGE algorithm of Teitz and Bart (1968) arbitrarily chooses a starting location pattern and then interchanges locations in this pattern, one by one, with vacant sites to minimize the objective function. The procedure is terminated when the objective function's value cannot be decreased further. Although the GREEDY algorithm is the superior of the three based on the worst-case performance criterion (Cornuejols, Fisher, and Nemhauser 1977), experience and experimentation have shown that, on the whole, INTERCHANGE algorithms are more efficient and robust. Most practical applications have tended to use the INTERCHANGE algorithm or to start with a GREEDY algorithm and then to apply the INTERCHANGE algorithm to the locational configuration obtained by the incremental search procedure. Mirchandani and Reilly (chap. 8 herein) provide a schematic outline of the steps in an interchange algorithm.

By dropping the integrality constraints on the decision variables, the network p-median problem can also be solved using linear programming. The optimal solution to the relaxed problem is often all integer and therefore provides an exact solution to the p-median problem (ReVelle and Swain, 1970). In cases where one or more of the decision variables are fractional, a branch-and-bound procedure can be used to find the optimal all-integer solution. Unfortunately, because of the large number of decision variables and constraints needed to represent typical location-allocation problems, the use of linear programming methods is relatively inefficient. An alternative procedure, recently reported to yield considerable success, is the Lagrangian relaxation method, which uses an iterative subgradient optimization technique to find the optimal solution (Cornuejols, Fisher, and Nemhauser, 1977; Erlenkotter, 1978; Narula, Ogbu, and Samuelsson, 1977).

Development of Model Applications

Because of its focus on maximizing systemwide accessibility, the p-median problem was applied in a number of different planning contexts in both the private and public sectors. As the application of location-allocation models progressed, the need to better represent the goals of particular systems in many cases required modification of the objective of minimizing total travel distance. This requirement led to a variety of extensions to the location-allocation problem. While these reformulations are too numerous to list exhaustively, in general the motivation underlying many of them was to introduce notions of equity in access.

The p-median problem's goal of systemwide efficiency often led to a spatial configuration in which individual accessibility to the system varied widely. To reduce such variability, constraints and alternative objective functions were suggested. One simple idea, for example, was to impose a constraint on the maximum distance separating any individual from a facility. An even stronger emphasis on reducing variability led to the suggestion that the objective should be to minimize the variance in accessibility. A third example is the minimax problem, which focuses on individual access rather than systemwide access and attempts to minimize the maximum distance separating any user from the closest facility (see Morrill and Symons, 1977). This objective is often referred to as the Rawls criteria.

Application of location-allocation models to emergency facilities such as ambulances and fire stations similarly led to the development of models particularly suited for the accessibility needs of such service delivery systems. The most popular among these models are the two covering problems: the set covering problem (see, for example, Toregas and ReVelle, 1972) and the maximal covering location problem (see, for example, Church and ReVelle, 1974). The covering models focus on locating facilities such that individual consumers are within a critical distance from their nearest facility. The set covering problem, for example, determines the minimum number of facilities and the optimal locations that together will place all individuals within the critical distance from a center. The maximal covering problem, on the other hand, restricts the number of centers to be located but finds their locations such that the maximum number of people can be covered by them. An individual is considered to be covered if located within the critical distance from the facility. (For mathematical formulations of the covering problems see, herein, Mirchandani and Reilly, chap. 8; Daskin, chap. 9).

While these different problem formulations developed somewhat independently and were proposed as alternatives to the p-median model, they are similar in structure to that model (Church and ReVelle, 1976; Hillsman, 1984). The solutions to these problems have the similar property that each demand node is allocated to only one facility, which is the closest one. Moreover, the models generally assume that the organization of service provision has no impact on the behavior of the users. Finally, these models implicitly take it that the environment in which the model is applied is completely stable and predictable. From a computational perspective, too, the models are similar in that they all can generally be solved as p-median problems with appropriate editing of the elements in the objective function (see Hillsman, 1984). Indeed, the general linear model proposed by Hillsman may be viewed as representing the final synthesis of these classic location-allocation models.

LOCATION-ALLOCATION MODELING: THE CONTEMPORARY PHASE

Location-allocation modeling in the classic phase, then, focused on normative and prescriptive analyses of well-defined location problems in which deterministic spatial allocations could be made. The contemporary period, however, is characterized by the development of models that can capture more complex behavior patterns of both producers and consumers. In addition, these models represent the environment more realistically. The contemporary phase has also brought new areas of application and interested people with new concerns. The critical readers of the classic location-allocation literature were predominantly operations researchers who, not unnaturally, were concerned with algorithmic efficiency and with the mathematical properties of optimal solutions (Cornuejols, Fisher, and Nemhauser, 1977; Francis and White, 1974; Handler and Mirchandani, 1979). The readers of the new literature have included, in addition to operations researchers, individuals with more diverse interests and experiences. This diversity has often led to the questioning of assumptions long accepted and to the development of even newer areas of application and research. The developments in the contemporary phase have brought greater realism to location-allocation problems because they have focused on better representing both producer and consumer behavior as well as the environment in which the model is applied.

This quest for better representation has led to the development of new location-allocation literatures in four areas. First, the representation of the decision-making behavior of the providers of services received attention. Later, a literature developed that emphasized more accurate representation of the process of consumer spatial choice. About the same time, it also became clear that in representing the locational characteristics of the environment in location-allocation models, some methods could lead to invalid conclusions, and a third literature began to develop on the representation of the geography of demand and of the distances, time, or transportation costs between places. Awareness also grew of the potential for error arising out of model misspecifications. The result was an increased interest in understanding and modeling the uncertainities in the environment in which a service is to be provided. Uncertainties arise from the stochastic nature of demand, travel time, and other service elements, or from potential changes in the location of other, often competitive service providers.

Representation of Decision Criteria

Multiple Objectives.   Most applications of location-allocation models in the classic phase used a single, systemwide objective to evaluate alternative location patterns. In reality, however, multiple criteria often apply in choosing location patterns. During the past decade, many researchers have focused on methods for solving multiple-objective location-allocation problems (for a review, see Cohon, 1978). In an example of such an application, Schilling (1980) used the criteria of property damage and personal casualty in determining the locational deployment of fire fighting units in Baltimore. Another study by ReVelle, Cohon, and Shobrys (1981) considers seven criteria of interest and illustrates the use of *value paths* to compare the performance of alternative solutions on individual criteria. In this volume, Daskin (chap. 9) presents a fire engine location and routing model that considers both the arrival time of vehicles and the coverage of demand. In the same genre, albeit in a different context, Goodchild and Noronha (chap. 5 herein) use multiobjective programming to determine the optimal location of gasoline service stations in London, Ontario. In contrast to earlier models of retail location, which looked only at planned purchases, Goodchild and Noronha take into account the demand generated by impulse as well as planned purchases.

An alternative to finding solutions that maximize outcomes along different objectives is the use of goal programming methods to find solutions that have minimum deviations from a set of stated goals (see, for example, Sutcliffe, Board, and Cheshire, 1984). Many decision makers are more comfortable evaluating planning alternatives that have been designed to minimize the discrepancy from predefined levels of selected criteria (goals) than alternatives designed either to optimize weighted sets of criteria or to meet arbitrary binding constraints.

The challenge inherent in any multiple-objective formulation is that a solution is rarely simultaneously optimal with respect to all the different criteria to be considered. The final solution, therefore, depends on the weights that are assigned to different criteria. Often, however, these weights are unknown. (If the weights were known a priori, the multiple objectives could be combined into a single one and the problem would reduce to the general linear model.) Consequently, many multiobjective procedures aim to identify the set of "noninferior" solutions from which the final choice can be made (Cohon, 1978). An alternative line of research has focused on determining the appropriate set of objective weights by studying decision makers' level of preference for various alternative plans that are presented to them (for a review of this approach, see Keeney and

Raiffa, 1976). Mirchandani and Reilly (chap. 8 herein) provide an illustration of *utility analysis* in which they ascertain the utility function of fire department officials in Albany, New York.

Multiple Types of Facilities. Classic location-allocation models all dealt with services that had only a single level of facility. Yet many service systems, health care delivery, for example, often include more than one type of facility and location modeling has then to coordinate activities among the different types. This requirement led to the development of hierarchical location-allocation models, which address the location of facilities in a manner that coordinates the activities at the different levels of the service delivery hierarchy. Church and Eaton (chap. 7 herein) provide a systematic review of the development of this literature. The authors also present two hierarchical location-allocation models that were applied to rural health service planning in Colombia and the Dominican Republic.

Ratick et al. (chap. 10 herein) provide another example of a location-allocation model that demands the coordination of activities among different types of facilities. They develop an interperiod network storage location-allocation (INSLA) model for designing commodity distribution systems. The interperiod model considers the possibility of storing commodities and thereby allows rationalization of potential cyclicity in demand and supply levels. The availability of storage and the optimal location of storage facilities is a key element that distinguishes INSLA problems from other location and transhipment problems.

Representing Consumer Spatial Choice

Location-allocation models developed in the classic period assumed that consumers patronize only a single facility, the one closest to them. Consumers may not behave in this fashion for a number of reasons. First, consumers may have a set of activities that they wish to perform within some time or space constraints. This situation gives rise to the phenomenon of multipurpose trips. A number of empirical studies have found multipurpose trips to constitute an important part of consumer travel (see, for example, O'Kelly, 1981). Organizing a multipurpose trip logically and efficiently will often make a facility most convenient to a consumer given the other activities to be performed, even though it is not the closest one (Mulligan, 1983; Ghosh and McLafferty, 1984).

A second reason why consumers might bypass the closest facility is that the facility itself may not give the same level of services at all times. For instance, when congestion occurs at the closest facility, a consumer may choose some other facility, expecting that service will be more convenient there (Berman and Larson, 1982). In other cases, decision makers may plan to constrain services in locations where demand is low. Tien and El-Tell (1984) model such a system for primary health care; the objective is to maximize the availability of physician-provided services as well as the general ability of people to reach the services. A further reason for not always visiting the closest facility and for patronizing more than one facility is that in many situations, especially those involving retail stores, individual facilities may vary in the level of services they provide, in their images, and in the prices they charge.

The realization that consumers may not always visit their closest facility, as they were assumed to do in the traditional location-allocation models, led a number of researchers to suggest that consumer choice should be viewed as a trade-off between distance and the in situ attractiveness of facilities. Two early papers by Lakshmanan and Hansen (1965) and Huff (1966), for example, used a probabilistic store-choice model to determine the optimal location of a new retail store. These early approaches were limited, however, in that they considered only the single-store-location problem. It was not until a number of years later that researchers demonstrated the advantage of using a location-allocation framework for solving the multiple-facility problem with probabilistic consumer choice (see, for example, Coehlo and Wilson, 1976; Hodgson, 1978; Achabal, Gorr, and Mahajan, 1982; Ghosh and McLafferty, 1982; Leonardi, 1983).

Incorporating probabilistic allocation rules in the location-allocation framework, of course, raised new questions regarding the formulation of models and algorithms to solve them. O'Kelly (chap. 11 herein) provides detailed discussion of this literature's development. He then demonstrates how stochastic models of consumer spatial behavior can be incorporated in location-allocation models, and he presents experiences with an algorithm designed to solve probabilistic location-allocation problems.

Representing the Environment

The Aggregation Problem.   In all location-allocation problems, demand data is usually aggregated according to some arbitary spatial unit. This unit, often a census area or some other administrative area, usually varies in size and shape, and the true geographic center of demand in the area is often unknown. Although a number of different methods are available for estimating the centroid for analysis purposes, the results from any location-allocation algorithm are highly sensitive to accuracy in data representation. In a pioneering study, Gould, Norbeck, and Rystedt (1971) used simulation experiments to illustrate this sensitivity. The experiments were conducted by computing the optimal multifacility locations for a range of geographical scales and levels of simulated errors in the spatial representation of demands. Knowing that the results might be sensitive to the spatial scale of the data did not, however, allow prediction of the consequences of inadequately representing the geographical distribution of demand and of selecting inappropriate scales at which to aggregate data. These have only recently been systematically explored (Goodchild 1979; Hillsman and Rhoda, 1978).

Casillas (chap. 12 herein) has found that optimal locations are stable over a wide range of degrees of spatial aggregation of data but that measures of the objective function are increasingly biased with higher degrees of spatial aggregation. His conclusions, although valuable for practical applications of location-allocation models, can be applied only to cases having no systematic bias in the spatial variation of levels of aggregation. The accuracy of this assumption in empirical applications is unknown. As a result of this and future work on the subject, more accurate methods for disaggregating spatial information will be used to represent data in location-allocation models. Some models for this purpose already exist (Bennett, Haining, and Griffith, 1984; Tobler, 1979), and their potential for use in location-allocation modeling will likely lead to renewed interest in their development.

Interplace Distances.   Discrete space applications of location-allocation modeling assume knowledge of time, distance, or costs of interactions between all places. In reality, such interplace distances or travel times must often be estimated, giving rise to another potential source of error in the data. In an early study, Nordbeck (1964) investigated the nature of this error by computing interplace distances from coordinates and comparing them with their counterparts measured along shortest-path routes. He then calibrated a *distance function* relating the two sets of distances and measured the errors associated with the function. Timbers (1967) computed a similar distance function between places in the United Kingdom and showed that systematic variations in the parameters of the distance function could be ascribed to town size, region of the country, and length of trip (see also Love and Morris, 1972, 1979). One technique for improving distance functions is to calibrate the models separately for distances between different regions (Fildes and Westwood, 1978; Nordbeck, 1964). The price of the resulting increase in the accuracy of distance estimates is, of course, loss of generality of the distance function for use in areas other than where it was calibrated.

In a reversal of the error-estimating pattern of the work described above, Ginsburgh and Hansen (1974) use the distances computed from coordinates to estimate the probability that a given interplace distance figured from a shortest-path algorithm is in error. They point out that practitioners commonly report difficulties, as a result of data encoding errors, in computing error-free distance matrices. They note that one piece of erroneous data used as input to a shortest-path algorithm will likely generate a large number of false distances.

In a manner similar to the estimation of distance functions, time-distance functions have been calibrated to estimate travel times from distances. Kolesar, Walker, and Hausner (1975) conducted experiments with ladder companies and battalion chiefs' cars in the New York City Fire Department; and similar studies, described and compared by Kolesar (1979, pp. 173–177), computed parameters for the same time-distance function in five U.S. cities. Scott, Factor and Gorry (1978, p. 411) computed the probability density function of the speed of ambulances en route to an emergency in Houston, Texas.

These studies used special geographic encoding procedures to compute route distances. No location-allocation studies have been identified that have used the U.S. Census geographic encoding scheme GBF/DIME System (U.S. Department of Commerce, 1978), which is available for most metropolitan areas of the United States. This system is currently being revised to develop an automated geographic data base known as the TIGER system (Marx, 1983). As a result of improvements in estimation procedures for distance and travel time and in geographical information systems, many location-allocation analyses are representing the environment far more realistically than in the classic period, when problems of data representation were rarely discussed.

Dealing with Uncertainty

A number of causes might raise uncertainty in the environment in which location-allocation models are applied. In some applications the underlying process that generates demand for service may itself be stochastic. This is the case, for example, when the locations of emergency services are being planned. Situations may arise when the closest emergency facility may not be available to attend an incident because it is engaged in servicing another incident (Mirchandani and Reilly, chap. 8 herein). Similarly, if the travel times along transportation links are stochastic, the response times of emergency vehicles can be estimated only probabilistically. In either case it would be inappropriate to simply use the distance to the closest emergency unit to evaluate the level of service provided. This condition led to the study of dynamic facility location models that consider uncertainty in the availability of units and account for stochastic travel times.

Much of the literature on dynamic location-allocation models was motivated by the early work done by the Rand Institute. Between 1969 and 1975, researchers at the institute undertook detailed analyses of emergency-service delivery systems and developed location models for planning such systems. The latter included both analytical location models, such as the Hypercube Queuing Model (Larson 1974, 1975), and simulation models (see, for example, Carter, Ignall, and Walker, 1975). In this volume, Daskin (chap. 9) and Mirchandani and Reilly (chap. 8) provide a detailed discussion of the literature dealing with environmental uncertainty and its impact on locations. Both chapters also present alternative location-allocation models that consider with great realism the complex pattern of needs for fire service and the pattern of response times of multiple fire fighting units.

Much of the progress in location-allocation modeling can be ascribed to the interaction of decision makers and analysts in defining the nature of the problem and in formulating the objective function that must be optimized. Many new and interesting objective functions were first identified through such interaction in the context of applied studies. This interaction is increasing as decision makers become more aware that identifying optimal locations is extremely sensitive to the choice of objective criteria and the relative weights assigned to them. An unresolved issue (see Rushton, chap. 14 herein) is the extent to which decision makers are able to define their preferences for alternative criteria independent of the context in which the problem exists. Do decision makers alter their preferences as a result of analyses that reveal to them, often for the first time, the degree to which some objective can easily be attained and others cannot? Analyses will often reveal unanticipated consequences of implementing a particular preference function. This argument leads Rushton to conclude that in the future, decision makers will use location-allocation models to create decision-support systems allowing them to explore systematically their decision space and the interaction between model results and model formulations.

Another task of future research will be to develop models that explicitly consider service outcomes as the objective criteria rather than relying on arbitrary distance standards (see, for example, McLafferty and Broe, 1985). However, this task will require a better understanding of the relationship between distance and outcome. In Seattle, for example, Mayer (1979) examined a sample of the three-tiered response system's response calls to reported life-threatening health conditions. He found that the response times of the first two tiers (fire engine companies and emergency medical technicians) were uncorrelated with death probabilities but that the response time of mobile intensive care units (MICU) significantly affected the probability of survival. In particular, death probabilities increased markedly when MICU response time exceeded nine minutes. Eighty five percent of MICU response times were within this limit. Such research results give essential information for estimating the improvements in outcome that may result from improvements in response time, and such knowledge relating response time and outcome provides the objective criteria for evaluating alternative service delivery configurations.

LOCATION-ALLOCATION MODELS AND LOCATION THEORY

The ability to better represent the behavior of producers and consumers and the environment in which the models are applied not only provided better facility location plans; it also allowed the use of location-allocation models in novel ways. For example, although in the past most applications of location-allocation models had been prescriptive in nature, a number of authors noted the usefulness of these models for developing and testing alternative theories of spatial organization. The location-allocation approach provides a powerful tool with which to systematically generate alternate spatial landscapes by varying different assumptions regarding producer and consumer behavior and the organization of the environment. Insofar as both location-allocation models and central place theories search for facility locations that best serve a dispersed population, the connection between these two research streams is potentially fruitful.

Beaumont (chap. 1 herein) demonstrates how location-allocation models can provide the tools with which to operationalize the underlying concepts of central place theories. Doing so allows one to study implications of those constructs in diverse spatial environments—a goal that eluded the classic theorists. As a consequence, location theorists need no longer restrict their investigations to simplified environments, and the richness of the classic theories is only now being fully demonstrated.

Location-allocation models also allow consideration of diversity in producer and consumer behavior patterns. For example, McLafferty and Ghosh (chap. 2 herein) explore the implications of multipurpose shopping for the structure of retail centers. They show that introducing multipurpose shopping in the axioms of central place theory provides a strong rationale for the agglomeration of low- and high-order firms and the development of retail hierarchies. They study the nature of this hierarchy using location-allocation models.

Ginsburg, Pestieau, and Thisse (chap. 4 herein) apply location-allocation models to location theory in yet another interesting way. They focus on the multiple objectives that political parties may try to fulfill in choosing their political platforms, and they formulate the problem of choosing the political platform as a location-allocation model. The result greatly enriches the traditional models of electoral competition based on Hotelling's (1929) principle of minimum differentiation and offers a theory that represents more accurately the complexities of the real world.

Location-allocation models provide a valuable tool in understanding and developing alternative theories of spatial structure. As the examples in this volume demonstrate, the implications of the postulates of central place theory and the theory of spatial competition can be more fully explored by formulating location-allocation models that incorporate the postulated process of spatial organization.

EVALUATING PAST LOCATION DECISIONS

In addition to furthering our understanding of location theory by creating alternative economic landscapes, location-allocation models can be used to evaluate past location decisions. Such applications use location-allocation models not as prescriptive tools but as descriptive models for understanding past patterns of locational decision making. Such modeling may have one or more of the following purposes: to evaluate the effectiveness of previous decisions in comparison with optimal location patterns; to infer the objectives that decision makers may have used in locating services; and, in systems that evolved as a result of individual providers' locational decisions, to assess whether the forces of competition led to efficiencies in the spatial organization (see, for example, Fisher and Rushton, 1979; McLafferty and Ghosh, 1982; Ghosh, 1979). This line of research, although comparatively new, is likely to be fruitful.

Tewari and Jena (chap. 6 herein) provide an illustration of how location-allocation models can be used to understand the behavior of decision makers. The authors analyze a case in which a government formalized the criteria for locating high schools and specified a process that administrative authorities should follow to locate new schools. Tewari and Jena use a location-allocation model to evaluate alternative locations that meet the stated criteria and to compute optimal locations. They then compare their optimal locations' performance on the stated criteria with the performance of locations selected by the educational authorities. They are able not only to demonstrate that the optimal locations are considerably superior on the stated criteria, but also to comment on aspects of the process of decision making about school locations that appear to be related to the poor performance of the selected locations. They are able, also, to comment on how the government could restate its performance measures so that a major part of the discrepancy between optimal and selected locations would not occur in the future.

A number of other reasons exist for evaluating past location decisions. One may ask, for example, whether changes in the location pattern over time lead to more efficient service delivery systems. This question is especially relevant when the delivery system itself results from decisions of individual providers rather than of a few centralized decision makers (Ghosh, 1979). Similarly, this approach can be used to test alternative hypotheses regarding the process of decision making. A particularly interesting application of this line of research, illustrated by Bell and Church (chap. 3 herein), is in archaeological studies, where knowledge about past decisions must be inferred from artifacts. The assumption underlying Bell and Church's study is that the principles a society (or decision makers) may have followed in organizing space can be discerned by studying the resulting pattern. The hypothesized principles can be corroborated by determining whether the spatial organization agrees with what would be optimal using a particular objective function that embodies those principles. Some may criticize such inferences on the ground that several alternative objective functions might lead to location patterns equally similar to the observed pattern. Bell and Church argue for the use of several alternative models to reach any conclusion and, just as importantly, for the use of location-allocation models in conjunction with other, more traditional evidence. Others may argue that the correspondence between the observed and optimal patterns results merely from chance. Such a hypothesis may be tested by using location-allocation modeling within a randomization framework. Indeed, as McLafferty and Ghosh (1982) show, location-allocation models may be more appropriate in this respect than traditional statistical procedures.

CONCLUSION

About twenty years have passed since location-allocation methods were first developed. At first restricted in their application to problems that involved finding locations that were optimal with respect to single criterion in relatively small areas, their development rapidly led to the solving of problems involving multiple criteria and for large areas. The performance characteristics of a diverse set of solution methods, now well-known, allow users a range of options from which to select the method most appropriate to the objective, data characteristics, computational facilities and the decision-making environment of the particular problem before them. Nevertheless, as the complexity of real problems is explored, new objective functions are recognized that can then be seen to be relevant in other problem contexts. The stochastic nature of the network of communication links, of the geographic pattern of demand, and of system performance itself, have challenged researchers in location-allocation analysis to further develop solution methods.

The typical goals of location-allocation analyses have changed markedly in the past decade from an exclusive concern to compute an optimal location pattern for a given activity to a range of concerns that suit the needs of an increasingly diverse group of users. These increasingly include persons who have little interest in the solution procedures per se, but are interested in improving the performance of a service-delivery system over which they exert some control. With the increasing use of location-allocation models by decision-makers has come the need for flexible analysis systems where decision-makers can explore alternative location patterns, use optimizing algorithms to compute interesting alternatives, and evaluate alternative options against criteria of interest some of which are highly qualitative in character. The availability of location-allocation methods have permitted a retrospective examination of past location decisions which, in turn, permit inferences to be made about the quality of the process for making location decisions. Other recent applications of these methods have included the development of location theory, including the modeling of competitive spatial behavior and the modeling of hierarchical location patterns.

The interest in behavior within systems has led to the forging of links between research on the relationship between the spatial organization of an activity and the behavior of people and the traditional location-allocation modeller who was an expert at selecting solution procedures that could find locations that were optimal with respect to a well-defined objective. The union between these two research groups, each with quite different traditions of methods and research objectives, is faced with a large number of unsolved questions. Particularly important is the recent recognition that behavior itself is sensitive to the geographical organization of many activities. Developing appropriate methods and finding appropriate contexts where these problems can be solved provides an interesting and large research agenda for the future.

REFERENCES

Achabal, D., W. R. Gorr, and V. Mahajan, 1982, MULTILOC: a multiple store location model, *Journal of Retailing* **58**:5–25.

Bennett, R. J., R. P. Haining, and D. A. Griffith, 1984, The problem of missing data on spatial surfaces, *Association of American Geographers Annals* **74**:138–156.

Berman, O. and R. C. Larson, 1982, The median problem with congestion, *Operations Research* **9**:119–126.

Bindschedler, A. E. and J. M. Moore, 1961, Optimal location of new machines in existing plant layouts, *Journal of Industrial Engineering* **12**:41–48.

Carter, G. M., E. J. Ignall, and W. Walker, 1975, A simulation model of the New York City Fire Department: Its use in deployment analysis. P-5110–1, *The Rand Corporation,* Santa Monica, California.

Christaller, W., 1933, *Central Places in Southern Germany,* trans. C. W. Baskin, 1966, Prentice-Hall, New Jersey.

Church, R. and C. ReVelle, 1974, The maximal covering location problem, *Regional Science Association Papers* **32**:101–118.

Church, R. and C. ReVelle, 1976, Theoretical and computational links between the *p*-median and the maximal covering location problems, *Geographical Analysis* **8**:406–415.

Coehlo, J. D. and A. G. Wilson, 1976, The optimum location and size of shopping centers, *Regional Studies* **10**:413–421.

Cohon, J. L., 1978, *Multiobjective Programming and Planning,* Academic Press, New York.

Cooper, L., 1963, Location-allocation problems, *Operations Research* **11**:331–343.

Cooper, L., 1964, Heuristic methods for location-allocation problems, *SIAM Review* **6**:37–53.

Cooper, L., 1967, Solution of generalized locational equilibrium problems, *Journal of Regional Science* **7**:1–18.

Cornuejols, G., M. L. Fisher and G. L. Nemhauser, 1977, Location of bank accounts to optimize float: An analytic study of exact and approximate algorithms, *Management Science* **23**:789–810.

Erlenkotter, D., 1978, A dual-based procedure for uncapacitated facility location, *Operations Research* **26**:992–1009.

Feldman, E., F. A. Lehrer, and T. L. Ray, 1966, Warehouse locations under continuous economies of scale, *Management Science* **12:**670–684.

Fildes, R. A. and J. B. Westwood, 1978, The development of linear distance functions for distribution analysis, *Operations Research Society Journal* **5:**92.

Fisher, H. B., and G. Rushton, 1979, Spatial efficiency of service locations and the regional development process, *Regional Science Association Papers* **42:**83–97.

Francis, R. L. and J. A. White, 1974, *Facility Layout and Locations: An Analytic Approach*, Prentice Hall, New Jersey.

Friedrich, C. J., 1929, *Alfred Weber's Theory of the Location of Industries*, University of Chicago Press, Chicago.

Ghosh, A., 1979, Changes in the spatial efficiency of periodic markets, *Tijdschrift Voor Economische en Sociale Geografie* **70:**293–299.

Ghosh, A. and S. L. McLafferty, 1982, Locating retail stoes in uncertain environments: a scenario planning approach, *Journal of Retailing* **58:**5–22.

Ghosh, A. and S. L. McLafferty, 1984, A model of consumer propensity for multipurpose shopping, *Geographical Analysis* **16:**244–249.

Ginsburgh, V. and P. Hansen, 1974, Procedures for the reduction of errors in road network data, *Operations Research Quarterly* **25:**321–322.

Goodchild, M. F., 1979, The aggregration problem in location-allocation, *Geographical Analysis* **11:**240–255.

Gould, P., S. Nordbeck, and B. Rystedt, 1971, Data sensitivity and scale experiments in locating multiple facilities, in *Multiple Location Analysis*, G. Tornquist, S. Nordbeck, B. Rystedt, and P. Gould, eds., Lund Studies in Geography, ser. **C(12)**, Sweden, pp. 67–86.

Hakimi, S. L., 1965, Optimum distributions of switching centers in a communication network and some related graph theoretic problems, *Operations Research* **13:**462–475.

Handler, G. Y. and P. B. Mirchandani, 1979, *Location on Networks, Theory and Algorithms*, MIT Press, Massachusetts.

Hansen, P., D. Peeters, and J. F. Thisse, 1983, Public facility location models: a selective survey, in *Locational Analysis of Public Facilities*, J. F. Thisse and H. G. Zoller, eds., North-Holland, New York, pp. 223–262.

Hillsman, E. L., 1984, The *p*-median structure as a unified linear model for location-allocation analysis, *Environment and Planning, A* **16:**305–318.

Hillsman, E. L., and R. Rhoda, 1978, Errors in measuring distances from populations to service centers, *Annals of Regional Science* **12:**74–88.

Hodgart, R. L., 1978, Optimizing access to public services: a review of problems, models and methods of locating central facilities, *Progress in Human Geography* **2:**17–48.

Hodgson, M. J., 1978, Towards more realistic allocation in location-allocation models: an interaction approach, *Environment and Planning, A* **10:**1273–1285.

Hotelling, H., 1929, Stability in competition, *Economic Journal* **39:**41–57.

Huff, D. A., 1966, A programmed solution for approximating an optimal retail location, *Land Economics* **42:**284–295.

Isard, W., 1956, *Location and Space-Economy*, M.I.T. Press, Massachusetts.

Kariv, O. and S. L. Hakimi, 1979, An algorithmic approach to network location problems, II: The p-medians, *SIAM Journal on Applied Mathematics* **37**:539–560.

Keeney, R. L. and H. Raiffa, 1976, *Decisions With Multiple Objectives: Preferences and Value Tradeoffs,* Wiley, New York.

Kolesar, P. J., 1979, Travel time and travel distance models, in *Fire Department Deployment Analysis,* W. E. Walker, J. M. Chaiken, and E. J. Ignall, eds., North-Holland, New York, pp. 157–201.

Kolesar, P. J., W. Walker, and J. Hausner, 1975, Determining the relation between fire engine travel times and travel distances in New York City, *Operations Research* **23**:615–627.

Kuehn, A. A. and M. J. Hamburger, 1963, A heuristic program for locating warehouses, *Management Science* **9**:643–666.

Kuhn, H. W., and R. E. Kuenne, 1962, An efficient algorithm for the commercial solution of the generalized Weber problem in spatial economics, *Journal of Regional Science* **4**:21–33.

Lakshmanan, T. R. and W. G. Hansen, 1965, A retail market potential model, *American Institute of Planners Journal* **31**:134–143.

Larson, R. C., 1974, A hypercube queueing model for facility location and redistricting in urban emergency services, *Computers and Operations Research* **1**:67–95.

Larson, R. C., 1975, Approximating the performance of urban emergency service systems, *Operations Research* **23**:845–868.

Launhardt, W., 1882, Die bestinmmuns des eweckmaessigsten standortes einer siewerblichen anlage, *Zeitschrift des Vereines Deutscher Ingenieure* **26**:106–115.

Lea, A., 1973, An annotated bibliography of location-allocation, Discussion Paper #13, Department of Geography, University of Toronto.

Leonardi, G., 1981, A unifying framework for public facility location problems—part 1: A critical overview and some unsolved problems, *Environment and Planning, A* **13**:1001–1028.

Leonardi, G., 1983, The use of random-utility theory in building location allocation models, in *Locational Analysis of Public Facilities,* J. F. Thisse and H. G. Zoller, eds., North-Holland, Amsterdam, pp. 357–383.

Losch, A., 1954, *The Economics of Location,* trans. W. H. Woglom and W. F. Stopler, Yale University Press, New Haven.

Love, R. F. and J. G. Morris, 1972, Modelling inter-city road distances by mathematical functions, *Operational Research Quarterly* **23**:61–71.

Love, R. F., and J. G. Morris, 1979, Mathematical models of road travel distances, *Management Science* **25**:130–139.

Maranzana, F. E., 1964, On the location of supply points to minimize transport costs, *Operational Research Quarterly* **15**:261–270.

Marx, R. W. 1983, Automating census geography, *American Demographics* **5**:30–33.

Mayer, J. D., 1979, Emergency medical service delays, response time and survival, *Medical Care* **17**:818–827.

McLafferty, S. L. and A. Ghosh, 1982, Issues in measuring differential access to public services, *Urban Studies* **19**:383–389.

McLafferty, S. L. and D. Broe, 1985, Locating medical services to maximize health outcomes, paper presented at the Association of American Geographers Annual Meetings, Detroit, April 1985.

Morrill, R. L. and J. Symons, 1977, Efficiency and equity aspects of optimum location, *Geographical Analysis* **9:**215–225.

Mulligan, G., 1983, Consumer demand and multipurpose shopping behavior, *Geographical Analysis* **15:**76–80.

Narula, S. C., U. I. Ogbu, and H. M. Samuelsson, 1977, An algorithm for the *p*-median problem, *Operations Research* **25:**709–713.

Nordbeck, S., 1964, Computing distances in road nets, *Regional Science Association Papers* **12:**207–220.

O'Kelly, M., 1981, A model of the demand for retail facilities incorporating multistop, multipurpose trips, *Geographical Analysis* **3:**140–156.

ReVelle, C. and R. Swain, 1970, Central facilities location, *Geographical Analysis* **2:**30–42.

ReVelle, C., D. Marks, and J. C. Liebman, 1970, An analysis of private and public sector location models, *Management Science* **16:**692–699.

ReVelle, C., D. Bigman, D. Schilling, J. Cohon, and R. Church, 1977, Facility location: a review of context-free and EMS models, *Health Services Research* **12:**129–146.

ReVelle, C., J. Cohon, and D. Shobrys, 1981, Multiple objective facility location, *Sistemi Urbani* **3:**319–343.

Rushton, G., M. F. Goodchild, and L. M. Ostresh, eds., 1973, *Computer Programs for Location-Allocation Problems,* Monograph 6, The Department of Geography, University of Iowa, Iowa City.

Schilling, D. A., 1980, Some models for fire protection locational decisions, *European Journal of Operations Research* **5:**1–7.

Schilling, B. A., A. McGarity, and C. ReVelle, 1982, Hidden attributes and the display of information in multiobjective analysis, *Management Science* **28:**236–242.

Scott, A. J., 1970, Location-allocation systems: a review, *Geographical Analysis* **2:**95–119.

Scott, D. W., L. E. Factor, and G. Gorry, 1978, Predicting the response time of an urban ambulance system, *Health Services Research* **13:**404–417.

Sutcliffe, C., J. Board, and P. Cheshire, 1984, Goal programming and allocating children to secondary schools in Reading, *Operations Research Society Journal* **35:**719–730.

Teitz, M. and P. Bart, 1968, Heuristic methods for estimated generalized vertex median of a weighted graph, *Operations Research* **16:**955–961.

Tien, J. M., and K. El-Tell, 1984, A quasihierarchical location-allocation model for primary health care planning, *IEEE Transactions on Systems, Man, and Cybernetics* **SMC-14:**373–380.

Timbers, J. A., 1967, Route factors in road networks, *Traffic Engineering and Control* **9:**392–394.

Tobler, W. R., 1979, Smooth pychnophylactic interpolation for geographical regions, *American Statictics Association Journal* **74(367):**519–530.

Toregas, C. and C. ReVelle, 1972, Location under time or distance constraints, *Regional Science Association Papers* **28:**133–142.

U.S. Department of Commerce, 1978, *GBF/DIME System: Description and uses,* Bureau of the Census, Washington, D.C.

Part I

LOCATION THEORY AND LOCATION-ALLOCATION MODELS

1

LOCATION-ALLOCATION MODELS AND CENTRAL PLACE THEORY

John R. Beaumont
University of London

Central place theory, which deduces the spatial pattern of settlements, has provided the foundation for many empirical studies of spatial organization. However, in applying the principles of the theory in a normative framework, such as the location of public facilities, a critical requirement is the ability to operationalize the underlying concepts of the theory. This paper argues and demonstrates that location-allocation models provide a suitable tool for operationalizing the concepts of central place theory. Location-allocation models, which jointly optimize the location of centers and the allocation of consumers to those centers, have been widely used for facility planning in both the private and public sectors (for a review see Leonardi, 1981; Beaumont, 1981). With the exception of locating emergency services, the majority of problems involve, as in central place systems, providing services for which consumers must travel to a facility, rather than services that are delivered to consumers. A basic tenet underlying this paper is that alternative formulations of location-allocation problems should be compared and evaluated to gain new insights into central place theory and to generate operational models of the theory.

As a foundation for the discussion in this chapter, the next section provides an overview of some selected features of central place theory. The discussion of location-allocation models starts with a presentation of a multifacility generalization of the Weber problem—the fundamental form of location-allocation

Acknowledgment: The author and this chapter have benefited from helpful comments by Avijit Ghosh, Gerry Rushton, and an anonymous referee.

problems. Alternative models that incorporate spatial interaction and price elasticity of consumer demand are developed. With these concepts as background, we examine alternative assumptions regarding producer and consumer behavior, and we develop a number of both single-level and hierarchical location-allocation models that operationalize aspects of central place theory. Finally, some suggestions regarding dynamic models are outlined. Throughout, the discussion demonstrates the use of location-allocation models to operationalize the basic concepts and structural assumptions of central place theory.

CENTRAL PLACE THEORY

An extensive examination of Christaller's (1966) and Losch's (1954) well-known central place theories is not essential to the argument here. However, at the outset, it is appropriate to examine some of the basic concepts and behavioral and structural characteristics in detail, specifically those directly relevant to modeling using a location-allocation framework.

Central to both Christaller's and Losch's theories are assumptions regarding the distribution of population. Christaller, for instance, assumed a uniform distribution in his theoretical deductions, although he recognized the importance of relaxing this assumption. Indeed, once centers develop, this assumption of uniform density is undermined because they have an indigneous population. However, it should be stressed that Christaller distinguished clearly between absolute importance and relative importance (where nodality is measured by absolute importance). This distinction relates to a division of the total demand for a center's functions into two components: the internally generated demand from the center itself, and the externally generated demand. An analysis of centrality is concerned with the externally generated demand (and in this sense, the distribution of the population is spatially uniform).

Losch considered uniform spatial distributions of population, both continuous and discontinuous, in some detail. However, although he had demonstrated that a set of hexagonal market areas would be the equilibrium for a continuous system of markets, his economic landscape was based on a discontinuous distribution assuming that "the population may be equally but not continuously distributed" (1954, p. 114). That is, he assumed that farms were distributed over a plain in a pattern that formed a basic lattice of equilateral triangles. This spatial backcloth is what constrains the potential central place structures that can be derived.

The concepts of range and threshold of a commodity are also basic to central place theory. The *range*, the maximum distance that a consumer will travel to purchase a specific good or service, relates to aspects of the spatial extent of centers, a characteristic that has been partly considered in location-allocation modeling by examinations of market-area delimitation and spatial coverage.

The *threshold* is a minimum bound on the range, a lower limit. The threshold of a commodity, implicit in Christaller's account, is the total effective demand required to support a particular function. Thus, the threshold can be associated with entrepreneurial decision making, whereas the range is primarily influenced by consumer decision making. Central place theory addresses the spatial interplay between the range and threshold. The minimum supply capacity, the threshold, is related to (normal) profitable production (which is not necessarily the maximum profit situation). Therefore, it is directly associated with consumer demand and the producer's revenue.

In a number of studies—such as Berry and Garrison (1958), Rushton (1972), and Puryear (1975)—exogenous, hypothetical threshold values have been employed to ascertain the number of central places. This derivation involves dividing the total demand by the threshold level to determine the maximum number of centers. It has been argued that the profitability/feasibility of service provision should be viewed as an endogenous factor resulting from the complex interplay of supply and demand (Beaumont and Clarke, 1980). However, in terms of such population threshold models, location-allocation models have been applied to specify and to solve types of issues not recognized earlier. Using a location-allocation framework, Goodchild and Massam (1969) discuss the impossibility of exactly solving the threshold-location problem on a surface of unequal population density; the Berry-Garrison formulation missed this point. Additional problems exist upon relaxing the assumption that consumers patronize their nearest center; the incorporation of spatial-interaction models within location-allocation formulations can provide useful insights on this matter.

At this stage, it is important to note some differences between Christaller's and Losch's uses of the threshold concept, since these differences have significant implications for the spatial structure of their deduced systems. In Christaller's system a center will offer a commodity of a given order only if demand for it is sufficient to meet the threshold requirement and only if the center already offers all the lower-order commodities. In contrast, in Losch's system the availability of lower-order commodities need not be available in a specific center for it to offer a higher-order commodity. This situation relates to Losch's superimposition of different hierarchical levels which, clearly, is unlikely to generate complete coincidence of centers. The formulation of hierarchical location-allocation problems should give special attention to this difference. A hierarchical spatial structure is an important component of central place systems and a phenomenon that has been recorded by many empirical investigations.

Thus, the nature of the hierarchies deduced by Christaller and Losch are different. Christaller assumed a much more rigid hierarchy in which a central place of a specific hierarchical order, n, offers not only those commodities applicable to its order, but also all the other commodities of a lower order, $n - 1, \ldots, 1$. Although this *successively inclusive* hierarchy can be thought of as a special case, it may be appropriate at the interurban level; however, it is inappropriate at the intraurban level. Interestingly, Parr (1979, p. 32) confines his analysis of structural change to successively inclusive hierarchies, because "without such a restriction, the range of possible structures associated with temporal changes would be virtually without limit." This fact should be borne in mind with regard to the formulation of hierarchical and dynamic location-allocation models.

The construction of hierarchically organized central place systems was approached from different directions by Christaller and Losch. Christaller built his system from the top order downward, whereas Losch built his system from the bottom order upward. A number of important differences can be attached to the alternative viewpoints, but neither Christaller nor Losch stressed the principle of sequential, hierarchical development of a central place system. The appearance of new hierarchies or the disappearance of existing orders, however, is a fundamental feature of structural change in a central place system, and it should be accommodated.

Finally, given the optimization nature of location-allocation mathematical programming models (Beaumont, 1982a), it should be noted that Christaller was interested in explaining the empirical distribution of central places, rather than in deriving an optimal spatial pattern:

When this . . . theory is confronted with reality, it becomes clear to what extent reality corresponds to theory, to what extent it is explained by it, and in what respect reality does not correspond with the theory and is therefore not explained by it (1966, pp. 4–5).

In contrast, it can be suggested that Losch was more interested in normative spatial configurations of centers. For instance, after deriving the optimal, so-called Loschian landscape, he prescribed a methodology to test the rationality of reality rather than to attempt to verify the theory by studying reality: "Comparison now has to be drawn no longer to test the theory, but to test reality! Now it must be determined whether reality is rational" (1954, p. 363).

Interpreting central place theory requires care, because no uniform interpretation and agreement exist. Beavon (1977), for example, highlights discrepancies between Christaller's original exposition and its various interpretations and the general confusion about Losch's work, which has not been given the same attention as Christaller's works. In relation to the location-allocation framework, Saey (1973) argues that Christaller's objective function was a set covering objective, rather than a minimizing of aggregate travel cost subject to the constraint of equal population around each center. Berry and Garrison (1958) had formulated the latter as the restated central place theory (and that formulation was first implemented as a location-allocation model and solved algorithmically by Goodchild and Massam (1969), using Cooper's (1964) alternating algorithm with the transportation problem of linear programming substituted for proximal assignment in the spatial-assignment phase of the algorithm).

Thus, at the outset, it must be appreciated that various commentators on Christaller have redefined his locational objective and modified his aggregation constraints to generate alternative patterns of spatial organization and efficiency.

LOCATION-ALLOCATION MODELS
AND CENTRAL PLACE THEORY

In developing location-allocation models of central place systems, it is important to first examine alternative optimizing frameworks. One possibility is to follow Christaller's so-called marketing principle and minimize the number of centers required to serve the population. For example, Christaller stated that "the primary and chief law of the distribution of central places organises central place systems so that . . . all parts of the region are supplied from the minimum possible number of functioning central places" (1966, p. 92).

Minimizing the number of centers may be appropriate when the demand for goods is exogenously fixed. More generally, however, the objective must be to maximize demand by optimally locating the centers, as indeed Getis and Getis (1966) suggest. To maximize demand requires incorporating the price elasticity of consumer demand in the location-allocation model. Thus the objective could be to maximize demand given the number of centers to be located. Alternatively, if the cost of establishing centers is available, this information can also be incorporated in the model. Kohsaka (1983), for example, introduces a budget constraint to reflect the upper limit on the total investment for the construction of centers.

As stated above, it is important to note that the demand for a good or service is affected by distance. Christaller's concept of range indicates that consumers are unwilling to travel further than a certain distance to purchase a good or service. Losch's analysis of the demand cone directly portrays the impact of distance on demand. Inherent in central place theory, too, is the assumption that consumers patronize the nearest center. Such spatial allocation can be modeled by minimizing the aggregated weighted distance (or transport cost), and it is called the *median* location-allocation problem. However, observed spatial behavior does not reflect such patterns, and it is important to incorporate spatial-interaction formulations in the model.

The following sections suggest a range of different location-allocation models to represent the noted characteristics of consumers' and producers' behavior. It is important to remember that determining the optimal location of centers and allocating consumers to them is an integrated process. In practice, however, it is possible to separate the stages because different spatial actors are involved:

center location involves decisions by producers, and allocation to them involves decisions by consumers. Kohsaka (1983) offers an alternative viewpoint by proposing a two-level location-allocation system in which the levels are distinguished by spatial scale. For example, in an intraurban-facility location problem, producers undertake demand-maximization behavior at a city level and then consumers undertake behavior involving the nearest-center assumption at a community level. This chapter, however, argues that although such multilevel control provides an appropriate framework for representing the hierarchical structure of a central place model, considering the location and the allocation stages separately allows both consumers' and producers' behavior to be studied explicitly.

SOME FUNDAMENTAL LOCATION-ALLOCATION PROBLEMS

The simplest type of location-allocation problem is the Weber problem (Friedrich, 1929), which involves locating a production center so as to minimize aggregate weighted distance from the various raw material sources. The extension of the Weber problem is the well-known p-median location-allocation model. The p-median problem is concerned with the optimal location of a set of p uncapacitated supply centers to minimize the total weighted distance between them and n demand locations. In continuous space, the p-median problem can be expressed as,

$$\text{Min } C = \sum_{i=1}^{n} \sum_{j=1}^{p} O_i \lambda_{ij} c_{ij} \tag{1.1}$$

subject to

$$\sum_{j=1}^{p} \lambda_{ij} = 1; \qquad i = 1, \ldots, n \tag{1.2}$$

and

$$\lambda_{ij} \begin{cases} 1 & i = 1, \ldots, n \\ 0 & j = 1, \ldots, p \end{cases} \tag{1.3}$$

where O_i is the quantity demanded at location i whose coordinates are (x_i, y_i); and λ_{ij} is a binary variable that is assigned a value of 1 if demand point i is allocated to center j, and zero otherwise. Note that in this formulation consumers are allocated to their nearest center. Constraints (1.2) and (1.3) ensure this all-or-nothing allocation procedure.

The distance (or the generalized transport cost, which is assumed to be proportional to distance) between the demand point i and the supply point j is represented by c_{ij} ($i = 1, \ldots, n; j = 1, \ldots, p$). The Euclidean metric,

$$c_{ij} = \sqrt{(x_i - x_j)^2 + (y_i - y_j)^2} \qquad (1.4)$$

is usually used to calculate this distance. With Euclidean distances, exact solution procedures for this minimization problem, with is neither convex nor concave (Cooper, 1967), have been developed by applying a branch-and-bound algorithm (Kuenne and Soland, 1972). A similar algorithm has been programmed by Ostresh (1973).

In addition, a number of heuristic solution procedures for this problem exist. Cooper's approach (1964, 1967) is considered here in more detail, because other location-allocation models can also be solved by following this procedure. Simply, it involves a sequence of allocating demand to centers and relocating centers until some convergence criterion is achieved. This so-called alternating procedure, therefore, captures the fundamental conceptualization of location-allocation problems. As stated earlier, these stages involve different spatial actors: producers make the location decision, and consumers make the allocation decision. For the p-median problem, the alternating procedure involves iterating through the following equations,

$$x_j = \frac{\displaystyle\sum_{i=1}^{n} O_i \lambda_{ij} x_i / c_{ij}}{\displaystyle\sum_{i=1}^{n} O_i \lambda_{ij} / c_{ij}} \qquad (1.5)$$

$$y_j = \frac{\displaystyle\sum_{i=1}^{n} O_i \lambda_{ij} y_i / c_{ij}}{\displaystyle\sum_{i=1}^{n} O_i \lambda_{ij} / c_{ij}} \qquad (1.6)$$

Equations (1.5) and (1.6) are derived by differentiating the objective function, equation (1.1) with respect to x_j and y_j, and setting the partial derivatives equal to zero; it is noted that

$$\frac{\partial c_{ij}}{\partial x_j} = \frac{-(x_i - x_j)}{c_{ij}} \quad \text{and} \quad \frac{\partial c_{ij}}{\partial y_j} = \frac{-(y_i - y_j)}{c_{ij}} \qquad (1.7)$$

One difficulty in using the alternating procedure is that the solution may be a local optimum, which is not necessarily a global optimum. Global optimality cannot be guaranteed. This fact may be of concern, since the spatial configuration of a local and a global optimum may be very different for a specific problem. As a rule, repeated runs using numerous starting values should be undertaken. Although there is no guarantee that a solution obtained repeatedly is the global optimum, it is highly probable that it is the desired answer.

It is possible to extend the previous model to include the unit cost of a commodity at center j. The objective would be to minimize total (transport and purchase) costs,

$$\text{Min } c = \sum_{i=1}^{n} \sum_{j=1}^{p} O_i \lambda_{ij} (r_j + c_{ij}) \tag{1.8}$$

where r_j is the cost of the good at center j. This formulation does permit an examination of the effect of differential prices, so both locational and price decisions of suppliers can be considered. If prices are spatially uniform, r_j is a constant in the optimization problem and the optimal pattern will be identical to that generated by the previous model. Thus it is possible to justify the omission of this specific center cost by specifying the number of centers required exogenously; that is, given a total budget constraint and a center's costs, the maximum number of centers is obtained directly. An alternative justification relates to the objective of maximizing profit (that is, aggregate revenue minus aggregate cost). In the p-median formulation, it is assumed that the level of demand is independent of the spatial organization of the centers and that the price of each good is constant; therefore, the total revenue is known before a model is used, because all demand is satisfied. Thus, in this case, the objective of profit maximization is identical to the objective of minimizing the total (transport) costs. An obvious extension would be to relax the assumptions of fixed demand and then to maximize profit rather than minimize total transport costs. This problem is considered in a later section.

As stated earlier, in addition to continuous-space location-allocation models a range of problems dealing with locations on a network has also been developed. The p-median formulation proposed by Hakimi (1964), which has been the basis for a number of extensions, should be examined in this context. Simply stated, the problem involves locating p centers on a network of demand points so that the total aggregate (or mean) distance of all consumers is at a minimum. As in the preceding case, every consumer is allocated to his or her nearest center. The problem can be formulated as a zero-one programming problem.

$$\text{Min } z = \sum_{i=1}^{n} \sum_{j=1}^{p} O_i \lambda_{ij} c_{ij} \tag{1.9}$$

subject to

$$\sum_{i=1}^{P} \lambda_{ij} = 1 \qquad (i = 1, 2, \ldots, n) \qquad \textbf{(1.10)}$$

$$\lambda_{ij} - \lambda_{jj} \geq 0 \qquad (i = 1, 2, \ldots, n; j = 1, 2, \ldots, p) \qquad \textbf{(1.11)}$$

$$\sum_{j=1}^{P} \lambda_{jj} = p \qquad \textbf{(1.12)}$$

$$\lambda_{ij} = \begin{cases} 1 & \text{if } C_{ij} = \min C_{i\kappa}/\kappa\epsilon\{s\} \\ 0 & \text{otherwise} \end{cases} \qquad (i = 1, 2, \ldots, n; j = 1, 2, \ldots, P) \qquad \textbf{(1.13)}$$

where n is the total number of demand nodes and p is the total number of centers. Using this p-median formulation, it is possible to examine the relationship between the location of centers and the minimum and maximum distances traveled by consumers. The relationship provides a basis for extending the discussion to consider the number of centers as well as their locations and size.

INCORPORATION OF SPATIAL-INTERACTION MODELS

The restrictive assumption that consumers patronize the nearest center is an inaccurate model of observed behavior (Beaumont, 1980; Hodgson, 1978; Sheppard, 1980). To relax this assumption, or at least to treat it as a special case, a spatial-interaction shopping model can be incorporated into the p-median problem. The spatial-interaction model is

$$S_{ij} = \frac{O_i W_j^{\alpha} e^{-\beta c_{ij}}}{\sum\limits_{j=1}^{p} W_j^{\alpha} e^{-\beta c_{ij}}} \qquad \textbf{(1.14)}$$

When consumers travel to the nearest center and market areas do not overlap, the parameter β is infinite. As choices, tastes, and other nontransportation factors increase in importance, β tends toward zero and transport costs are relatively less crucial in consumer behavior. Transport costs are no longer a contributor to consumer choice when β equals zero.

At first sight, the natural generalization of the p-median problem would be given by

$$\text{Min } Z = \sum_{i=1}^{n} \sum_{j=1}^{p} S_{ij} c_{ij} \qquad \textbf{(1.15)}$$

This is an unconstrained optimization problem, because the all-or-nothing nature of the p-median model is relaxed. This problem can be solved by an alternate location-allocation heuristic procedure (Beaumont, 1980).

LOCATIONAL SURPLUS MAXIMIZATION

In this section, the narrow emphasis of most location-allocation formulations on the aggregate transport-cost criterion is questioned when spatial-interaction models have been incorporated. Minimizing transport cost is considered an inappropriate criterion for measuring user benefit when consumer choice is embedded, and the section outlines an alternàtive welfare function that is consistent with the behavioral model. Furthermore, the alternative formulation is shown to be consistent with the nearest-center hypothesis when no dispersion is exhibited in the choice between alternative centers. Simply stated, it is a more general formulation.

Neuberger (1971) and Williams (1976), for instance, both demonstrate that for evaluation of transport and land-use, aggregate travel cost is, in general, an unsuitable benefit criterion when consumer choice is included. It is only appropriate when there is no differentiation between centers and consumers and when it is assumed that everyone patronizes the nearest center. More fundamentally, transport cost is untenable with respect to the theoretical foundation of the location decision-making process because it disregards the spatial-choice process generating trip patterns.

Neuberger (1971) suggested that a suitably defined locational surplus function, which is based on Marshall's microeconomic concept of consumers' surplus, is an appropriate criterion for measuring user benefit. To derive the locational surplus welfare criterion, the change in benefit perceived by consumers when the configuration of supply points is changed from an initial state—$s^\circ = (x_1^\circ, y_1^\circ, \ldots, x_p^\circ, y_p^\circ)$, with a corresponding cost matrix, $c^\circ = (c_{11}^\circ, \ldots, c_{np}^\circ)$—to a final state, $s' = (x_1', y_1', \ldots, x_p', y_p')$, is considered. This alteration in benefit, ΔLS, is given by the path integral

$$\Delta LS = - \sum_{i=1}^{n} \sum_{j=1}^{p} \int_{\rho} dc\, s_{ij}(c) \tag{1.16}$$

where ρ is some path in cost-space between the initial and final states, s° and s', respectively (Williams and Senior, 1977). As the function satisfies integrability conditions (Williams, 1976), the integral is independent of path, and the change in locational surplus, ΔLS, is given by

$$\Delta LS = \frac{1}{\beta} \sum_{i=1}^{n} O_i \log \frac{\sum\limits_{j=1}^{p} W_j^\alpha e^{-\beta c'_{ij}}}{\sum\limits_{j=1}^{p} W_j^\alpha e^{-\beta c^\circ_{ij}}} \tag{1.17}$$

The denominator in this equation is a constant in the context of optimizing the supply locations. A continuous-space location-allocation problem for maximizing locational surplus can be written as

$$\text{Max } LS = \frac{1}{\beta} \sum_{i=1}^{n} O_i \log \sum_{j=1}^{p} W_j^{\alpha} e^{-\beta c_{ij}} \qquad (1.18)$$

Here optimization involves locating p centers (each of which is assumed to be of an adequate capacity to accommodate any demand that may be placed upon it, that is, to be uncapacitated) to maximize locational surplus of consumers at n exogenous locations. Consumer trend patterns are specified by the spatial-interaction model of equation (1.14). This problem can be solved by employing an alternate location-allocation heuristic (Beaumont, 1980).

Varying the value of the β parameter portrays that the nearest-center hypothesis can be thought of as merely one member of a family of more general models. At the limit, when β is infinite, the maximization of locational surplus is equivalent to minimizing total transport costs. The optimal solutions to the locational surplus problem are dependent on the value of the β parameter (Beaumont, 1980). The degree of market-area overlap depends on the value of the β parameter; a small value of β means relatively large overlapping market areas through longer trips, and a large value of β implies little market-area overlap and restricted travel. Although this model is concerned with a single level, note that the nearest-center hypothesis, producing nonoverlapping market areas, is necessary for a rigid hierarchical nesting.

Thus, this problem, in addition to demonstrating consistency with the p-median problem, derives a suitable measure for analyzing user benefit. However, it should be noted that in this definition of locational surplus, only transport costs are allowed to change; the in situ attractiveness of centers, W_j, is held constant. As Coehlo and Williams (1978) demonstrate, it is possible to incorporate both types of changes simultaneously by considering the utility of trip decision making.

For completeness, let us present an alternative mathematical programming model, which Leonardi (1978) has employed in his analysis of multifacility optimal location and size problems. The spatial-interaction benefits, described by equation (1.18), are defined as locational surplus, although given the conventional Hansen-type (1959) definition of physical accessibility

$$\sum_{j} W_j^{\alpha} e^{-\beta c_{ij}}$$

it is possible also to interpret them as accessibility benefits (see, for example, Williams and Senior, 1978).

Leonardi (1978) applied a so-called log-accessibility objective function subject to a constraint on total center size W. This application is formally stated as

$$\underset{\{W_j\}}{\text{Max }} LA = \sum_{i=1}^{n} O_i \log \sum_{j=1}^{n} W_j e^{-\beta c_{ij}} \qquad (1.19)$$

subject to

$$\sum_{j=1}^{n} W_j = W \quad \text{and} \quad W_j \geqslant 0, \forall_j \qquad (1.20)$$

This constrained problem can be translated into an unconstrained one by using the associated Lagrangian expression, L:

$$\underset{\{W_j\}}{\text{Max }} L = \sum_{i=1}^{n} O_i \log \sum_{j=1}^{n} W_j e^{-\beta c_{ij}} + \lambda(W - \sum_{j=1}^{n} W_j) + \sum_{j} v_j W_j \qquad (1.21)$$

where λ and v_j are the Lagrangian multipliers associated with constraints (1.20) and (1.21). The necessary optimality conditions can be ascertained by differentiating the Lagrangian with respect to W_j, λ and v_j, and by equating them to zero (assuming W_j, λ and v_j are nonzero). That is,

$$\frac{\partial L}{\partial W_j} = 0, \frac{\partial L}{\partial \lambda} = 0, \text{ and } \frac{\partial L}{\partial v_j} = 0 \qquad (1.22)$$

Now,

$$W_j \frac{\partial L}{\partial W_j} = \sum_{i=1}^{n} \frac{O_i e^{-\beta c_{ij}}}{\sum\limits_{j=1}^{n} W_j e^{-\beta c_{ij}}} - \lambda \qquad (1.23)$$

or

$$\frac{\partial L}{\partial W_j} = \frac{1}{W_j} \sum_{i=1}^{n} \frac{O_i W_j e^{-\beta c_{ij}}}{\sum\limits_{j=1}^{n} W_j e^{-\beta c_{ij}}} - \lambda \qquad (1.24)$$

It is possible to rewrite right-hand side of the equation as

$$\frac{\sum\limits_{i=1}^{n} S_{ij}}{W_j} - \lambda \qquad (1.25)$$

where S_{ij} is the gravity-type, shopping model [note that for consistency with Leonardi's (1978) formulation, the α parameter is dropped]. It is, therefore, possible to write the optimality condition as

$$\frac{\sum_{i=1}^{n} S_{ij}}{W_j} - \lambda = 0 \tag{1.26}$$

or as

$$\sum_{i=1}^{n} S_{ij} = \lambda W_j \tag{1.27}$$

which is a balancing condition, equating the flow of revenue accruing to a center j to its functional size (Harris and Wilson, 1978). This optimality condition means that center size, W_j, is allocated in proportion to the level of demand,

$$\sum_i S_{ij}$$

and consequently, proportional alterations in center size would occur if the constraint on total center size was modified. As a result, there would be no change in the relative structure of a central place system if there was overall system growth. However, if it is assumed that the aggregate production costs involve fixed and variable costs that vary over space, a combinatorial element of selecting a set of centers is present (in a similar way to the usual exogenous specification of the number of centers to be located). If this feature was incorporated into the above formulation, in any expansion (or contraction) of a central place system, changes in centers' sizes would take account of those changes' relative cost advantages.

LOCATION-ALLOCATION MODELS AND CENTRAL PLACE CONCEPTS

The previous sections have presented a number of extensions to the p-median problem. This section formulates a variety of simple location-allocation models that operationalize aspects of central place theory. Christaller (1966), for instance, had to be content with verbally describing spatial interaction and demand elasticity concepts without formally capturing them in any model.

As with the models in the previous section, the models considered in this section are all static and associated with one level. They do provide a foundation, however, to develop hierarchical and dynamic models, which are the topics of later sections.

The previous discussion of a demand cone around each center is now replaced by the idea of a similar cone around each consumer to capture price elasticity of consumer demand. One result of this characteristic is that the aggregate level of satisfied demand (and profit) depends on the spatial structure of the centers in relation to the distribution of consumers. A natural extension, presented in this section, is a location-allocation model representing the profit maximization motive of individual centers, one of Losch's (1954) optimizing principles. Finally, links with the basic concepts, threshold and range, are made through a discussion of capacitated and covering-type problems, respectively.

Price Elasticity of Consumer Demands

The existing emphasis on perfectly price inelastic demand formulations of the location-allocation problem is misplaced in certain circumstances in view of the evidence that, for particular types of services, the level of individual consumer demand depends on a consumer's proximity to a center. This evidence contradicts the usual assumption in many location-allocation models that the level of aggregate service utilization is independent of the pattern of facilities. Away from the perfectly inelastic situation, the level of actual demand will be a function of the distance required to travel to a center. Following Sheppard (1981, p. 436), two kinds of demand elasticity can be recognized:

First, the number of trips made from i will be inversely related to the accessibility of (centers) to demand points . . . Second, the fraction of trips from i that terminate at a given destination . . . will depend on the attractiveness of j relative to other destinations.

The latter has been considered already in terms of the derivation of a locational surplus function. The former can be conceptualized in terms of a demand cone for each consumer, rather than for each supplier. The level of expenditure, O_i, available to purchase a commodity decreases the further a consumer is from the center at i (see Fig. 1.1). Beyond a certain distance, the user range, transport costs will be so restrictive that no purchase is possible.

No restrictive assumptions are placed on the spatial pattern of demand, although it is assumed that every uncapacitated center offers the same goods and services at a standard price and that the consumers incur the cost of transporting them. The analyses preclude directly examining the possibility of a combination of price, quantity, and locational changes as a response to alterations in the models' parameters; spatial patterns and the level of aggregate demand are the topics of interest. A number of possible representations are considered.

If a center is indicated at j, the distance effect on the actual demand level from i, $O_{i(j)}$, can be given as,

$$O_{i(j)} = O_i^* e^{-v c_{ij}} \tag{1.28}$$

Demand

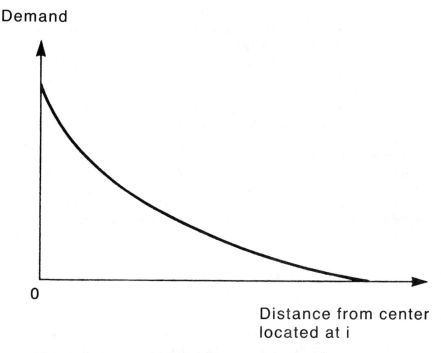

Figure 1.1.   Demand varies with distance from the center.

where O_i^* is the potential level of demand for a consumer located at i if a center is also located at i. Since, by definition, c_{ij} is equal to zero, $O_{i(i)}$ is equal to O_i. Thus, the parameter ν is directly related to the effect of distance on the demand level. When ν is zero, demand is perfectly inelastic. For this discussion we assume ν to be positive; that is, demand is assumed to fall at a negative exponential rate with distance. This assumption could be easily changed, if necessary.

In this discussion of demand elasticity, no account of competing centers and consumer choice has been included. The spatial-interaction model, which was applied previously, is a perfectly inelastic demand formulation. Harris and Tanner (1974) consider an alternative spatial-interaction model that in the notation used here, can be written as

$$S_{ij} = \frac{O_i W_j^\alpha e^{-\beta c_{ij}}}{e^{-\beta\eta} + \sum_{j=1}^{p} W_j^\alpha e^{-\beta c_{ij}}}.$$ **(1.29)**

where η can be thought of as a parameter that is associated with price elasticity of consumer demand. Clearly, for a given value of β, equation (1.29) is the perfectly inelastic demand formulation, represented by equation (1.14), when η is $+ \infty$ (disregarding the β parameter). Alternatively, as η tends towards $- \infty$, the term $e^{-\beta\eta}$ tends toward $+ \infty$ and, in so doing, reduces the level of spatial interaction to zero.

In fact, the modifications in consumers' behavior, specifically their demand, is not only related to distance. The concept of accessibility is much more pertinent because it reflects both the effect of transport costs and the related in situ attractiveness of centers.

Because consumer behavior is affected not only by distance but also by the in situ attractiveness of centers, an additional formulation is proposed, attempting to portray the effect of both distance and the relative attractiveness of centers on the level of demand. This formulation can be formally written as

$$S_{ij} = O_i e^{-\alpha \log \sum_{j=1}^{p} W_j e^{-\beta c_{ij}}} \frac{W_j e^{-\beta c_{ij}}}{\sum_{j=1}^{p} W_j e^{-\beta c_{ij}}} \qquad (1.30)$$

In equation (1.30), actual level of demand is related explicitly to the log-accessibility of the centers to a consumer (see Hansen, 1959) and α is a scaling parameter that relates to elasticity of demand. When α equals zero, the spatial configuration of centers is identical to the one generated by equation (1.14), a characteristic that depends on the value of the dispersion parameter, β. Thus, with the incorporation of elastic travel demand, trips converge to zero at the limit of α, rather than necessarily to the pattern of consumers going to their nearest center.

In relation to this discussion about consumers' demand, it is appropriate to allude to the conventional concepts of neoclassical economics: the substitution and income effects. For example, *ceteris paribus,* if the distance to all the centers increased in identical proportions for consumers in one zone, the pattern of trip distribution would remain the same but the level of effective demand would decline—the income effect. At another extreme, if a large new center was located in zone i, consumers in zone i would obviously alter their patterns of patronage because of its relative attractiveness in comparison to its competing centers—the substitution effect. In practice, when assessing overall changes in the pattern of travel demand, it is important to explicitly distinguish both these effects.

For completeness, following Neuberger's (1971) approach to consumers' surplus, a locational surplus objective function, which is associated with the above travel demand model, can be derived as follows:

$$\text{Max } LS = \frac{1}{\alpha\beta} \sum_{i=1}^{n} O_i e^{\alpha \log \sum_{j=1}^{p} W_j e^{-\beta c_{ij}}} \tag{1.31}$$

Although the simulation of individual components to determine aggregate consumer surplus is a very useful property, such welfare functions should take account of distributional issues. Arguably, it might be more appropriate to apply an objective function concerned with the equalization of consumers' accessibility to centers. With regard to this spatial-equity issue, attention is drawn to the family of so-called center problems, which minimize the maximum of the weighted distances between supply and demand points. It should also be remembered that models formulated to maximize aggregate demand usually fail to take account of this important distributional issue.

Finally, and analogous to the reflection of elastic travel demand through the incorporation of accessibility to all destinations in the definition of an origin's level of trip generation, it would be possible to consider aspects of multipurpose trips by extending the definition of in situ attractions (see also Fotheringham's, 1983, recent proposal for a new set of spatial-interaction models based on the theory of competing destinations). Specifically, if the in situ attractiveness of j is extended to indicate the center's accessibility to other centers, it should be possible to consider some multipurpose-trip effects (see McLafferty and Ghosh, chap. 2 herein).

Maximizing Demand

As the location of centers affects the overall level of demand, profit is not independent of spatial structure. Maximizing the demand for a particular center means that centers compete with each other rather than cooperate as in the common p-median problem. The profit maximizing behavior is presented in a game theoretical mould by individual centers attempting to position themselves at a location that achieves for them as extensive a market as is tenable. The unconstrained optimization problem is a continuous-space location-allocation problem which, for the generalized situation with n exogenous discrete demand points and p competing centers, can be formally represented by

$$\text{Max } D_j = \sum_{i=1}^{n} S_{ij} \tag{1.32}$$

where D_j is the level of demand for a center located at j. This general format can, obviously, be used for a number of the preceding price elastic formulations. In numerical experiments with perfectly inelastic demand, Beaumont (1980) replicates the well-known result of spatial clustering when spatial competition exists (see also Hotelling, 1929; Smithies, 1941). However, when elasticity of demand is introduced, ". . . spatial clustering is an exception rather than the rule" (Ghosh and Craig, 1983, p. 20). In a similar way, although for a location problem on a network, Holmes, Williams, and Brown (1972) proposed a model to maximize the utilization of centers. The model incorporates Christaller's concept of an upper range, S, beyond which demand is zero. Formally, the demand function for node i using center j, V_{ij}, is stated as

$$V_{ij} = \begin{cases} O_i - md_{ij} & d_{ij} < S \\ O & d_{ij} \geqslant S \end{cases} \tag{1.33}$$

where O_i is the level of demand if a center is located at i; m is a parameter reflecting the decline in demand per unit of distance; and d_{ij} is the distance between node i and node j. Following the structure of a p-median problem, the objective function of this model can be formulated as

$$\text{Max } D = \sum_{i=1}^{n} \sum_{j=1}^{n} V_{ij} \lambda_{ij} \tag{1.34}$$

In examining similarities and differences among models, it is important to note what ReVelle and Schilling (1975) demonstrated: if there are sufficient centers in the system to ensure that the nearest center to each demand node is within a distance of S units (that is, that the demand level does not decrease to zero for any demand node), the optimal pattern for this model is identical to that derived by the more conventional p-median problem.

LOCATION, SIZE, AND NUMBER OF CENTERS

The original objective of Christaller's 1966 study was to find ". . . a general explanation for the sizes, number and distribution of towns" (p. 2). In all the location-allocation formulations discussed so far, the number of centers is an exogenous input. This subsection presents a number of models that determine the number of centers endogenously.

In the first model, total system costs, TSC (the sum of travel and operating costs), is analyzed in two stages (see also Dokmeci, 1973). First, given n demand points and their level of demand O_i, the p-median problem (or one of its derivative models) is solved for various values of p. This solution determines the location and the necessary size of the centers, W_j (assuming a direct relationship between demand and size). In addition, it generates the value of the minimum aggregate transport costs, ATC. The feasible sets of locations for different values of p are thus determined, and in the second stage of the analysis, center size is introduced explicitly on the supply side. A simple assumption is that fixed and marginal costs of providing a service are spatially uniform and are determined by the size of the center. For instance, total costs of supply for a center located at TCS_j^*, can be written as

$$TCS_j^* = a + bW_j \tag{1.35}$$

where a is the fixed cost and b is a constant marginal cost parameter. Alternatively, functions reflecting both economies and diseconomies of scale can be used. The total cost of supply, TCS, for a system of p centers is therefore represented by

$$TCS = pa + b \sum_{j=1}^{p} W_j \tag{1.36}$$

A system's total cost for p centers is simply the total travel cost plus the total supply cost. The optimal number of facilities (and their locations and sizes) is found by selecting the value of p that minimizes total system costs, that is,

$$\underset{\{p\}}{\text{Min}}\, TSC = ATC + TCS \tag{1.37}$$

The simplicity of the model is attractive, and the framework permits various formulations. Separation of the two facets of behavior—consumer behavior and supplier behavior—is especially useful; however, it goes without saying that the interrelationship between supply and demand is fundamentally significant.

A second model relates to White and Case's (1974) discussion of the total covering problem which, like the "marketing principle," minimizes the number of centers required to completely satisfy the demand of a set of consumers. Formally, this model can be represented as a zero-one, uncapacitated programming problem,

$$\text{Min}\, N = \sum_{j=1}^{p} \ell_j \tag{1.38}$$

subject to

$$\sum_{j=1}^{p} a_{ij}\ell_j \geq 1 \qquad i = 1, \ldots, n \tag{1.39}$$

$$a_{ij};\, \ell_j = 0 \text{ or } 1 \qquad i = 1, \ldots, n;\, j = 1, \ldots, p \tag{1.40}$$

where ℓ_j is a binary variable set equal to one if a center is assigned to location j, and to zero otherwise. Similarly, the covering coefficient, a_{ij}, has a value of one if consumers at location i are allocated to a center at location j, and it has a value of zero otherwise.

Both these models assume that the centers are uncapacitated, that is, that there is no upper limit to their possible size. Therefore, it is likely that the optimal pattern will have an allocation of demand that results in large variation in the size of the centers. In certain circumstances, such as the provision of standardized facilities (Beaumont and Sixsmith, 1984), this situation is inefficient and additional constraints must be introduced to overcome the problem.

In conjunction with central place theory, and specifically with the concept of a threshold or lower capacity, it is interesting that when examining central place theory in a location-allocation framework, Rushton (1972) applied the threshold of a service to determine the number of centers (see also Goodchild and Massam, 1969; Morrill and Kelley, 1970). More satisfactory is the endogenous determination of the threshold level for profitable supply, which is found in the spatial-allocation model proposed by Beaumont and Clarke (1980). It is also possible to add a constraint to the location-allocation models that explicitly requires the minimum center size to be above an exogenously given level, and such a formulation is considered now (see also Puryear, 1975; Kohsaka, 1983).

To date, attention has focused primarily on the demand side of location-allocation models, and clearly, the very essence of the problems relates to the complex interactions between supply and demand over space. By modifying Leonardi's (1981) so-called accessibility-congestion sensitive allocation model, Fig. 1.2 presents the structure of a capacity-constrained location-allocation model. While there are often benefits for consumers in patronizing large centers, disadvantages do occur if large numbers of other consumers use the same center—a congestion effect.

Inputs into the model include the location of all possible sites; the upper capacity of the centers, U; the lower capacity (threshold) constraint, M; and a parameter, σ, which reflects the attractiveness of unused capacity. The exogenous minimum size constraint is necessary; otherwise all the original feasible sites would probably have a small allocation of customers. Anyway, in practice, a simulation procedure (see Goodchild, 1978; Rushton, McLafferty, and Ghosh, 1981) that embeds an indication of a relative lack of congestion as a measure of attractiveness in the allocation process can be used to generate a pattern of centers from the set of possible sites. At the beginning of the simulation, for convenience, it would be assumed that all the possible sites were equally attractive (unless additional information was available to indicate otherwise). Thus it is unnecessary to give the number of centers required; this characteristic and their location are given as part of the results. Note that although the iterative structure may give an impression of pseudodynamics, this formulation is static; it produces

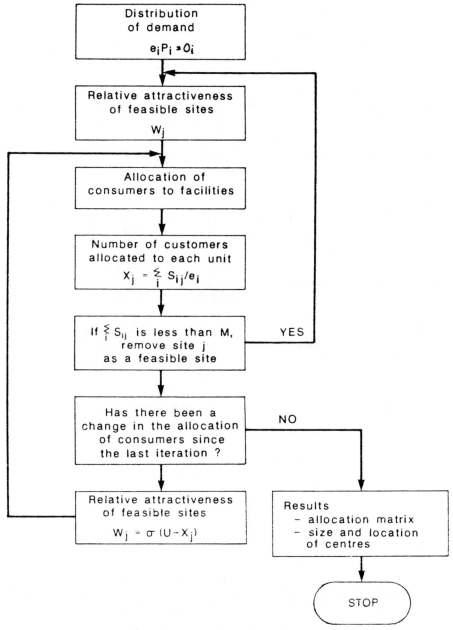

Figure 1.2. A capacity-constrained, location-allocation model.

a pattern of centers that is consistent with the level and distribution of demand at one point in time.

In contrast to using the threshold as a basis for determining the number of centers, it is also possible, by a simple extension of the p-median problem, to use the concept of the range. The solution of the p-median problem allocates consumers to their nearest center, but the distances that they must travel often vary greatly. To insert an element of equity into the formulation, a constraint on the maximum distance between a consumer and his or her nearest center can be introduced. This addition results in the p-median problem with a maximum distance constraint (Khumawala, 1973; Hillsman and Rushton, 1975). However, the way this problem is specified, p centers would not always be able to cater to all the customers within the maximum distance constraint. Consequently, an alternative formulation, the *location-set covering* problem, has been proposed (Toregas and ReVelle, 1972). This formulation involves determining the minimum number and the optimal location of centers so that the aggregate weighted distance is minimized subject to a maximum distance constraint. Therefore, the formulation is very similar to Christaller's marketing principle. Moreover, the maximum distance constraint can be thought to represent a service's range— the maximum distance a consumer will travel to patronize a center. Formally, in the critical distance, or range of a service, one center must be located within d^{crit} distance units of each demand location. For each demand point i, a set of possible center locations, N_i, that satisfy the maximum distance constraints can be defined as

$$N_i = \{j | c_{ij} \le d^{\text{crit}}\}; \qquad i = 1, \ldots, n \qquad \textbf{(1.41)}$$

The optimization problem can be stated as

$$\text{Min } ND = \sum_{j=1}^{p} \ell_j \qquad \textbf{(1.42)}$$

subject to

$$\sum_{j \in N_i} \ell_j \ge 1; \quad i = 1, \ldots, n \qquad \textbf{(1.43)}$$

Thus, the location-set covering problem provides the smallest number of centers needed to ensure an exhaustive cover (that is, to ensure that all demand locations have a service center within the critical distance). In terms of central place theory, it directly incorporates the concept of the range. An alternative but related formulation involves determining the largest number of centers, given that a minimum (threshold) center size is satisfied. Using this type of model, which Bahrenberg (1982) terms a "maximum supply dispersion" model, in conjunction with the location set-covering model, it is possible to operationalize the two basic principles of central place theory for the locational planning of the services.

Hierarchical Location-Allocation Models

Hierarchical location-allocation models have received relatively little attention from analysts. Location-allocation models usually take account of interactions between supply and demand at one level but fail to capture the interactions between centers of different orders. This lack of concern for multiple orders is to some extent surprising, because the hierarchical organization of various services is important for their coordinated provision. More specifically, it is the major characteristic that must be addressed if location-allocation models are to be employed to operationalize central place concepts. This section outlines simple hierarchical location-allocation models and suggests some extensions.

The most straightforward way to generate a hierarchical location-allocation model is to superimpose various levels (with different numbers of centers or varying upper distance constraints) on top of each other. For example, both Christaller (1966) and Losch (1954) superimposed the spatial patterns of centers providing each commodity to arrive at the location pattern of multiple goods. Indeed, the successively inclusive principle of a Christallerian hierarchy has been applied in location-allocation modeling (see, for example, Calvo and Marks, 1973; Tien, El-Tell, and Simons, 1983). Some have argued, however, that both Christaller and Losch failed to incorporate important functional dependencies between orders.

The single-level problem has also been extended to provide the basis of hierarchical location-allocation models; Narula and Ogbu's (1979) two-hierarchy location-allocation problem, for instance, reduces to the p-median problem. In discussing the location of emergency medical service vehicles, Daskin and Stern (1981) extend the traditional set covering problem by using hierarchical and multiobjective programming. They formulate a hierarchical objective covering problem to determine the minimum number of vehicles needed to cover all zones while simultaneously maximizing multiple coverage of zones.

An alternative approach to these single one-step models is to formulate a hierarchical problem as a series of single-level p-median problems (see, for example, Fisher and Rushton, 1979), which can be solved from either the top down or the bottom up. Dokmeci's (1973) hierarchical location-allocation model minimizes total system costs (the sum of supply and transport costs) by building hierarchy of centers starting with the lowest order, which contains the most centers, rather than proceeding from the top downward. Although this difference is not really significant, the method of nesting hierarchies is interesting and deserves brief discussion. Given the location of the demand points, the configuration of the first-order centers that minimizes aggregate weighted transport cost is readily determined (as in the p-median problem). To ascertain the locations of the second-order centers, both are original demand locations and the first-order centers' locations are assumed to be demand locations, and ". . . to determine those of third level facilities, the demand, first, and second levels are considered as demand locations" (Dokmeci, 1973, p. 443).

Some argue, however, that Dokmeci's hierarchy is founded on invalid assumptions. In general, a particular order of centers does not itself demand services from a higher-order level; only the original demand points require satisfying. Even if it is suggested that the population associated with the supply of a particular commodity bundle also demands commodities, no mechanism dictates that they demand commodities of the order they are concerned with and all lower orders. (An iterative, converging process, as described in relation to population size, may be appropriate here.)

Obviously, a number of variations can be suggested integrating features of the other nonhierarchical models described in the preceding sections. For instance, it may be deemed appropriate to explicitly take account of the interactions between centers of different orders as well as of the interactions between demand points and centers. That is, one approach would be to extend the p-median problem to include both the aggregate weighted distances between demand and supply points and between supply points of different orders (assuming centers of the same level provide identical services). Formally, given N levels $(k = 1, \ldots, N)$, the problem can be stated as

$$\operatorname*{Min}_{\{s(k)\}} AWC = \sum_{k=1}^{N} \sum_{i=1}^{n} \sum_{j(k)=1}^{p(k)} 0(k)_i A_{ij(k)} C_{ij(k)}$$

$$+ \sum_{k=1}^{N-1} \sum_{j(k)=1}^{p(k)} \sum_{\ell(k+1)=1}^{p(k+1)} V_{j(k)\ell(k+1)} C_{j(k)\ell(k+1)}$$

(1.44)

where there are n demand points and $j(k)$ center of order k. The spatial-allocation component between consumers and centers is $A_{ij(k)}$. The probability of a consumer at i patronizing a center at j, for a good k, could be defined by a conventional spatial-interaction model. The value $V_{k\ell}$ allocates each center of order k to the nearest center of order ℓ. This model per se is not explored in any further detail here, but its characteristics are incorporated in the models described below.

An alternative model is based on a hierarchy of N levels in which the highest order is depicted as level N and the hierarchical structure is built downward as an incremental type of location-allocation problem. Following Christaller (1966) a successively inclusive hierarchy is assumed, although it can be relaxed easily. The first stage can be thought of as a usual single-level location-allocation model that involves locating optimally (according to some specified criterion) a set of $p(N)$ centers of order N in relation to the spatial distribution of n demand points, where $n > p(N)$. Once this problem has been solved, it provides the information for the first incremental problem. Specifically, given the optimal pattern of centers of order N (and the original distribution of demand points), the next stage is to locate optimally a set of $p(N - 1)$ centers of order $N - 1$, where $p(N - 1) > p(N)$. As order $N - 1$ services are provided by the highest-order centers (at least, under the assumption of a successively inclusive hierarchy), it is necessary to locate only $[p(N - 1) - p(N)]$ centers of order $N - 1$, given the location of the centers of order N. As Fig. 1.3 shows schematically, this procedure is repeated until the complete spatial hierarchy is determined.

The foregoing formulations are continuous-space location-allocation problems, and it would be appropriate to consider a similar problem based on a network. In fact, using Hillsman's (1980, p. 9) ALLOC V algorithm, it is possible to compare the effects of either building up or building down a spatial hierarchy for a p-median heuristic for a hierarchical network problem. Constraining some centers at specified locations allows the construction of different spatial hierarchies. Interestingly, "The top-down approach usually gives a different pattern of centers in the hierarchy than does the bottom-up approach" (p. 9).

These preliminary suggestions on hierarchical location-allocation models demonstrate that although much more research is required in this direction, such models offer a framework that permits the operationalization of central place concepts. The various extensions suggested to the nonhierarchical p-median problem, such as distance constraints, capacity constraints, and differential attraction/interaction-based allocation rules, would be a useful starting place. An additional avenue would be to view hierarchical development as a temporal process, incorporating it into a dynamic location-allocation model. This topic is briefly considered in the next section.

DYNAMIC LOCATION-ALLOCATION MODELS

Up to this point, all the formulations have been atemporal. The models generate optimal, static spatial patterns in which all the activity takes place at one instance. This section undertakes a short examination of simple dynamic location-allocation models. This treatment provides a basis from which to discuss hierarchical development over time—evolution of a central place system (Beaumont, 1982b, 1984). Christaller (1966, p. 84) recognized the importance of time as a controlling principle in spatial organization, and Berry (1967) suggested the possible collapse of Christaller's hierarchy over time resulting in areas of specialization.

Scott (1971, 1975) and Sheppard (1974) offer some general structural frameworks for the integration of the spatial and discrete temporal dimensions in center location models. Their dimensions include the timing of establishing

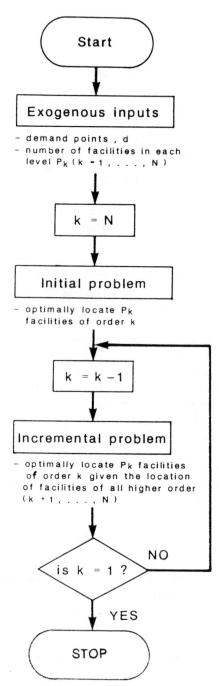

Figure 1.3. A hierarchical location-allocation model: A successively inclusive hierarchy.

new centers. Rao and Rutenberg (1977) present a model with a continuous time dimension (although it is restricted to examining a single center's change of size).

A static formulation is sufficient if neither the level nor the location of demand alters over time. Particularly in the competitive environment of the private sector, a center location found to be optimal at the outset could become undesirable as new, competing centers develop. Consequently, in considering a pattern's short-term attractiveness, it is essential to also consider its ability to protect its position from future competition. Given this potential instability of short-term solutions, Ghosh and Craig (1983) suggest that, using the game theoretic concept of maximin strategies, an apposite objective for a center would be to maximize, over a long-term planning horizon, the discounted net present value of the investment. (Other potentially useful directions to pursue include the literature on adaptive decisions and on decision making under uncertainty.)

If temporal variation does occur, it is possible to extend the static problem. For example, Scott (1975) has applied a modified Weber problem to minimize aggregate weighted transport costs over T time periods, during which time the number (n_t), level (O_{it}), and location (x_{it}, y_{it}) of demand points change. The problem is to locate, in the first time period, one center that takes into account these future variations. Formally, this can be represented as

$$\operatorname*{Min}_{\{x_{jt}, y_{jt}\}} Z = \sum_{t=1}^{T} \sum_{i=1}^{n_t} O_{it} c_{ijt} \tag{1.45}$$

and the subscript t represents the time period. This problem can be solved by a simple heuristic procedure.

Scott (1975) compared the resulting optimal center location for the dynamic Weber problem with those for the static formulation at different time periods. If the locations were greatly different, the center would be likely to relocate at some time, and costs of relocation must be included in the model. Otherwise the problem reduces to a series of static problems. Scott (1975) offers an alternative model in which an existing center may close and a new one open. It is assumed that when the center relocates it incurs a fixed cost, a. For simplicity, it is also assumed that a is independent of time and the distance over which relocation occurs, although of course this assumption is easily relaxed. The problem can be written as

$$\operatorname*{Min}_{\{x_{jt}, y_{jt}\}} Z = a_1 + \sum_{i=1}^{n_1} O_{i1} c_{ij1} + \sum_{t=2}^{T} \left(a\lambda_t + \sum_{i=1}^{n_t} O_{it} c_{ijt} \right) \tag{1.46}$$

subject to

$$\lambda_t \text{ is } 0 \text{ or } 1 \qquad t = 2, \ldots, T \qquad \qquad (1.47)$$

The value a_1 is the cost of establishing the center in the first time period. Constraint (1.47) concerns the binary variable that relates to whether relocation occurs or not: if λ_t equals 1, relocation occurs, and if it is 0, no relocation occurs. Scott (1975) demonstrated how the set of binary variables (λ_t; $t = 1, \ldots, T$) assist in solving the problem through a numerical evaluation of all the possible combinations (although such an evaluation is probably inefficient over a large number of time periods).

Extending aspects of these models allows the replacement of a truly dynamic model by a series of static problems, thus outlining a multilayer approach for developing a central place structure. This approach works on the problem of optimal facility location by dividing the time horizon of interest into different layers, each of which is concerned with a different time interval (see Fig. 1.4). The highest layer covers the longest time interval, and, while a static model would be solved for each layer, the specific models employed in different

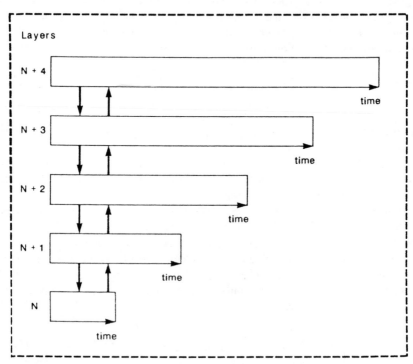

Figure 1.4. A multilayer framework for dynamic location-allocation problems.

layers would probably be different because of the different amounts of available information and uncertainty involved. In this approach's simplest form, the models would be solved for the demand at the end of each time interval a. Consequently, the definition of the layers is fundamentally important. It would be possible, however, to introduce some kind of discounting process to take account of variations within each period. (In such a situation, stochastic location-allocation models would be useful to assess the variability in the optimal solutions.) Although the presented framework is simple, it would be straightforward to extend the feedback links between layers. As it stands, however, each layer is viewed separately as a static location-allocation problem, but the solution of one layer, say $N + 1$, is consistent with an incremental problem based on the optimal solution for the layer N. Particular attention should be given to core center locations; that is, optimal locations that are recurrent solutions for different layers. Simple information feedbacks are also included in Fig. 1.4 to indicate the need to reexamine a lower layer after the results of a higher layer are known. This stance is much more satisfactory than assuming, at least implicitly, that all the pertinent information is known. In fact, in model development, the generation of different results for comparison would likely be especially instructive. A specific problem would have a particular framework; its nature would determine the desirable number of layers and the lengths of the different time intervals.

These dynamic models are rather simple, but they do provide a foundation, like the static p-median problem, from which to generate a variety of extensions and reformulations. One possibility is to include time-dependent constraints, such as maximum distance or maximum capacity constraints, to reflect changes in consumer and producer behavior over time. Alternatively, hierarchical development and structural change could be represented. For instance, hierarchical development could be related to aspects of center relocation and expansion (or contraction), which would be combined with an endogenous, dynamic attractiveness function dependent on changing capacity levels and demand characteristics.

For completeness, a framework for a dynamic, successively inclusive hierarchical location-allocation model can be outlined. The development process is assumed to be primarily incremental in nature, taking account of existing infrastructure. The model suggests no specific optimizing criterion. However, if a consumers' surplus objective function is used, it would be necessary to include changes in benefits related to changes in origin and destination characteristics as well as those related to changes in transport costs. For example, extra trips may be beneficial because of the increased attractiveness/accessibility of some centers, and this benefit should not be considered solely in terms of extra transport costs.

Fig. 1.5 portrays a dynamic, incremental (successively inclusive), hierarchical location-allocation model. Conceptually, given the need to explicitly take account

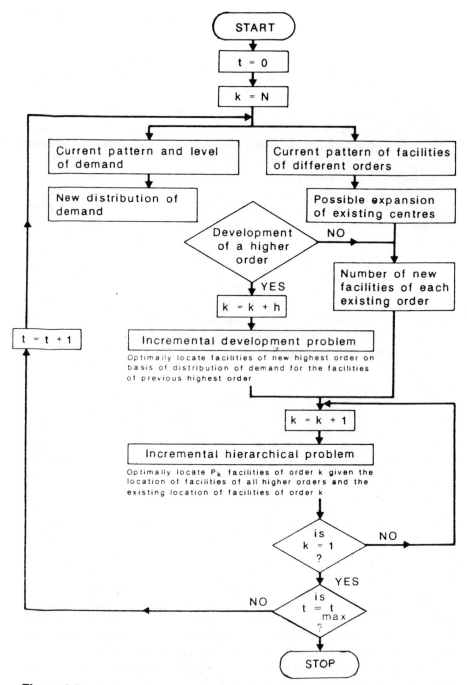

Figure 1.5. A dynamic (successively inclusive) hierarchical location-allocation model.

of existing stock, a benefit of the framework is its explicit and combined representation of present and future distribution of centers of different orders. Indeed,
the present configuration of centers is probably the most important factor determining the future central place structure. Future development is, therefore,
realistically mirrored as part of an ongoing process. Decisions relating to center
extension or emergence must take account not only of the aspects related to
consumers' behavior, such as the level of demand, but also of the aspects
related to producers' behavior, such as operating costs, scale economies or
diseconomies, and capital costs. (Indeed, in general, insufficient attention has
been given to providing a sound economic basis for cost functions; this neglect
is in stark contrast to, say, the analyses of consumers' spatial-choice behavior.)
Assuming an elastic demand formulation, it would be possible by successively
analyzing problems with different numbers of centers to consider the impact
of center attractiveness, congestion, and accessibility on the overall level of
utilization. Particular attention should be given in these various analyses to
whether there is a common core of optimal locations.

Such a framework is, therefore, very flexible and, although it is necessary
to specify clearly the actual formulation required in particular situations, could
be operationalized easily. The framework's position in relation to the planning
process has been examined elsewhere (Beaumont, 1983), although the solutions
may not be optimal at all time periods.

REFERENCES

Beaumont, J. R., 1980, Spatial interaction models and the location-allocation problem, *Journal of Regional Science* **20(1)**:37–50.
Beaumont, J. R., 1981, Location-allocation problems in a plane: a review of some models, *Socio-Economic Planning Sciences* **15(5)**:217–229.
Beaumont, J. R., 1982*a*, Mathematical programming in human geography, *Mathematical Social Sciences* **2**:213–243.
Beaumont, J. R., 1982*b*, Towards a conceptualisation of evolution in environmental systems, *International Journal of Man-Machine Studies* **16**:113–145.
Beaumont, J. R., 1983, Characteristics of a comprehensive framework for location-allocation models, *Sistemi Urbani* **2(3)**:349–364.
Beaumont, J. R., 1984, A description of structural change in a central place system: a speculation using *Q*-analysis, *International Journal of Man-Machine Studies* **20**:567–594.
Beaumont, J. R., and M. C. Clarke, 1980, improving supply side representations in urban models with special reference to central place and Lowry models, *Sistemi Urbani* **2(1)**:3–12.
Beaumont, J. R. and A. J. Sixsmith, 1984, Elderly severely mentally infirm (ESMI) units in Lancashire: an assessment of resource allocation over space, in *Planning and Analysis in Health Care Systems*, M. Clarke, ed., Pion, London, pp. 163–193.
Beavon, K. S. O., 1977, *Central Place Theory: A Reinterpretation*, Longman, London.

Berry, B. J. L., 1967, *Geography of market centers and retail distribution*, Prentice-Hall, New Jersey.

Berry, B. J. L., and W. L. Garrison, 1958, Recent developments of central place theory, *Regional Science Association Papers and Proceedings* **4:**107–120.

Calvo, A. B. and D. H. Marks, 1973, Location of health care facilities: an analytical approach, *Socio-Economic Planning Sciences* **7:**407–422.

Christaller, W., 1966, *Central Places in Southern Germany*, Prentice-Hall, New Jersey.

Coehlo, J. D., and H. C. W. L. Williams, 1978, On the design of land use plans through locational surplus maximisation, *Regional Science Association Papers* **40:**71–85.

Cooper, L., 1964, Heuristic methods for location-allocation problems, *SIAM Review* **6:**37–52.

Cooper, L., 1967, Solutions of generalised equilibrium models, *Journal of Regional Science* **7:**1–18.

Daskin, M. S., and E. H. Stern, 1981, A hierarchical objective set covering model for emergency medical service vehicle deployment, *Transportation Science* **15:**137–152.

Dokmeci, V. F., 1973, An optimisation model for a hierarchical spatial system, *Journal of Regional Science* **13:**439–451.

Fisher, H. B. and G. Rushton, 1979, Spatial efficiency of service locations and the regional development process, *Papers of the Regional Science Association Papers* **42:**83–97.

Fotheringham, A. S., 1983, A new set of spatial interaction models: the theory of competing destinations, *Environment and Planning, A* **15:**15–36.

Friedrich, C. J., 1929, *Alfred Weber's Theory of the Location of Industries*, University of Chicago Press, Chicago.

Getis, A. and J. Getis, 1966, Christaller's central place theory, *Journal of Geography* **65:**200–226

Ghosh, A. and C. S. Craig, 1983, Facility planning in competitive environments: a location-allocation approach, Graduate School of Business Administration, New York University, mimeo.

Goodchild, M. F., 1978, Spatial choice in location-allocation problems: the role of endogenous attraction, *Geographical Analysis* **10:**65–72.

Goodchild, M. F., and B. H. Massam, 1969, Some least cost models of spatial administrative systems in Southern Ontario, *Geografiska Annaler* **51B:**86–94.

Hakimi, L. S., 1964, Optimum locations of switching centers and the absolute centers and medians of a graph, *Operations Research* **12:**450–459.

Hansen, W. G., 1959, How accessibility shapes land use, *American Institute of Planners Journal* **25:**73–76.

Harris, A. J. and J. C. Tanner, 1974, Transport demand models based on personal characteristics, *TRRL Supplementary Report* **65UC**, TRRL, Crowthorne.

Harris, B., and A. G. Wilson, 1978, Equilibrium values and dynamics of attractiveness terms in production-constrained spatial interaction models, *Environment and Planning, A* **10:**371–388.

Hillsman, E., 1980, *Heuristic Solutions to Location-Allocation Problems*, Monograph 7, Department of Geography, University of Iowa, Iowa city.

Hillsman, E. and G. Rushton, 1975, The p-median problem with maximum distance constraints: a comment, *Geographical Analysis* **7**:85–90.

Hodgson, M. J., 1978, Towards more realistic allocation in location-allocation models: an interaction approach, *Environment and Planning, A* **10**:1273–1285.

Holmes, J., F. Williams, and L. Brown, 1972, Facility location under a maximum travel restriction: an example using day care facilities, *Geographical Analysis* **4**:258–266.

Hotelling, H., 1929, Stability in competition, *Economic Journal* **39**:41–57.

Khumawala, B. M., 1973, An efficient algorithm for the p-median with maximum distance constraints, *Geographical Analysis* **5**:309–321.

Kohsaka, H., 1983, A central-place model as a two-level location-allocation system, *Environment and Planning, A* **15**:5–14.

Kuenne, R. E., and R. M. Soland, 1972, Exact and approximate solutions to the multi-source Weber problem, *Mathematical Programming* **3**:193–209.

Leonardi, G., 1978, Optimum facility by accessibility maximising, *Environment and Planning, A* **10**:1287–1306.

Leonardi, G., 1981, A unifying framework for public facility location problems—part 1: a critical overview and some unsolved problems, *Environment and Planning, A.* **13**:1001–1028.

Losch, A., 1954, *The Economics of Location,* Yale University Press, Connecticut.

Morill, R. L. and M. B. Kelly, 1970, The simulation of hospital use and estimation of location efficiency, *Geographical Analysis* **2**:283–300.

Narula, S. C. and U. I. Ogbu, 1979, An hierarchical location-allocation problem, *Omega* **7**:137–143.

Neuberger, H. L. T., 1971, User benefit in the evaluation of transport and land use plans, *Journal of Transport Economics and Policy* **5**:52–75.

Ostresh, L. M., 1973, Exact solutions to the m-centre location-allocation problem, in *Computer Programs for Location-Allocation Problems,* G. Rushton, M. F. Goodchild, and L. M. Ostresh, eds., Monograph 6, Department of Geography, University of Iowa, Iowa City.

Parr, J. B., 1979, Regional economic change and regional spatial structure: some interrelationships, *Environment and Planning, A.* **11**:825–837.

Puryear, D., 1975, A programming model of central place theory, *Journal of Regional Science* **15**:307–316.

Rao, R. C. and D. P. Rutenberg, 1977, Multilocation plant sizing and timing, *Management Science* **23**:1187–1198.

Revelle, C., R. Church, and D. Schilling, 1975, A note on the location model of Holmes, Williams and Brown, *Geographical Analysis* **7**:457–459.

Rushton, G., 1972, Map transformations of point patterns: central place patterns in areas of variable population density, *Regional Science Association Papers* **28**:111–129.

Rushton, G., S. L. McLafferty, and A. Ghosh, 1981, Equilibrium locations for public service: individual preferences and social choice, *Geographical Analysis* **13**:196–202.

Saey, P., 1973, Three fallacies in the literature on central place theory, *Tijdschrift voor Economische en Sociale Geografie* **64**:181–194.

Scott, A. J., 1971, Dynamic location-allocation systems: some basic planning strategies, *Environment and Planning* **3**:73–82.

Scott, A. J., 1975, Discrete dynamic locational systems: notes on a structural framework, in *Dynamic Allocation of Urban Space,* A. Karlquist, L. Lundquist, and F. Snickars, eds., Saxon House, Farnborough, pp. 121–158.

Sheppard, E. S., 1974, A conceptual framework for dynamic location-allocation analysis, *Environment and Planning, A* **6**:547–564.

Sheppard, E. S., 1980, Location and the demand for travel, *Geographical Analysis* **12**:111–127.

Sheppard, E. S., 1981, Public facility location with elastic demand: users' benefits and redistribution issues, *Sistemi Urbani* **3**:435–454.

Smithies, A., 1941, Optimum location in spatial competition, *Journal of Political Economy* **49**:423–459.

Tien, J. M., E. El-Tell, and G. R. Simons, 1983, Improved formulations to the hierarchical health facility location-allocation problem, *IEEE Transactions on Systems, Man and Cybernetics* **SMC-13**:1128–1132.

Toregas, C. and C. ReVelle, 1972, Optimal location under time or distance constraints, *Regional Science Association Papers and Proceedings* **28**:133–143.

White, J. A. and K. E. Case, 1974, On covering problems and the central facilities location problem, *Geographical Analysis* **6**:281–293.

Williams, H. C. W. L., 1976, Travel demand models, duality relations and user benefit analysis, *Journal of Regional Science* **16**(2):147–166.

Williams, H. C. W. L. and M. L. Senior, 1977, A retail location model with overlapping market areas: Hotelling's problem revisited, *Urban Studies* **14**:203–205.

Williams, H. C. W. L. and M. L. Senior, 1978, Accessibility, spatial interaction and the evaluation of land use transportation plans, in *Spatial Interaction Theory and Planning Models,* A. Karlquist, L. Lundquist, F. Snickars, and J. W. Weibull, eds., North-Holland, Amsterdam, pp. 253–287.

2

OPTIMAL LOCATION AND ALLOCATION WITH MULTIPURPOSE SHOPPING

Sara McLafferty
Columbia University

Avijit Ghosh
New York University

Since their introduction in the 1960s, location-allocation procedures and their application to facility location problems have attracted growing interest. These models allow systematic evaluation of a large number of possible locational configurations in terms of some stated criteria, and this feature makes them uniquely suitable for spatial planning. Although the majority of applications of location-allocation models to date have been in planning situations, such models also have a great and largely untapped potential in the development of theories of spatial organization. In particular, insofar as both location-allocation models and central place theory search for facility locations that best serve a dispersed population, the connections between these two research streams are potentially fruitful. Yet only a few recent studies have used location-allocation modeling to develop and test alternative central place theories (Beaumont, 1980; Kohsaka, 1983, 1984; Puryear, 1975).

This chapter develops a location-allocation model to study the spatial organization of retail firms in a multipurpose-trip environment. Classic theories of location start with the postulate that consumers make only single-purpose trips to the nearest store. A number of recent papers, however, have drawn attention to the phenomenon of multipurpose shopping, wherein consumers bypass their closest outlet and visit more distant ones to buy two or more goods at the same place. Since multipurpose shopping is known to be common in practice,

Acknowledgment: The authors wish to thank two anonymous reviewers for their comments on an earlier draft of this paper. Financial support provided by NSF Grant Number SES 8301010 is gratefully acknowledged.

its effect on the growth and development of retail centers deserves attention. The first section of this chapter postulates a model of consumer behavior that determines both the optimal propensity for single- and multipurpose shopping and the frequency of trips made by individual consumers. This model becomes an integral part of the location-allocation procedure developed in the chapter's second section. With respect to retail firms, the location-allocation model finds the locations that minimize consumers' total shopping costs. The result is a two-stage iterative location model that simultaneously determines the optimal shopping itinerary and travel pattern for each consumer and the optimal locations for retail firms. The location-allocation model is used to study the nature and properties of optimal location patterns in multipurpose-trip environments. The chapter's third section develops a simulation procedure to study the impact of changes in price and in cost of operations on the locational configuration of retail firms.

Our ultimate goal is to improve upon classic theories of spatial organization by introducing more realistic postulates of consumer behavior. In this aim the study is similar to a number of previous ones that have also attempted to incorporate more realistic assumptions into classic central place theory (see, for example, Curry, 1967; Eaton and Lipsey, 1982; Kohsaka, 1984; Rushton, 1971; Mulligan, 1983, 1984; White, 1977). Rushton (1971) and White (1977) analyzed the effect of a gravity-type spatial-interaction model on the size and spacing of central places. More recent studies by Eaton and Lipsey (1982), Mulligan (1983), O'Kelly (1983), and Kohsaka (1984) explicitly consider multipurpose travel and its impact on central place systems. These studies indicate that multipurpose shopping increases the revenues and expands the market areas for lower-order firms that agglomerate with higher-order sellers. As a result, the spatial organization of firms may be different from that proposed in classic central place theories such as Christaller, (1966) and Losch, (1954).

TRIP PATTERNS AND THE FREQUENCY OF TRIPS

Underlying any theory of spatial organization are postulates regarding the behavior of consumers and producers. The theories demonstrate the locational consequences of those postulates (Rushton, 1971). A theory of location should consider both the spatial and the temporal aspects of consumer behavior. As noted by Curry (1967), "It is evident in the shopping problem that activities operating in the time domain such as inventory management must have spatial implications and cannot be divorced from geographic analysis" (p. 222). In the past, researchers have devoted much attention to the spatial aspects of shopping behavior while rather neglecting the temporal dimension. The present model considers both the spatial and temporal dimensions. Only by considering both simultaneously is it possible to fully understand consumer spatial behavior.

To incorporate the temporal dimension of shopping in our consumer model, we give explicit consideration to three components of consumer shopping costs: (a) cost of travel, (b) cost of the goods, and (c) cost of holding inventory. Together these costs determine the spatial pattern of shopping trips as well as the frequency of such trips. Although the effect of transportation cost on consumer travel is well recognized in location theory, the impact of inventory cost has been largely ignored. Curry (1967) acknowledged the importance of such costs but concentrated only on producers' inventory. An early study by Reinhardt (1973) and more recent studies by Bacon (1983) and Lentnek, Harwitz, and Narula (1981) also note the effect of inventory costs on consumer shopping decisions. It is costly to store goods, owing to the opportunity cost of holding stock as well as the cost of physical storage. The risk of perishability also increases with storage and imposes an additional cost on the consumer. Although the different components of inventory cost can be identified separately, we refer to inventory cost as the sum of all costs associated with storing goods. Our proposed consumer model postulates that consumers are rational, and that when faced with an array of shopping tasks, they choose the shopping itinerary that minimizes total shopping costs—the sum of travel costs, cost of goods, and inventory costs. The attention to total shopping costs allows us to determine each consumer's shopping frequency for different goods, the travel pattern, and the mix of single- and multipurpose trips.

The explicit consideration of multipurpose travel is another important feature of our model. The importance of multipurpose trips in consumer shopping is now well established. Empirical studies have found that between 30% and 50% of all shopping trips are multipurpose (Hanson 1980; O'Kelly 1981). Such shopping trips are a rational behavior pattern that reduces the time and cost of travel. Although the possibility of multipurpose trips by consumers has long been recognized, only recently has the effect of such trips on spatial organization been investigated (Eaton and Lipsey, 1982; Mulligan, 1983, 1984; O'Kelly, 1983; Kohsaka, 1984). Eaton and Lipsey (1982) argue that multipurpose shopping provides a fundamental economic rationale for the agglomeration of low- and high-order firms in a central place system. For a linear market, they show that the presence of multipurpose shopping enhances the revenue of low-order firms located at high-order places. There is, therefore, an incentive for low-order firms to agglomerate with the higher-order firms. Thus Eaton and Lipsey are able to deduce the hierarchical structure of central place systems, rather than assuming it, as Christaller and Losch did.

McLafferty (1982) presents a normative facility location model that incorporates multipurpose shopping. The model reduces consumer travel by locating facilities so as to increase the potential for multipurpose trips. In this model, as well as in Eaton and Lipsey (1982), individual consumers' propensities for multipurpose shopping are determined based on the frequencies of single-purpose trips for various goods. Such a procedure is valid as long as shopping frequencies are determined by the perishability of goods or by restrictions on supply (e.g., when goods are provided by itinerant traders). In general, however, the shopping frequency that is optimal when only single-purpose trips are considered may not be so when multipurpose trips are allowed. The opportunity for multipurpose shopping may result in radical changes in shopping patterns as consumers adjust the frequency of travel to reduce shopping cost.

Mulligan (1984) analyzes the impact of multipurpose shopping on the population sizes of the urban centers. His study uses the shopping model developed earlier by the author (Mulligan, 1983) and, to generate consumer trip patterns, specifies a set of multipurpose-shopping propensities at different levels of the central place hierarchy. Using numerical simulations Mulligan shows that multipurpose shopping leads to differential growth of centers at different levels of the hierarchy, and that "different types of central places on the same level of the hierarchy have different populations" (p. 53). One drawback of the study, however, is that the multipurpose-shopping propensities are specified exogenous to the model. As the author himself notes, "It would be more elegant to identify the multipurpose shopping propensities on an endogenous basis" (p. 54).

In a recent paper, Kohsaka (1984) studies the effect of multipurpose-shopping trips on a central place system. He develops a demand-generation model in which the probability that demand for a good will occur within a time period follows a Poisson distribution and the interval of demand fits a negative exponential distribution. Based on these assumptions Kohsaka deduces the maximum and minimum levels of multipurpose shopping. These levels are derived from the parameters of the Poisson and negative exponential distributions, which are spatially invariant. One would expect, however, that these parameters would differ as a function of the consumer's location and available shopping opportunities.

O'Kelly (1983) models the impact of multipurpose shopping on the size of retail facilities. The model incorporates a spatial-interaction function to determine the effect of facility size on destination choice. Since facility size is endogenous to the model, the trip pattern is related to the spatial structure, and the propensity for multipurpose shopping varies spatially. The focus of O'Kelly's study, however, is on multistop shopping trips in which consumers visit a number of locations on a single trip. Further, the locations of facilities are assumed to be fixed, and only their sizes are allowed to vary. (The contextual demand model of Lentnek, Harwitz, and Narula, 1981, also considers multistop trips.)

Previous studies of the impact of multipurpose trips on spatial organization have made important theoretical and methodological advances in analyzing central place systems, and such studies provide a motivation for the research described here. Our point of departure, however, is the specification of the consumer model that directly links spatial structure and multipurpose-trip behavior. In an earlier paper (Ghosh and McLafferty, 1984) we showed that the rate of multipurpose shopping varies with the consumer's location relative to shopping opportunities. In this chapter, we analyze the implications of such a finding for the spatial organization of firms. Consumers in our model act to minimize total shopping cost by choosing the optimal frequency of single- and multipurpose trips and the appropriate amount of goods to buy during each type of trip. These choices involve consideration of storage costs, cost of travel, and the price of goods at different places. The details of the model are presented next.

THE SHOPPING MODEL

For expositional clarity, we initially consider a two-level retail hierarchy. The extension to a multilevel system is discussed briefly later. The following assumptions are made regarding the behavior of the consumers and the nature of the spatial system:

1. There is a planning horizon during which consumers demand fixed amounts of two goods j and k. While both the length of the planning horizon and amounts of goods demanded during this period may vary among consumers, for simplicity we assume that each consumer purchases D_j units of good j and D_k units of good k during the planning horizon. Good k is a higher-order good ($D_k < D_j$). We consider the goods to be essential to the consumers, so that the consumers purchase the same quantity of the good irrespective of the price. Thus demands for the goods are completely inelastic. The consumer's goal is to minimize the total shopping cost during the planning horizon.

2. Place j is a low-order center that sells good j. Place k, on the other hand, is a high-order center that offers both good j and good k.

3. T_j is the cost of a round trip to the nearest j place, and T_k is the cost of a round trip to the nearest k place.

4. Each good is consumed at a constant rate through time. We ignore the possibility of random fluctuations in consumption rates. This stance is similar to the assumption of constant stock depletion adopted by Lentnek, Harwitz, and Narula (1981).

5. The cost of holding inventory, $S\{\cdot\}$, is a monotonically increasing function of the monetary value of the average level of the stock held by the consumer. Thus, an inventory cost of 25% implies that the cost of holding a unit in inventory for the entire planning horizon is 25% of the purchase price of the unit. By convention, $S\{0\} = 0$ and there is an upper limit to the amount of good that can be stored at any time.

6. P_{jj} is the mill price of good j at place j, and P_{jk} is the price of the same good at place k. P_{kk} is the price of good k at place k.

Since we assume that consumers minimize total shopping costs, if there are no restrictions on the amount of goods that can be bought on a single trip and there are no inventory holding costs, a consumer will purchase all the required goods on a single trip. When inventory costs are significant, however, a rational consumer will trade off the cost of transportation with inventory holding cost. This trade-off determines the frequency of trips and the amount of goods bought on each trip.

Initially, consider the case when only single-purpose trips are allowed. Let N_j and N_k be the number of trips made to places j and k respectively and P_{jj} and P_{kk} the prices of goods j and k, respectively. Since each good is consumed at a constant rate through time, the average inventory of good j is $D_j/2N_j$, and that of good k is $D_k/2N_k$. The total shopping cost can then be written as

$$TC = N_jT_j + S\{(D_j/2N_j)P_j\} + P_jD_j$$
$$+ N_kT_k + S\{(D_k/2N_k)P_k\} + P_kD_k \qquad (2.1)$$

In this equation total transportation costs increase with the number of trips and storage costs decrease correspondingly. If the mill prices of the goods are spatially stationary, the consumer will trade off transportation costs with storage costs and choose the optimal number of trips that minimizes total shopping costs. The optimal values of N_j and N_k, obtained by minimizing equation (2.1), determine the optimal frequencies of trips and the amount of goods bought on each trip. Equation (2.1) thus provides a basic model of consumer shopping.

Now if there is a seller of good j at place k whose mill price is less than the delivered price at place j, there will always be the possibility of multipurpose shopping (Mulligan, 1983). We define x as the proportion of good j that is bought on multipurpose trips to place k ($0 < x < 1$). The optimal value of x depends on the relationship between the cost of goods, storage costs, and the relative costs of transportation to places of different order. Since transportation costs depend on an individual's location in relation to shopping opportunities, the value of x will be spatially nonstationary. To determine the value of x we develop the multipurpose-shopping model.

Shopping and consumption patterns for nonimpulse items tend to follow a regular pattern. On each visit to a particular type of place, therefore, equal amounts of goods are purchased. That is, if xD_j and D_k units of goods j and k, respectively, are bought on N_k trips to place k, then (xD_j/N_k) units of good j and (D_k/N_k) units of good k are purchased on each trip. Similarly, the amount of good j bought on each trip to place j is $[(1 - x)D_j/N_j]$. The average inventory of good j, therefore, is $(xD_j/2N_k)$ units during x portion of the planning horizon and $[(1 - x)D_j/2N_j]$ units during the rest of the planning horizon. The average inventory of good k is $(D_k/2N_k)$. The total shopping cost is

$$TC = N_jT_j + S\{(1-x)(D_j/2N_j)P_{jj}\}(1-x) + (1-x)P_{jj}D_j$$
$$+ N_kT_k + S\{x(D_j/2N_k)P_{jk}\}x + P_{jk}D_j$$
$$+ S\{(D_k/2N_k)P_{kk}\} + P_{kk}D_k \qquad (2.2)$$

The integers N_j and N_k are nonnegative. The parameter x determines the rate of multipurpose shopping. Equation (2.2) is minimized over values of N_j, N_k, and x to determine the optimal shopping strategy for each consumer. For the general case where there are m types of goods, the multipurpose-shopping model can be written as follows:

$$\text{Min } TC = \sum_{j=1}^{m} N_j T_j + \sum_{j=1}^{m} \left(\sum_{k=j}^{m} X_{jk} D_j P_{jk} \right)$$
$$+ \sum_{j=1}^{m} \left[\sum_{k=j}^{m} S \left(\frac{x_{jk}D_j}{2N_k} x_{jk} P_{jk} \right) \right]$$

subject to $\qquad\qquad\qquad\qquad\qquad\qquad\qquad\qquad\qquad\qquad$ (2.3)

$$\sum_{k=1}^{m} x_{jk} = 1 \quad \text{for all } j$$

$aN_j = N_k \quad$ for $k = j + 1, \ldots, m$ and all j
N_j, N_k, a are all nonnegative integers

The nature of consumer demand generation and trip patterns is critical to determining the optimal location of firms serving a spatially dispersed population. As Papageorgiou and Brummel (1975) note, the relationship between trip patterns and purchase frequencies is central in explaining shopping behavior. We use the shopping model to determine the optimal frequency of trips and the rate of multipurpose shopping, rather than specifying them exogenously. Thus, we make explicit why observed regularities in shopping behavior arise and provide a model for predicting such behavior. In contrast to earlier studies, the rate of multipurpose shopping varies spatially as a function of the consumer's location relative to alternative shopping opportunities.

THE LOCATION-ALLOCATION MODEL

To study the spatial organization of firms operating in multipurpose-shopping environments, this section develops a location-allocation procedure in which the shopping model presented in the previous section is used to determine consumer patronage of alternative locations. Thus, embedded in the location-allocation procedure is an optimization model for calculating each consumer's travel and shopping pattern. Since the shopping model predicts consumer behavior in multipurpose-trip environments, the location-allocation procedure is used to investigate the impact of such consumer behavior on the spatial organization of firms.

Previous studies investigating the effect on central place systems of different postulates about consumer behavior have typically started with the classic central place network and analyzed the growth and decline of centers caused by changes in the consumer allocation rule (see, for example, Kohsaka, 1984; Mulligan, 1984; Rushton, 1971). We, however, follow McLafferty (1982) in that rather than assuming a Christallerian network as the starting point of our analyses, we begin by finding the optimal locations of firms in the presence of multipurpose shopping. But whereas McLafferty (1982) assumed that all consumers have an equal propensity for multipurpose shopping, we use the shopping model to determine the optimal shopping behavior of each consumer.

To illustrate our model we use a two-dimensional region consisting of 81 points evenly spaced on a square lattice. Each demand point has the same population and is a feasible location for a retail firm. For clarity, we use the index m ($m = 1, 2, \ldots, 81$) to refer to population nodes and the index n ($n = 1, 2, \ldots, 81$) to refer to them as potential center locations. (Nodes are numbered consecutively from left to right, top to bottom). We consider a two-level retail hierarchy with one high-order firm already established in the region. The optimal locations of lower-order firms are determined by the model. The overall objective is to minimize the total shopping cost of all consumers, $\sum_i TC_i$,

where TC_i ($i = 1, 2, \ldots, n$) is the shopping cost of consumer i given a particular locational configuration. For any locational configuration of firms, the optimal values of N_j, N_k and x for each consumer are determined by solving problem (2.3). These values enable us to calculate both the shopping cost of the consumers and also the total revenues of each firm. Given a particular locational configuration, the sales revenue of firm j can be calculated as follows:

if j is low-order place

$$\sum_{i \epsilon X_j} (1 - x_i^*) D_j P_{jj} \tag{2.4}$$

if j is high-order place

$$\sum_{i \epsilon X_j} (1 - x_i^*) D_j P_{jj} + \sum_{i=1}^{n} x_i^* D_j P_{jk}$$

In equation (2.4), x_i^* is the optimal value of x_i for each location i found by solving problem (2.3) for each demand node. The set of demand points for which the outlet at j is the closest low-order firm is x_j. If a node is equidistant from two or more low-order firms, the revenue is divided equally among those firms. A vertex substitution algorithm is used to find the locational configuration that minimizes total shopping cost for all consumers.

To define the shopping model, the following parameters must be specified: transportation costs from any node m to a center n; the cost of holding inventory; and the demands for the two goods and their prices. For all results presented in this section, the cost of a round trip from any node m to a j-level place is calculated as $(d_{mj} + 1.0)*1.5$. The cost of a multipurpose trip to a k-level store is $(d_{mk} + 1.0)*8.0$, reflecting the extra cost of purchasing multiple goods. The distance between places m and j, d_{mj} is calculated using the Euclidean metric. The demand for the low-order good is 400 units and that for the high-order good is 40 units, and their retail prices are \$10 and \$40, respectively. Further, the cost of holding a unit of a good in inventory for the entire planning horizon is 40% of the monetary cost of that unit. It is not possible to store more than 30 units of any good in inventory, and there is a minimum lot size constraint of 5 units. To determine the optimal number of low-order firms, we assume that the minimum threshold revenue for each firm is \$30,000.

Table 2.1 shows the optimal locational configuration and the revenue for each firm in the patterns for 5, 6, and 7 centers. In all cases, one of the low-order firms agglomerates with the high-order firm at node 41. In the 7-center case, however, three of the firms fail to earn the threshold revenue. Since at equilibrium all firms must generate at least the threshold level of revenue, the maximum number of low-order firms the region can support is six. With six firms at their optimal locations the total shopping cost is \$478,466, of which

Table 2.1. Optimal Locational Configuration for Different Numbers of Firms

5 Firms		6 Firms		7 Firms	
Location	Sales	Location	Sales	Location	Sales
41	114933.3	41	112866.7	41	108133.3
12	52268.7	20	42100.0	38	29700.0
26	52268.7	26	42100.0	26	42100.0
70	52268.7	71	43466.7	71	43466.7
56	52268.7	66	52266.7	66	42100.0
		14	31200.0	14	29700.0
				11	29800.0
Transport cost	12642.0		12525.3		12358.7
Storage cost	12363.2		12340.9		12335.3
Multipurpose trips	.2467		.2403		.2340

$12,341 and $12,525 are spent on transportation and on inventory costs, respectively.

In most cases there were multiple optimal solutions to the location problems. Thus, in addition to the equilibrium configurations shown in the text, there were a number of other solutions with the same properties. One reason for the multiple optima is the symmetry of the square lattice used in the simulations. The use of discrete space also gives rise to assymetries in the equilibrium patterns.

The optimal travel patterns of consumers at different nodes and their rates or multipurpose shopping are shown in Figs. 2.1 and 2.2. Consumers at node 41, for example, being closest to both types of stores, make the maximum number of trips—18 to the low-order place and 6 to the high-order one (see Fig. 2.1). They purchase 25% of the low-order good on multipurpose trips. On each multipurpose trip these consumers purchase 6.67 units of the high-order good and 16.67 units of the low-order one. The schedule of single- and multipurpose trips over the planning horizon for consumers at nodes 41 and 40 are shown in Fig. 2.3.

The optimal travel pattern and rate of multipurpose shopping in the region exhibit significant spatial variation. Consumers at node 41, who make 18 single-purpose trips and 6 multipurpose trips, purchase 25% of the low-order good on multipurpose trips. Consumers in the neighboring nodes, on the other hand, make 10 single-purpose and 5 multipurpose trips and buy 33% of the low-order good during multipurpose shopping. The impact of travel cost on the optimal amount of multipurpose shopping is evident from Fig. 2.1. At the center of the region (near the high-order place) the number of multipurpose trips is 6. The number of multipurpose trips decreases as one moves away

9,3	9,3	9,3	12,3	12,3	12,3	9,3	9,3	9,3
12,3	12,3	12,3	12,3	16,4	12,3	12,3	12,3	12,3
12,3	12,3	12,3	12,4	12,4	12,4	12,3	12,3	12,3
12,3	12,3	12,4	10,5	10,5	10,5	12,4	12,3	12,3
9,3	9,3	8,4	10,5	18,6	10,5	8,4	9,3	9,3
9,3	9,3	8,4	10,5	10,5	10,5	8,4	9,3	9,3
9,3	12,3	12,3	12,4	8,4	8,4	12,3	12,3	12,3
9,3	12,3	12,3	12,3	9,3	9,3	12,3	12,3	12,3
9,3	12,3	12,3	12,3	9,3	9,3	12,3	12,3	12,3

☐ Location of high order firm

── Location of low order firm

Figure 2.1. Optimal numbers of single- and multipurpose trips for each demand node. The first figure is the number of single-purpose trips and the second one the number of multipurpose trips.

0.25	0.25	0.25	0.20	0.20	0.20	0.25	0.25	0.25
0.20	0.20	0.20	0.20	<u>0.20</u>	0.20	0.20	0.20	0.20
0.20	<u>0.20</u>	0.20	0.25	0.25	0.25	0.20	<u>0.20</u>	0.20
0.20	0.20	0.25	0.33	0.33	0.33	0.25	0.20	0.20
0.25	0.25	0.33	0.33	0.25	0.33	0.33	0.25	0.25
0.25	0.25	0.33	0.33	0.33	0.33	0.33	0.25	0.25
0.25	0.20	0.20	0.25	0.33	0.33	0.20	0.20	0.20
0.25	0.20	<u>0.20</u>	0.20	0.25	0.25	0.20	<u>0.20</u>	0.20
0.25	0.20	0.20	0.20	0.25	0.25	0.20	0.20	0.20

☐ Location of high order firm

__ Location of low order firm

Figure 2.2. Optimal rate of multipurpose shopping.

from the center of the region. Consumers near the edges of the region, for example, make only 3 multipurpose trips. Similarly, the distance to the nearest low-order place affects the number of trips made to it. It is the relationship between costs of travel to low- and high-order places, however, that influences the final trip frequencies and the optimal level of multipurpose shopping. Figs. 2.1 and 2.2 also show that consumers having the same propensity for multipurpose shopping (i.e., having the same value of x) may indeed have markedly different trip frequencies. Compare, for example, the travel pattern of consumers at node 1 with that of consumers at node 41. Although both groups purchase 25% of the low-order good on multipurpose-shopping trips, consumers at node 1 make only 9 single-purpose trips and 3 multipurpose trips, in contrast to 18

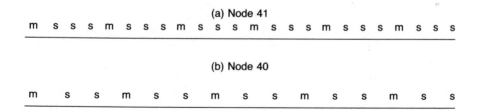

(a) Node 41

m s s s m s s s m s s s m s s s m s s s m s s s

(b) Node 40

m   s   s   m   s   s   m   s   s   m   s   s   m   s   s

Figure 2.3. Schedule of single- and multipurpose trips over the planning horizon; m = multipurpose trip; s = single-purpose trip.

single-purpose and 6 multipurpose trips by consumers at node 41. The latter consumers make more frequent trips to take advantage of their highly accessible location and to thereby reduce inventory costs.

The possibility of multipurpose shopping reduces total travel costs for all consumers. Without multipurpose shopping, consumers would need to make more frequent shopping trips for the low-order good. Consumers at nodes 1 and 41, for example, would make 12 and 24 trips, respectively, to the low-order place, in comparison to the 9 and 18 trips they make with multipurpose shopping. Total travel cost for all consumers in the region increases to $13,119 from $12,340, and inventory cost would increase to $12,883 from $12,525. Multipurpose shopping, since it reduces the frequency of travel, is an effective way to reduce total shopping cost.

Transportation costs and the cost of holding inventory both affect the optimal rate of multipurpose shopping. Inventory holding cost can change owing to a change in either inventory carrying cost or the price of the goods. Since interest payments are a major component of inventory carrying cost, variations in the interest rate influence the level of inventory holding cost. Table 2.2 shows how changes in the cost of goods affect optimal travel and shopping patterns. As the price of a good falls, it is less costly to store goods, and consequently consumers make fewer trips. For example, when the price of the low-order good decreases, consumers purchase more of that good per trip, since the cost of storing the good is lower. In all cases, this situation leads to fewer single-purpose trips, and in some cases even to fewer trips to the high-order place. The net impact is that greater amounts of the low-order good are purchased on multipurpose trips. Increases in the price of the low-order good, on the other hand, lead to more single-purpose trips, owing to the increased cost of holding inventory. The consumer at node 1, for example, makes 12 single-

Table 2.2. Impact of Price Changes on Multipurpose Shopping

P_{jj}[a]	P_{kk}	Storage Costs	Transport Costs	Level of Multipurpose Shopping
5	40	11,332	10,967	.3189[b]
10	40	12,341	12,525	.2403
20	40	14,333	14,528	.2025
30	40	16,261	16,000	.2000
10	10	8,477	9,260	.2000
10	20	10,008	10,545	.2089
10	60	14,264	14,217	.2761
10	80	15,749	15,727	.3107

[a] $P_{jj} = P_{jk}$.
[b] Overall proportion of good j bought on multipurpose trips.

purpose trips when the unit price of the low-order good is 30, 9 when it is 10, and only 6 when the price declines to 5. *Ceteris paribus,* the level of multipurpose shopping decreases from 32% to 20% as the price increases over that range. As one might expect, the impact of increasing the price of the higher-order good has the opposite effect. It increases the frequency of trips to the high-order place relative to those to the low-order place. This increase, in turn, provides added opportunities for multipurpose shopping. As seen in Table 2.2, the proportion of multipurpose shopping increases from 20% to more than 31% as the price of the higher-order good increases from 10 to 80.

An important feature of the optimal locational configuration in the presence of multipurpose shopping is the agglomeration of low-order firms with higher-order ones. The benefit of such agglomeration can be substantial, since the agglomerated low-order firm not only sells to consumers within its own market area but also benefits from multipurpose shopping by other consumers. All consumers in the region patronize the agglomerated low-order firm at node 41 to some extent. As a result, the sales of the agglomerated low-order firm are $112,866, compared to only $52,266 for the highest-selling nonagglomerated low-order firm. Since this agglomerated low order firm's substantial advantage is due to multipurpose shopping, the high-order firm may attempt to claim part of this excess revenue by negotiating rental subsidies or side payments. The maximum possible subsidy can be determined by comparing the present revenue of $112,866 units with the revenue the firm would earn in the absence of agglomeration. In that situation the maximum revenue a low-order firm can earn is $58,000. The difference of $54,866 is the maximum possible payment to the high-order firm. When there is more than one high-order firm, the benefit from agglomerating with any one of them depends on the level of shopping at the different high-order places. Therefore, the amount of payment a high-order firm can negotiate depends on its ability to attract customers and encourage multipurpose shopping.

In practice, this phenomenon is exemplified when real estate developers, recognizing the benefit of agglomeration, offer subsidized rates to high-order firms that will attract low-order firms. This arrangement often occurs in regional shopping malls where large anchor stores pay significantly lower rates than the smaller stores attracted to the mall because of the large stores' drawing power.

The agglomeration of low-order firms leads to the development of a hierarchical spatial organization of sellers (see Eaton and Lipsey, 1982; McLafferty and Ghosh, 1984). In this hierarchical system, high-order places sell high- and lower-order goods, while only low-order goods are available at low-order places. It is important to note that this hierarchy arises as part of the process formulated in the location model and is not exogenously specified as in the classic theories of Christaller and Losch. Similarly, Rushton (1971) and White

(1977) start by assuming the existence of a hierarchy rather than by deducing the hierarchical structures from the basic behavioral postulates of their models.

Another feature of the multipurpose-shopping equilibrium spatial pattern is that it contains fewer firms than without multipurpose shopping. This feature results because with multipurpose shopping the agglomerated firms cannibalize the sales of their nonagglomerated counterparts. As Table 2.1 shows, the number of firms at equilibrium in the multipurpose-shopping case is six. When only single-purpose trips are allowed, the number of low-order firms in the equilibrium configuration is nine, and each firm earns a revenue of $36,000. Thus multipurpose shopping leads to higher sales for agglomerated firms and consequently to fewer firms at equilibrium. Another characteristic of the location pattern is the possibility of early entry by low-order firms. Since the potential for agglomeration is limited in any region, there is an incentive for low-order firms to enter early to preempt desirable locations. The timing of such entry depends on the magnitude of the agglomeration benefit and the level of competitive threat.

EFFECT OF COST AND PRICE VARIATIONS ON LOCATIONS

Thus far we have assumed that the price of the low-order good is constant across the different stores. It is likely, however, that in the long run the agglomerated low-order firm, owing to its higher level of sales, can benefit from economies of scale and charge a lower price. This ability further increases its attractiveness over unagglomerated firms. In this section we analyze the impact of such price changes on the final locational configuration. The motivation for the analysis is to investigate the conditions under which the location patterns are stable. Since price reductions increase the drawing power of the agglomerated firm, it is possible that in some cases a series of price reductions may ultimately result in the demise of all nonagglomerated firms (see Goodchild, 1978).

The cost of operating a retail establishment comprises two elements: the cost of procuring the goods (or the cost of providing the service) and the cost of operating a store. We assume that both these costs depend on the volume of operation. First consider the cost of procuring goods. Assume that the retail firm is a reseller of goods purchased from a manufacturer (for simplicity, we ignore the cost of transportation between the manufacturer and the retailer). The price charged by the manufacturer (pc) depends on the total number of units sold by the retailer: $pc = 5.0$ if $v < 10,000$, and $pc = a - b(\log v)$ otherwise, where a and b are parameters of the cost function and v is the number of units sold. The manufacturer charges a base price of $5 for orders of less than 10,000. Orders above this level benefit from quantity discounts, as shown in the second expression in the equation. Quantity discounts by the manufacturer provide the retail firm with the flexibility to reduce its own price.

To model the impact of changes in procurement cost on the retailer's price, we assume that retailers follow the common practice of adding a fixed markup percentage to the procurement cost to obtain the final retail price. We use a markup percentage of 100% of cost.

In the initial 6-center equilibrium pattern, the agglomerated firm sells 11,286 units, making it eligible to receive a quantity discount. If the values of a and b are 25 and -5, respectively, the agglomerated firm can reduce its price to $9.47 per unit. This price reduction makes multipurpose shopping even more attractive and increases sales of the agglomerated firm to 19,316 units. With this price change, the overall proportion of low-order goods bought on multipurpose trips increases to 51% from 24% in the base case. As a result, sales of the agglomerated firm increase from $112,866 to $182,925 and the net revenue (after deducting the cost of procurement) increases from $56,430 to $96,580. As a consequence of the sales increase for the agglomerated low-order firm, the sales of four of the five nonagglomerated firms drop below the threshold level. Of the nonagglomerated low-order firms, only the one at site 66 is able to survive. The failure of four competitors increases sales of the agglomerated firm even further and subsequently results in the closure of the sole unagglomerated firm in the region. Thus, in this case the ultimate location pattern collapses to just one high-order and one agglomerated low-order firm.

The desirability of a price-reduction policy depends not only on sales but also on the cost of operating a store. As sales increase, operating cost rises owing to the need for additional labor, space, and operating expenses. Assume the fixed cost, fc, to be related to unit sales in the following manner: $fc = 15,000$ if $v < 5,000$, and $fc = (v - 5000)^y/1000$ if $v > 5000$. Total operating cost is fixed at $15,000 as long as sales are less than 5,000 units. The change in operating costs above this level depends on the value of the parameter y.

With the relationship between costs and volume defined, we can now investigate dynamic changes in the system. Let the values of a, b, and y be 11.667, -1.667, and 1.8, respectively. Initially, with all firms charging $10 for the low-order good, the agglomerated low-order firm sells 11,286 units of the low-order good, for a profit of $34,560. Since the firm sells more than 10,000 units, the manufacturer offers a quantity discount and the firm can reduce its retail price to $9.82 per unit. The price reduction makes multipurpose shopping even more attractive, and profits of the agglomerated firm increase to $38,748 from sales of 12,592 units. The proportion of low-order good purchased on multipurpose trips increases from 24.03% to 31.06%. As a result of increased sales for the agglomerated firm, the low-order firm at node 14 fails to break even. As shown in Table 2.3, this failure results in added sales for the five remaining firms.

As a consequence of the increased level of multipurpose shopping, the market share of the agglomerated firm increases further, creating the potential for even

Table 2.3. Changes in Optimal Configuration as Agglomerated Firm Takes Advantage of Quantity Discounts

	Iteration			
	1	2	3	4
Price (P_{jk})	10.00	9.82	9.82	9.55
Sales of agglomerated firm	112,867	125,295	134,247	187,533
Profit of agglomerated firm	34,560	38,748	41,093	51,731
Multipurpose-shopping rate	.2403	.3106	.3264	.5116
Number of firms	6	6	5	5

greater price discounts. With the price at $9.82 per unit, the agglomerated firm sells 13,670 units. Now, however, it can reduce its price further to $9.55—which increases multipurpose shopping to 51.16%—and both sales and profits of the agglomerated firm increase. Of the unagglomerated firms, the one located at node 71 incurs a loss of $1,325 and is forced to close. Sales are now concentrated among the four remaining firms. The additional sales allow the agglomerated firm to reduce its price even further to $8.90. Such a price reduction, while leading to increased sales, does not result in additional profit for the agglomerated firm. The additional operating costs necessary to serve the increased number of users is greater than the additional revenue generated. There is no incentive, therefore, for the firm to adopt such a strategy. Thus, the final equilibrium pattern consists of four of the low-order firms, and 51.15% of the low-order good is bought during multipurpose shopping. Fig. 2.4 shows the optimal travel pattern of consumers for this locational configuration.

The nature of the final locational configuration depends on the parameters of the cost and profit equations. If the value of y is 1.7, for example, the location and price adjustments are similar, until the very last stage, to the previous case. With only four low-order firms remaining in the region, the agglomerated firm can increase its profit even further by reducing its price to $8.90. This reduction increases sales to $304,284. The dramatic increase in sales for the agglomerated firm results in all nonagglomerated firms failing to break even. The final locational pattern, therefore, consists of one low-order firm agglomerated with the high-order firm. On the other hand, the nature of the final equilibrium is very different when the value of y is 1.9. Owing to the sharp rise in fixed cost with volume increases, no benefit results from price reduction. As a consequence, the agglomerated firm maintains its price and there is no change in the locational configuration.

Apart from changes in the value of y, the locational configuration is also affected by changes in the quantity-discount rate. If the values of a and b are (a)$P_{jk} = 9.82$; Number of Firms = 6

0.25	0.25	0.25	0.25	0.20	0.25	0.25	0.25	0.25
0.25	0.20	0.25	0.34	0.20	0.34	0.25	0.20	0.25
0.20	0.20	0.34	0.34	0.50	0.34	0.34	0.20	0.20
0.25	0.34	0.34	0.50	0.50	0.50	0.34	0.34	0.25
0.25	0.34	0.50	0.50	0.34	0.50	0.50	0.34	0.25
0.34	0.34	0.50	0.50	0.50	0.50	0.50	0.34	0.25
0.25	0.25	0.34	0.34	0.50	0.50	0.34	0.20	0.25
0.25	0.20	0.20	0.34	0.34	0.34	0.20	0.20	0.20
0.25	0.25	0.20	0.25	0.25	0.25	0.25	0.20	0.25

(b)$P_{jk} = 9.82$; Number of Firms = 5

0.25	0.25	0.25	0.34	0.50	0.34	0.25	0.25	0.25
0.25	0.20	0.25	0.34	0.50	0.34	0.25	0.20	0.25
0.20	0.20	0.34	0.50	0.50	0.50	0.34	0.20	0.20
0.25	0.34	0.34	0.50	0.50	0.50	0.34	0.34	0.25
0.25	0.34	0.50	0.50	0.34	0.50	0.50	0.34	0.25
0.34	0.34	0.50	0.50	0.50	0.50	0.50	0.34	0.25
0.25	0.25	0.34	0.34	0.50	0.50	0.34	0.20	0.25
0.25	0.20	0.20	0.34	0.34	0.34	0.20	0.20	0.20
0.25	0.25	0.20	0.25	0.25	0.25	0.25	0.20	0.25

(c) $P_{jk} = 9.55$; Number of Firms = 5

0.51	0.51	0.51	0.51	0.51	0.51	0.51	0.51	0.51
0.51	0.51	0.51	0.51	0.51	0.51	0.51	0.51	0.51
0.51	0.51	0.51	0.51	0.51	0.51	0.51	0.51	0.51
0.51	0.51	0.51	0.51	0.51	0.51	0.51	0.51	0.51
0.51	0.51	0.51	0.51	0.51	0.51	0.51	0.51	0.51
0.51	0.51	0.51	0.51	0.51	0.51	0.51	0.51	0.51
0.51	0.51	0.51	0.51	0.51	0.51	0.51	0.51	0.51
0.51	0.51	0.51	0.51	0.51	0.51	0.51	0.51	0.51
0.51	0.51	0.51	0.51	0.51	0.51	0.51	0.51	0.51

☐ Location of high order firm
— Location of low order firm

Figure 2.4. Change in multipurpose shopping rate due to price change.

25 and −2.5, respectively, the agglomerated firm earns a substantial quantity discount on its initial sale of 11,286 units. It can therefore reduce its price to $9.47. This reduction increases shopping at node 41 to the extent that all nonagglomerated forms fail to break even; so for this quantity-discount schedule, the agglomerated firm can easily drive all its nonagglomerated competitors out

of business through predatory pricing. Thus, the benefit of agglomeration inherent in the hierarchical organization is further accentuated when the agglomerated firm can offer discriminatory price reductions owing to its high level of initial sales. Increases in the cost of operations, however, counterbalance this trend. The implication is that the nature of the final locational configuration depends on the shape of both the cost and quantity-discount functions.

CONCLUSION

This chapter has developed a location-allocation model to investigate the impact of multipurpose trips on the growth and spatial organization of central place systems. The model captures the process of mutual adjustment between patterns of consumer travel and shopping and facility locations by systematically varying locations and determining each consumer's optimal trip pattern and multipurpose-shopping rate. Thus, the consumer's decision calculus is embedded in the choice of optimal locations. This method allows us to analyze the spatial consequences of consumer behavior postulates that are more realistic than heretofore, and thus to probe into the origins and evolution of central place patterns.

The results of the numerical simulations allow a number of generalizations regarding central place systems in multipurpose-trip environments. It is shown that the optimal frequencies of single- and multipurpose trips depend on the consumer's location relative to shopping opportunities, as well as on cost and price factors. Thus the model derives an explicit link between spatial structure and spatial behavior. Even when all consumers in an area have equivalent tastes and incomes, the organization of space itself brings variability to shopping patterns.

The presence of multipurpose shopping has important effects on the equilibrium number and location of firms. Like previous studies (Eaton and Lipsey, 1982), our research provides a strong economic rationale for the agglomeration of low- and high-order firms. Agglomerated low-order firms cannibalize the market areas of their nonagglomerated competitors and thus earn excess profits. Revenues of the nonagglomerated firms decrease as a result, and over the long run such firms may drop out of the landscape. The equilibrium number of firms is usually less than it would have been under the standard assumption of no multipurpose shopping. A consequence of agglomeration is the development of a hierarchical spatial organization much like that described by Christaller (1933). The significance is that this hierarchy, rather than being exogenously specified, arises as a result of the consumer process postulated in the shopping model.

The simulation procedure allows us to investigate the comparative static of the spatial system. Specifically, we have looked at the impact of procurement and operating costs on the equilibrium number of firms. Economies in procurement costs resulting from quantity discounts may accentuate the benefit of agglomeration and lead to even fewer nonagglomerated firms. If such quantity discounts are high, agglomerated firms may adopt predatory pricing policies to drive their nonagglomerated competitors out of business. The benefit of price reductions must, however, be trade off with higher sales volume's effect on operating costs. The latter may arise due to discontinuities in technology or the negative effects of congestion. The nature of the final equilibrium depends on the interaction of these factors.

This study has provided some important insights into central place systems in multipurpose-trip environments, but a number of extensions remain on the agenda for future research. One is to investigate the implications for land rent. We have already indicated how high-order firms may negotiate subsidies from low-order firms in exchange for agglomeration rights. This situation helps determine the retail rent to be paid by different firms. One can also consider the pattern of consumer rents and land values. If we assume that all households buy only shopping goods, land/housing, and transportation, then the pattern of urban rents is related to the variability in shopping costs over the region. Thus, one can build a comprehensive model that considers both residential and commercial land values.

Future research should also focus on integrating other aspects of consumer shopping behavior into our model. For example, one might analyze comparison shopping and multipurpose shopping within a single model. Comparison shopping leads to the clustering of firms offering similar goods (Eaton and Lipsey 1979), while multipurpose shopping leads to hierarchical agglomeration of dissimilar firms. Thus the equilibrium pattern of central places reflects two types of agglomerative effects. Similarly, it would be interesting to consider a model in which consumers trade off the utility from shopping with that from nonshopping activities (Mulligan, 1983, 1984). This trade-off would lead to spatially elastic demand that might geographically limit the drawing power of the agglomerated firms.

Another direction for future research is toward the construction of hierarchical location-allocation models. A number of such models have been proposed in the literature (see for example, Dokmeci, 1977, 1980; Narula and Ogbu, 1979; Kohsaka, 1984; O'Kelly and Storbeck, 1984). However, as O'Kelly and Storbeck note, "Many previous studies have presented primarily structural elaborations of more basic single level locations problems" (p. 121). What has been lacking is a behavioral process that links consumer behavior at different levels of the hierarchy and allocates demand to different levels. The consumer shopping model presented here helps to solve these problems. The construction of a multilevel hierarchical location-allocation model incorporating multipurpose shopping is the subject of our ongoing research.

REFERENCES

Bacon, R. 1983, A model of the frequency of consumer purchase patterns: A contribution to the theory of retail locations, in *Economics of Distribution: Proceedings of the Second International Conference on the Economics of Distribution,* F. Angeli, ed., CESCOM, Milan, pp. 579–589.

Beaumont, J. R., 1980, Spatial interaction models and the location-allocation problem, *Journal of Regional Science* **20:**37–50.

Christaller, W., 1933, *Central Places in Southern Germany,* trans. C. W. Baskin, 1966, Prentice-Hall, New Jersey.

Curry, L., 1967, Central places in the random spatial economy, *Journal of Regional Science* **7:**217–238.

Dokmeci, V., 1973, An optimization model for a hierarchical spatial system, *Journal of Regional Science* **13:**439–451.

Dokmeci, V., 1980, A multiobjective model for the regional planning of health facilities, *Environment and Planning, A* **11:**517–525.

Eaton, B. and R. Lipsey, 1979, Comparison shopping and the clustering of homogeneous firms, *Journal of Regional Science* **19:**421–435.

Eaton, C. and R. Lipsey, 1982, An economic theory of central places. *Economic Journal* **92:**56–72.

Ghosh, A. and S. McLafferty, 1984, A model of consumer propensity for multipurpose shopping, *Geographical Analysis,* **16:**244–249.

Goodchild, M., 1978, Spatial choice in location allocation problems: the role of endogenous attraction, *Geographical Analysis,* **10:**65–72.

Hanson, S., 1980, Spatial diversification and multipurpose travel: implications for choice theory, *Geographical Analysis,* **12:**245–257.

Kohsaka, H., 1983, A central-place model as a two-level location-allocation system, *Environment and Planning, A* **15:**5–14.

Kohsaka, H., 1984, An optimization of the central place system in terms of the multipurpose shopping trip, *Geographical Analysis,* **16:**250–269.

Losch, A., 1954, *The Economics of Location,* Yale University Press, Connecticut.

Lentnek, B., M. Harwitz, and S. Narula, 1981, Spatial choice in consumer behavior: towards a contextual theory of demand, *Economic Geography* **57:**362–373.

McLafferty, S. 1982, Locating services to minimize consumer travel with multipurpose travel, paper presented at the International Conference on Improving Geographical Accessibility to Rural Services, Bangalore, India, July 1982.

McLafferty, S. and A. Ghosh, 1984, A simulation model of spatial competition with multipurpose trips, *Modelling and Simulation* **15**(part 1):477–482.

Mulligan, G., 1983, Consumer demand with multipurpose shopping behavior, *Geographical Analysis,* **15:**76–80.

Mulligan, G., 1984, Central place populations: some implications of consumer shopping behavior, *Association of American Geographers Annals* **74:**44–56.

Narula, S. and U. Ogbu, 1979, An hierarchical location-allocation problem, *Omega* **7:**137–143.

O'Kelly, M., 1981, A model of demand for retail facilities incorporating multistop, multipurpose trips, *Geographical Analysis,* **13:**134–148.

O'Kelly, M., 1983, Multipurpose shopping trips and the size of retail facilities, *Association of American Geographers Annals* **73**:231–239.

O'Kelly, M. and J. Storbeck, 1984, Hierarchical location models with probabilistic allocation, *Regional Studies* **18**:121–129.

Papageorgio, G. and A. Brummel, 1975, Crude inferences on spatial behavior, *Association of American Geographers Annals* **65**:1–13.

Puryear, D., 1975, A programming model of central place theory, *Journal of Regional Science* **15**:307–316.

Reinhardt, P. G., 1973, A theory of household grocery inventory holdings, *Kyklos* **26**:497–511.

Rushton, G., 1971, Postulates of central place theory and the properties of central place systems, *Geographical Analysis,* **13**:140–156.

White, R. W., 1977, Dynamic central place theory: results of a simulation approach, *Geographical Analysis,* **9**:226–243.

3

LOCATION-ALLOCATION MODELING
IN ARCHAEOLOGY

Thomas L. Bell

The University of Tennessee, Knoxville

Richard L. Church

University of California, Santa Barbara

Archaeologists deal with many aspects of past cultures. One of the most intriguing and potentially rewarding areas of research is the degree to which certain facets of a sociocultural organization can be ascertained by examining the settlement configuration of the culture of interest. It is a basic tenet of the ''new'' archaeology that settlement systems reflect broader societal goals and that analytically examining the artifacts of such systems can provide clues to social organization and evolution (Watson, 1972). Whether a society maintained an equalitarian social structure characteristic of simple hunter-gatherers or a highly stratified structure in which centralized elite groups extracted tribute from a peripheral populace, the settlement configuration will reflect the structure. Unfortunately, for settlements in most ancient cultures, archaeologists frequently are hampered by the lack of accurate data on the precise location, areal extent, population, and length of habitation. It is therefore often impossible to glean a complete picture of societal organization from an analytical examination of the often scant archaeological record. In such cases, archaeologists turn to other indicators of past cultural organization. Pottery and lithic artifacts, exotic trade items, organization of activities within single sites, and religious icons are but a few of the facets that have been examined to increase our understanding of past cultures.

Recently, the data necessary to conduct studies of past settlement systems

Acknowledgment: Funding for this research was provided by the Human Geography and Regional Science Branch of the National Science Foundation Grant Number SES79–23686. The authors wish to thank Larry Gorenflo, Avijit Ghosh, and an anonymous referee for their cogent and useful suggestions on an initial draft of this chapter.

through systematic field surveys have become more widely available. As a result, archaeologists have increasingly turned their attention to the systemic aspects of ancient cultures by examining entire settlement patterns. This examination has often been within the context of classic geographic location theory, especially central place theory (see Parsons, 1972; Smith, 1976; Clarke, 1977; Crumley, 1979). Few archaeologists have expanded their research methods to include recent advances in location-allocation modeling and optimization techniques. The few optimization models applied have been directed at rather simply organized societies such as hunter-gatherers, and then with no explicit focus on location (Reidhead, 1976, 1979, 1980; Keene, 1979, 1981). The objective of this chapter is to expand archaeological applications of location-allocation models to the study of more complex cultural systems.

This chapter presents a number of location-allocation models that have proved useful in the analysis of archaeological settlement patterns. We describe locational studies in two disparate cultural and environmental settings spanning thousands of years and two different continents: the upper Nile Valley during the Ramessid period (ca. 1317–1070 B.C.) and the Basin of Mexico during the Late Horizon (Aztec) period (A.D. 1350–1519). The models used in these two study areas were fashioned differently both to reflect place-to-place differences in the importance placed on settlement location criteria and to address different research questions.

The use of this family of models should not be interpreted as a complete break with earlier attempts to examine prehistoric settlement patterns. We view the application of location-allocation models to study settlement systems as an extension of earlier archaeological applications of central place theory (e.g., Flannery, 1972; Johnson, 1972, 1977; Marcus, 1973; Smith, 1979). Indeed, it can be shown that central place theory is but a limiting case of more encompassing and flexible hierarchical p-median location-allocation models (Rushton, personal communication; Bell, 1979; Hillsman, 1980). The value of the approaches employed in this chapter lies mainly in their flexibility, which allows one to address specific research questions in a more precise manner than before.

SETTLEMENT EFFICIENCY AND ARCHAEOLOGICAL SETTLEMENT PATTERNS

Central place patterns arise from the interplay of particular market economic forces and environmental conditions (Rushton, 1971). To examine the settlement patterns of complex prehistoric societies in a realistic manner, central place theory itself would need to be modified (Steponaitis, 1978, 1981).

Location-allocation models may, however, be used to test whether observed archaeological settlement patterns are similar to those that can be generated given certain assumptions about underlying locational forces and environmental circumstances. These models are not as restrictive in their assumptions as central place theory *per se*.

Settlement systems are, by their very nature, the produce of myriad sociopolitical, economic, and cultural forces. One strategy for modeling the evolution of settlement systems is to adopt a multiple-objective approach. Given the nature of such problems, single-objective models in many cases appear too simplistic; multiobjective models can be better used to reflect system complexity. This chapter discusses several possible multiobjective models. The goal in each case is to develop a flexible approach to settlement research—an approach that can incorporate multiple decision-making objectives and evolutionary forces and yet yield optimal, or close-to-optimal configurations against which to compare actual data.

In location-allocation modeling the emphasis is on the properties of the solution other than the deduced geometry. In contrast, most previous archaeological applications of central place theory have focused on the geometrical (i.e., hexagonal) arrangement of places at the expense of critically evaluating the forces that generated the geometric configurations. Although settlement configuration (form) is an important aspect of settlement studies, location-allocation modeling also offers insight into the processes underlying the development of spatial systems. Such processes may be deduced by the changing pattern of variable weights in a multiple-objective formulation that is applied to settlement data in a temporal sequence.

In this chapter the application of location-allocation modeling to archaeological settlement systems proceeds chronologically. Data on more recent archaeological settlement systems are more detailed. They are therefore more testable for contemporaneity of habitation and contain more material artifacts visible on the extent landscape. The location-allocation models developed for Nile Valley settlements are thus somewhat more simplistic than those applied to the more recent settlements in the Late Horizon Basin of Mexico.

POLITICAL CENTRALIZATION OF THE NILE VALLEY DURING THE RAMESSIDE PERIOD

As noted, location-allocation models can be useful in testing archaeological hypotheses, and the results of such applications may serve as independent confirmation of hypotheses derived from field archaeology. For example, some Egyptologists have concluded that evidence strongly suggests that a bureaucratic centralization of authority took place during the reign of the Ramessid pharaohs (ca. 1317–1070 B.C.). Their conclusions are based both on textual evidence (especially the Wilbour Papyrus) and on the assumed nature of tribute flow among upper Nile settlements (Butzer, 1976; O'Connor, 1972). This section tests a location model that is appropriate to a politically complex society in which there is evidence of bureaucratic centralization of authority (Kauffman, 1981; Kauffman, Church, and Bell, 1981). The data are from the region of the immediate Nile floodplain from the first cataract near Aswan to just south of Cairo, excluding the Faiyum Depression. They are used to test whether the settlement system corresponded to that expected for a centralized bureaucracy. The data set includes a total of 128 settlements contained within 23 distinct administrative units called nomes (Fig. 3.1).

We developed a location-allocation model to test the assumption that political centralization was an overarching goal of the pharaohs. Differences between competitive market and centrally administered settlement systems should be reflected by the nature of the relationship among centers of the same functional size. As a result of competition for consumers, market centers tend to become evenly spaced if the population being served is fairly evenly dispersed. In a system with strong central administration, on the other hand, the size and shape of a center's administrative district "remains unaffected by that center's nearness to other politically affiliated centers of equivalent order" (Steponaitis, 1978, p. 427). The location-allocation procedure permits one to assess whether the observed settlement system reflects the pattern of a centrally administered system.

The Maximal Covering Location Problem

The assumption that a central authority exerts maximal control over a population through regional administrative centers is analogous to the underlying premise of the so-called location set-covering and maximal covering location problems, which are members of the set of general location-allocation models (Church and ReVelle, 1974, 1976; Toregas and Revelle, 1982). Scott (1971) has noted that one of the economic systems most appropriately analyzed by location-allocation models "corresponds . . . to a system of complete centralization of decision-making, . . . It is in the nature of such systems to seek out cost-minimizing solutions" (p. 1).

It has been suggested that centralization of decision making was operative in close to its ideal form in Ramessid Egypt (O'Connor, 1972). The maximal covering location problem (MCLP), in particular, focuses on the supplier of services (i.e., the central decision maker). The problem may be stated as follows: maximize coverage (population served) within a desired service distance by locating a fixed number of facilities. In Ramessid Egypt, the main objective of settlement location is hypothesized to have been the maximization of administrative control—coverage of the populace by a central authority.

Because it may not be possible to achieve total population coverage given a specified number of facilities and a theoretical maximum covering distance, the next best thing would be to cover as much of the population as possible within that distance constraint; the problem is to maximize the number of people

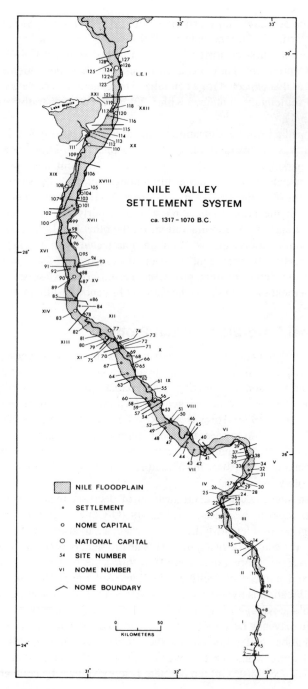

Figure 3.1. Approximate location of 128 settlements in the Upper Nile Valley during the Ramessid period.

covered, or controlled, within a theoretical maximum service distance by locating a fixed number of facilities (in this setting, the nome capitals, which are the administrative centers).

The MCLP was employed as a multiobjective simulation model; the goal was to reconstruct the primary decision-making criteria applied in locating nome capitals. It was assumed that at least some of these criteria might be at odds with each other and that actual locations may have been compromise solutions. The research analyzed the different combinations and weightings of these criteria, seeking the trade-off that would produce a spatial pattern of centers closely approximating the known archaeological situation. Good correspondence between actual and simulated patterns would suggest that the combination of objective-function criteria and weights that produced the comparable simulated pattern could have been those either consciously or unconsciously pursued by the Ramessid administrators.

Two measures of settlement importance were designated for each Nile Valley site. The first measure was a composite score developed by Butzer (1976) based on the value of selected socioeconomic attributes (e.g., number of villas and courtyards, presence of a mayor). The highest individual site scores were at Menfe and Weset, the national capitals, each with a score of 14 (Butzer, 1976). The total score for the 128 settlements was 509. The second measure was an ordinal scale of 1 to 4 based on score values, and it totaled 213 for the 128 settlements. These weights of settlement importance were used in the present research both to roughly measure site population and to evaluate the site's importance as an element of the settlement system. The MCLP was solved within a multiobjective trade-off setting in which the two decision-making criteria were assumed: (1) maximizing the importance of the settlements selected as administrative centers, as measured either by the composite score based on the socioeconomic attributes of the sites or by the ordinal rank of the sites (this criterion will be called preference), and (2) minimizing the number of sites left uncovered within a specified covering radius S (this criterion will be called coverage). The MCLP was used to analyze the efficiency of the configuration of Ramessid administrative centers according to the following formulation:

$$\text{Min } Z = \left[w_1 \left(\sum_i A_i \bar{Y}_i \right) - w_2 \left(\sum_j B_j X_j \right) \right] \tag{3.1}$$

subject to

$$\sum X_j + \bar{Y}_i > 1 \text{ for all } i \in I$$
$$\sum X_j = p$$
$$X_j = 0, 1 \text{ for all } j \in J$$
$$\bar{Y}_i = 0, 1 \text{ for all } i \in I$$

where

I = denotes the set of demand nodes (sites to be covered),
J = denotes the set of administrative centers,
S = the distance beyond which a demand node is considered uncovered,
D_{ij} = the shortest distance from node i to node j
X_j = 1 if an administrative center is allocated to j, 0 otherwise,
Y_i = 1 if the node is not covered, 0 otherwise
A_i = the population at demand node i,
B_j = measure of settlement importance (e.g., rank, score),
$N_i = j \in J \mid D_{ij} \leq S$,
p = the number of administrative centers to be located,
w_1 = objective-function weight for coverage (nonnegative),
w_2 = objective-function weight for preference (nonnegative).

In this formulation, N_i is the set of sites eligible to cover demand node i. A demand point is covered when the closest administrative center is at a distance less than or equal to S. Consequently, a demand node is uncovered when the closest administrative center is at a distance greater than S. The objective is to find the minimum number of people left uncovered if p administrative centers are located along the Nile River and its major canals.

Holding both covering distance and number of facilities constant, a trade-off curve may also be developed by varying the weights (w_1, w_2) given to the two components of the objective function.

Choosing an Appropriate Covering Distance

Selecting an appropriate maximal covering distance is important, and it is a choice for which there are no firm anthropological precedents for the setting in question. The first approximation of a maximal covering distance (19 km) was based on calculating the average linear spacing of 23 principal sites over the 873-km length of the region (Defense Mapping Agency, 1960). It is reasonable to assume that an administrative official in the course of duties or a peasant attending a religious or state ceremony in the capital might travel 19–20 km. Such journeys would not be frequent journeys, nor would they be made entirely on foot. Most travel of government personnel, as well as movement of grain and exotic goods, was by barge on the Nile (Kees, 1961). Correspondingly, the majority of nome capitals were at the time either located directly on the river or linked to it by canal (Butzer, 1976, p. 78).

A second estimation of covering distance, 22 km, was based on the mean river distance between capitals, 44 km. Half of this distance represents the estimated maximal covering radius of each capital. Two different covering radii were used to determine how sensitive the derived locational configurations were to changes in this parameter.

Implications of the Simulation Results for the Ramessid Settlement System

The first set of objective-function weights for each pairing of coverage and preference variables was always 1 for coverage and 0 for preference. This weighting is designated as the $p1$ solution set in Fig. 3.2 and can be interpreted as an approach that maximizes coverage of demand points at the expense of any importance placed on the functional value of the sites chosen as administrative centers.

The second set of objective-function weights calculated were diametrically opposed to the first, with coverage given a weight of 0 and preference a weight of 1. In these solutions, designated by the $p5$ points in Fig. 3.2, 23 facility centers with the highest functional score or rank are chosen from the set of sites first in an attempt to optimize the functional (or population) value of the facility centers.

Coverage and preference values for the $p1$ and $p5$ solution sets could then be plotted on a graph, with coverage values on the X axis and preference on the Y axis. The slope of the line between these two points was calculated and used as the objective-function weight for preference in the next $(p3)$ run (Cohon et al., 1979). The $p2$ objective-function weight is the slope between points $p1$ and $p3$, while the $p4$ objective-function weight is the slope between points $p2$ and $p5$ (Table 3.1).

The Ramessid Settlement Hierarchy

We are fortunate to have certain clues as to the locational organization of Egypt during Ramesside times. Textual sources reveal the hierarchy of administrative centers believed to have been operative. The most important towns in the system were the national capitals, Menfe (Memphis) and Weset (Luxor) which, respectively, dominated the northern and southern parts of Egypt. Menfe was the primary seat of government during the Ramessid period, although the delta towns of Tanis and Bubastis operated as temporary seats of government when the royal family was in residence.

Next in importance were the capitals of the nomes, or provinces, into which Egypt was divided from the Old Kingdom onward. Ptolemaic documents associated nome capitals with dominance in the administrative, economic, and religious activities of the nomes. Most of the known mayors, who are believed to have been key in provincial administration, were also associated with nome capitals.

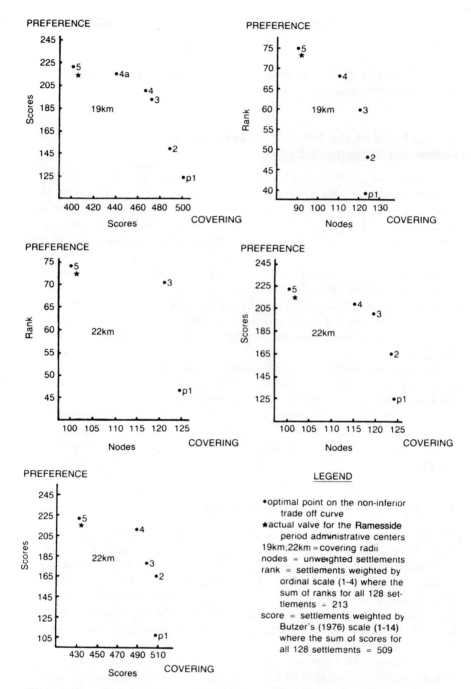

Figure 3.2. Trade-off curves for solutions to the maximal covering location problem for 23 administrative centers chosen from among 128 Nile Valley settlements during the Ramessid period.

84

Table 3.1. Summary Table of Data Displayed Graphically in Fig. 3.2

Solution Sets	Sum of Coverage Values	Sum of Preference Value for Admin. Ctrs.	No. of Nome Capitals in Solution	% Pop. Covered by Solution	% Nome Capitals in Solution
19 km: Coverage (nodes)[a]					
Preference (ranks)[b]					
p1	123	38	03	96.1	13.0
p2	123	49	08	96.1	34.8
p3	119	60	12	93.0	52.2
p4	111	68	16	86.7	69.6
p5	90	74	20	70.3	87.0
19 km: Coverage (scores)[c]					
Preference (scores)					
p1	495	126	07	97.2	30.4
p2	492	155	10	96.7	43.5
p3	471	196	15	92.5	65.2
p4	465	202	16	91.4	69.6
p4a	402	221	20	79.0	82.6
p5	436	219	19	85.7	87.0
22 km: Coverage (nodes)					
Preference (ranks)					
p1	124	46	06	96.9	26.1
p3	122	70	15	95.3	65.2
p5	100	74	19	78.1	82.6
22 km: Coverage (nodes)					
Preference (scores)					
p1	124	126	07	96.9	30.4
p2	124	163	11	96.9	47.8
p3	119	199	16	93.0	69.6
p4	115	210	18	89.8	78.3
p5	100	221	20	78.1	87.0
22 km: Coverage (scores)					
Preference (scores)					
p1	505	109	06	99.2	26.1
p2	505	161	11	99.2	47.8
p3	499	180	13	93.0	56.5
p4	485	213	16	89.8	69.6
p5	432	221	20	78.1	87.0

[a] Total sum of nodes = 128 (i.e., the 128 settlements are treated as having equal weight).
[b] Total sum of rank = 213.
[c] Total sum of scores = 509.

Other towns are known to have had mayors and to have collected taxes, but the Wilbour Papyrus suggests that the capitals were much more important than other towns in terms of the amount of land that their temples and officials controlled (Gardiner, 1947).

Results of the Application of the MCLP to the Ramessid Settlement Hierarchy

The application of the MCLP to the Nile Valley data, as depicted graphically in Fig. 3.2 and summarized in Table 3.1, shows good compatibility between the variables of coverage and preference. The percentage of population covered by the solutions ranges from a low of 70.3% (19 km nodes/ranks, $p5$) to a high of 99.2% (22 km scores/scores, $p1$). Although the $p1/p2$ solution sets generally do not include a high percentage of nome capitals, the sites chosen as facility locations in most cases have scores within the range of nome capital scores or are close neighbors of a capital.

Eight capitals are present in all but one of the $p1/p2$ solution sets: (4) Yebu, (12) Edjbo, (14) Nekhab, (2) Weset, (56) Khant Min, (75) Shashotep, (83) Kos, and (124) Menfe (numbers refer to Fig. 3.1). When $p3$ solution sets are taken into account, four more capitals are regularly included: (38) Gebtyu, (39) Inu, (41) He Sekhem, and (65) Djuka. These solution sets were generated with relatively little weight placed on the preference objective, which suggests that the hypothesis that nome capitals were located so as to maximize administrative control over the Nile Valley population is a plausible interpretation of Ramessid goal orientation.

Eleven capitals are present in 70% or more of the solution sets when taken together without regard for variation in covering distance or objective-function weights. With the exceptions of (39) Inu, and (110) Ninsu, all of these sites are present in the $p1$ through $p3$ solutions. One noncapital site, (8) Enboyet, is also present in a total of 83% of the solution sets. This site has a rather high preference score for a noncapital. It is also located almost equidistant from Yebu and Edjbo, the nearest capitals. Textual sources attribute a mayor to Enboyet. The results of the simulation procedure suggest that mayors may have been key administrative links in certain regions of the valley. This implication is especially strong in the numerically small but densely settled population of the extreme southern portion of the region, where the distance between capitals is two to three times that in the rest of the valley. The other area in which mayors may have played an important role was the more densely settled and numerically populous northern part of the valley, where they may have significantly supplemented the activities of nome capital personnel. In this region, Tjeni (52) and Mertum (116) are both designated as having had mayors and are present in approximately 40–50% of the solution sets. The presence of certain nome capitals and other important centers (especially those with a mayor) varies somewhat as a function of the coverage distance radius used. This result suggests that dividing the Nile Valley into meaningful subregions operating with reasonably homogeneous administrative goals should produce a more viable interpretive model of Ramessid administration.

Two nome capitals, Pi Nemty (77) and Hebnu (95), rarely are included in any solution set. Their continued status as capitals, even though they seem to be spatially inefficient, must be explained in terms of cultural, economic, or physical variables beyond the scope of the model. Pi Nemty is located in proximity to two other capitals, Shashotep (75) and Siyawti (80). It is possible that its survival as a capital was linked to very dense settlement in the area during Ramesside times. This density may now be masked both by the increased destruction, caused by modern farming practices, of archaeological sites in the area and by the significant eastward shift in the course of the Nile river (Butzer, 1960). These factors would alter the surviving record of the settlement pattern and hence the applicability of the model.

An alternative explanation for the viability of these three capital sites is their connection with important trade routes. Pi Nemty (77) controlled the entrance to a major alabaster and gold trade route across the Eastern Desert. Siyawti (80), on the opposite bank, was a primary port of trade for goods flowing between the Kharga Oasis in the Western Desert and the Nile Valley (Kees, 1961, p. 97). Hebnu (95), the other nome capital rarely included in any solution set, was not a port of entry for caravan routes through the desert, but it owed its continued prominence to its earlier role, during the Middle Kingdom, as the southern terminus of the northeastern frontier's defenses against Asia. Although its role as a defense command post declined considerably throughout the New Kingdom's periods of heightened national security and prosperity, the momentum it had gained as a center of population and political prominence was sufficient to carry it through (O'Connor, 1972).

O'Connor (1972) suggests that three designated nome capitals—H Nesu (104), Spermeru (108), and Shena Khen (119)—were relatively unimportant as administrative centers during the Ramessid period. These capitals appear in less than 40% of the solutions, a result to be expected if their inefficient locations were a prime factor in their fall from prominence. Perhaps the political importance of Spermeru and H Nesu was overshadowed at this time by that of Hardai (101), which had become both administratively and economically important to the Ramessids (O'Connor, 1972).

In a number of instances, the location-allocation process designated sites close to known capitals as administrative centers. The two with the highest rates of inclusion—Nekhen (13), 11 times, and Sako (102), 14 times—are among those that Butzer (1976) specifies as being alternate capitals. That three other neighboring noncapitals had mayors in residence underlines the continued importance of mayors as administratve links outside of, or in addition to, the formal nome capital network. It also supports the hypothesis that the spatial arrangement of mayors may have been functionally more important during the Ramessid era than that of the traditional nome structure.

We have formulated the working hypothesis that in choosing locations for regional administrative centers—nome capitals and/or mayors—the Ramessid bureaucracy's main objective was to maximize control of the Nile Valley population. The results of the maximal covering location model heighten and support this hypothesis. Caution is in order, however, when imputing the Ramessid pharaohs' influence on the siting of administrative centers. The set of administrative centers at any time is the result of evolutionary processes. Our models have captured only a snapshot of these dynamic forces, and even that is blurred by the lack of tight temporal control over the data set. Some settlements subsequently lost importance, while others enjoyed increased hierarchical status as their fortunes, population, and influence grew. Mayors and other administrators at larger sites probably did consciously attempt to maximize their power and influence in order to extract increased tribute from the populations of other settlements within their suzerainty. However, even centers located to maximize their administrative coverage were still subject to the ecological constraints affecting their administrative efficiency relative to other settlements within their hinterlands.

ANALYSIS OF LATE HORIZON (AZTEC) SETTLEMENT IN THE NORTHEASTERN BASIN OF MEXICO

This second application of location-allocation modeling examined Late Horizon (Aztec) settlement in the northeastern portion of the Basin of Mexico (the Temascalapa-Teotihuacan region). In particular, we focused on two locational principles appropriate to this ecological and anthropological setting: administration and subsistence (Gorenflo, Church, and Bell, 1983). To help place this study in context, we begin with a brief summary of its sociocultural setting (see Gibson, 1964, for a more complete treatment). During the first portion of the Late Horizon period (A.D. 1350–1430), immediately preceding the settlement data used in this analysis, the Basin of Mexico was characterized by the presence of politically separate, competing city-states called tlatoque, initially numbering 14 but increasing to 50 by the end of this 80-year phase. Numerous conflicts among these political units adjusted and altered the administrative landscape of the region. Administrative requirements increased greatly toward the end of the period, resulting both from the competition among city-states and from the steady increase in the basin population. Among other things, the forces of conflict and population growth, in conjunction with an increasingly centralized administrative order, escalated the emphasis on extracting as much tribute as possible from the basin's somewhat fragile, semiarid environment (Evans, 1980, pp. 44–46).

In A.D. 1430 the political setting of city-states ended abruptly when the Triple Alliance became the controlling political body for the entire basin. This

takeover by the Tenochtitlan-Texcoco armies added a new administrative tier
to the area's ruling order as the previously independent city-states were brought
under the power of a single political force. The takeover resulted in a new
link in tribute flow; Alliance members brought various portions of the basin
into their control and, if anything, placed an even greater strain on the already
heavily stressed system's ability to provide subsistence. Population growth as
a whole continued at a rapid rate, culminating somewhere between 1 and 1.2
million by the Spanish arrival in 1519 (Sanders, Parsons, and Santley, 1979,
p. 184).

Both archaeological and ethnohistorical evidence suggest the presence during
the Late Horizon period (A.D. 1430–1519) of a carefully organized administrative
hierarchy of settlements, a hierarchy reflected in the Aztec social order as well
as in the regional system. We are, therefore, interested in examining settlement
location for administrative purposes. The goal is to determine how the population
in the northeastern portion of the basin could best have been covered to control
corvée labor, tribute flow, and political unrest and to establish political hegemony
over subservient centers. During the Late Horizon (Aztec) period region-wide
population was extremely dense, especially considering the labor-intensive agri-
culture characteristic of this semiarid area. Thus, we are also concerned with
how settlement location enabled the extraction of subsistence products, specifically maize, in an efficient manner.

Modeling Late Horizon Location in the
Temascalapa-Teotihuacan Regions

We confined this application to a carefully prescribed area in the northeastern
portion of the Basin of Mexico. The area of interest comprises approximately
900 square kilometers in what are known as the Temascalapa and Teotihuacan
survey regions (Fig. 3.3), representing a portion of the basin covered by a
surface survey to locate all archaeological sites. There were 344 settlements
containing permanent population during the Late Horizon period in this region.
Supplementing the archaeological remains in the Temascalapa-Teotihuacan area,
the region's agricultural potential has been examined carefully by authors interested in various ecological facets of the area's prehistory, making it particularly
attractive for the present analysis (e.g., Sanders, 1965, 1976a, 1976b; Sanders,
Parsons, and Santley, 1979; Evans, 1980).

Ethnohistoric sources suggest that travel among settlements in this region
involved a series of major and minor connecting trails, an idea supported by
the few extant maps from the early postconquest period (see Paso y Troncoso,
1908; Linne, 1948). Unfortunately, we do not know the exact configuration of
such a network for the study area and, as a result, are forced to estimate
interaction. The location of the arcs connecting major nodal pairs was first

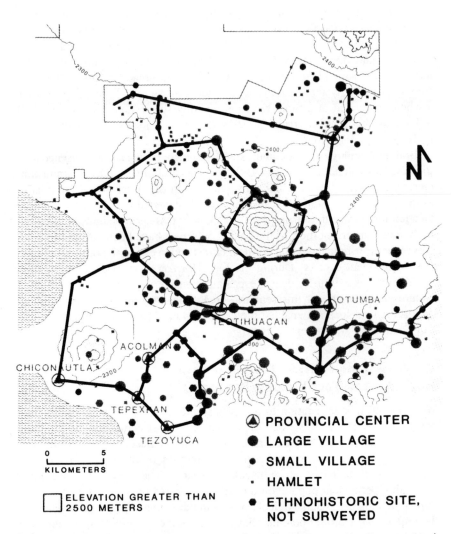

Figure 3.3.   Basic network configuration of the Temascalapa-Teotihuacan study area, used for determining site-interaction travel time.

approximated by considering the location of intermediate settlements and topographic barriers (Gorenflo, n.d.). The resulting estimated network is depicted in Fig. 3.3. The interaction measures expressed in terms of the travel time between nodes were calculated based on a formula developed by Tobler (personal communication) that relates walking speed to slope:

$$v = 6e^{-3.5|s+.05|}$$

where

v = velocity and s = slope.

For the purpose of examining location with respect to food production, we calculated the agricultural capability of each site based on the estimated number of food producers present and the type of agricultural practices possible in different parts of the study area. We simplified matters by examining only one crop, maize, which was the dominant agricultural product. Estimates of maize production were generated for the 344 sites based, among other factors, on Evans' (1980, pp. 145–150) estimate of farming productivity under different ecological conditions in the Teotihuacan Valley (Table 3.2) and Sanders and Santley's (1980) calculation of area cultivated by each family. About 50% of the inhabitants of regional centers and 10% of the village dwellers are assumed to have produced no food themselves (see Sanders, Parsons, and Santley, 1979, pp. 158–177). We extrapolated Evans' piedmont figure to the Temascapala piedmont region assuming that the latter would have been similar in productivity to the rainfall-irrigated piedmont of the Teotihuacan Valley.

Maximizing Coverage in the Northeastern Basin of Mexico

The model developed to examine settlement in the northeastern basin employs a desired-service-distance concept: facilities are located to serve administratively all users, or as many as possible, within a desired range. This concept is similar to the maximal covering location model of Church and ReVelle (1974). In

Table 3.2. Productivity Estimates for Maize in the Teotihuacan Valley During the Late Horizon Period (1350–1520 A.D.)

Type of Agriculture	kg of Maize/ha
Not arable	0
Rainfall piedmont	300
Floodwater terrace	800
Floodwater alluvium	1,000
Irrigated alluvium	1,400
Chinampa (man-made islands of reeds and mud)	3,000

Source: Evans, 1980.

addition to simulating administrative concerns, we wanted the model to be sensitive both to the distribution of food-production potential in the region and to site type. Thus, we developed a multiobjective maximal covering location model, with an associated preference function for settlement type, as follows:

$$\text{Max } Z = \sum_{i \in I} (w_1 \text{pop}_i + w_2 \text{prod}_i) Y_i + \sum_{j \in J} (w_3 v_j X_j) \qquad (3.2)$$

subject to

$$\sum_{j \in N_i} X_j \geqslant Y_i$$

$$\sum_{j \in J} X_j = p$$

$$X_j = (0, 1) \quad \text{for all } j \in J$$
$$Y_i = (0, 1) \quad \text{for all } i \in I$$

where

w_1, w_2, and w_3 = relative weightings for population, productivity, and site type, respectively,
Y_i = 1 if a settlement i is covered, 0 otherwise,
y_j = relative importance of site type,
pop_i = population to be served at settlement i,
prod_i = the maize productivity at settlement i.

This model is flexible, allowing a focus on either population or productivity separately or, through the weights, on a combination of the two. Moreover, we can examine settlement configurations so that any of the 344 site locations in the Temascalapa-Teotihuacan region is feasible; or, through the use of the preference function ($w_3 v_j$) we can "force" centers to be located at specific site types. This weighting procedure can be used to determine the level of efficiency for different coverage distances of the observed configuration of regional centers. We examined three different coverage distances S, beginning at 1.5 hours of travel (approximately 7.5 km over level ground), a value that enables nearly complete coverage with the arrangement of ten centers, and coverage in the 85% range for seven centers (the number of regional centers actually present in the area during the Aztec Late Horizon). With this as a foundation, we constrained the solutions in two ways. First, to analyze the efficiency possible within empirical limitations, we forced centers to be placed at known regional-center locations. Second, we examined each of two more-restricted coverage distances, 1.25 and 1.0 hours, to assess the possible efficiency of various site configurations with less covering ability. Because of space limitations, this chapter reports only a few of the models tested.

We began by examining administrative coverage within a 1.5-hour limit. We specified an objective function where $w_3 = 1.0$, $w_2 = 0$, and $w_3 = 0$. At first we allowed administrative facilities to be placed at any of the 344 sites present in our data set. If we located seven administrative centers to maximize the above objective, only one of the actual Aztec regional-center sites (Tepexpan) appeared in the most efficient solution (Fig. 3.4). There was a general emphasis on the use of larger sites, but to optimize coverage of population, alternatives to actual administrative centers were chosen.

Locating seven political centers in this manner produced an administrative coverage value of about 85%. Moreover, although we placed no emphasis on productivity coverage, we find a robustness in the locational patterns, as approximately 86% of the regional productivity also is located within 1.5 hours of these sites. Some redundancy is to be expected between population coverage and agricultural coverage, for the presence of more individuals at a site generally enables greater food production. But it should be recalled that productivity estimates incorporate the different agricultural practices of which various portions of a region are capable and also account for different numbers of nonfood producers in the various sorts of sites. The relationship between administrative coverage and productivity coverage therefore departs from linearity. In the end, however, there appears to be very little trade-off between covering for administrative purposes and covering to maximize productivity. For all intents and purposes, covering for administrative purposes in the northeastern Basin of Mexico means covering for agricultural productivity as well.

We obtain different configurations if we weight the ethnohistorically defined regional centers very heavily, thereby adding to their attractiveness for inclusion in the covering model solution. By using various combinations of weightings, different cases can be examined. If, for example, we maximize an objective with a slight weight on population ($w_1 = 0.25$), no weight or agricultural productivity ($w_2 = 0$), and a heavy weight on the preference for use of an actual Aztec regional center ($w_3 = 5000$), the seven-site solution contains all of the actual regional centers (Fig. 3.5). As expected, because of the inclusion of these inefficiently located administrative centers, coverage drops to 58% for population and 53% for productivity.

Conclusions Regarding the Aztec Settlement System

Although other factors were no doubt important in the evolution of settlement patterns in the northeastern Basin of Mexico, one can argue that the coverage of administration and productivity would have been fundamental to the success

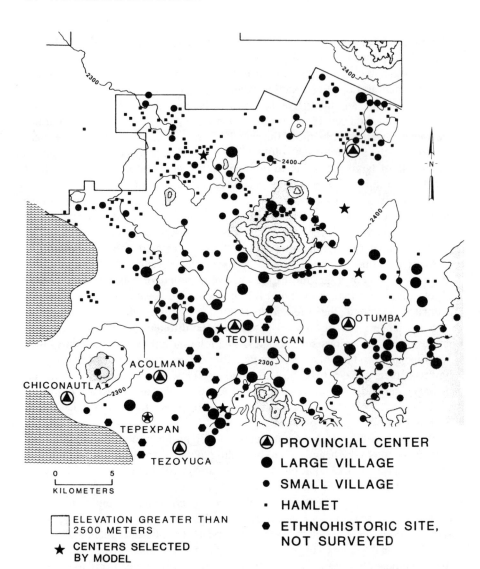

Figure 3.4. Seven-site solution; $S = 1.5$, $w_1 = 1.0$, other weights $= 0$.

of such a complex sociopolitical system—especially if that system were under heavy population stress.

The study suggests that a reasonably efficient spatial organization of sites for the purpose of covering population and productivity was possible with a relatively small service time-distance. The degree of simultaneous coverage

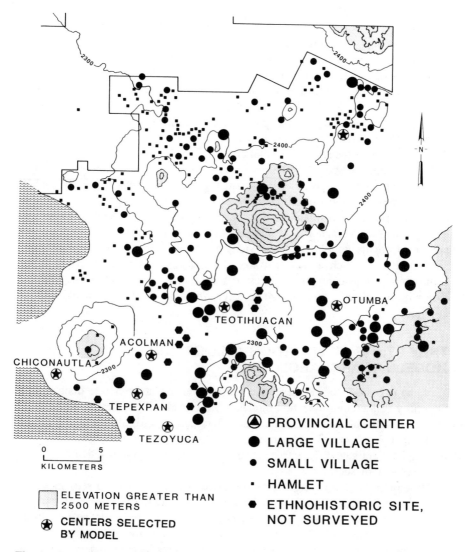

Figure 3.5. Seven-site solution, stressing known center locations ($S = 1.5$, $w_1 = 0.25$, $w_2 = 0$, $w_3 = 5000$).

for both variables in the seven-center solution was approximately 85% at $S = 1.5$. Yet if we evaluate the efficiency of the seven ethnohistorically defined regional centers, the level of coverage at the same service distance drops by approximately 30%. That actual location patterns depart so greatly from the most efficient solution is a matter that deserves further anthropological and geographical attention. Several possible reasons might be considered. The leaders at this time *imposed* a great deal of spatial and political organization. One of the leaders of the Triple Alliance, shortly after its rise to power, intertwined hinterlands of the six centers in the Teotihuacan Valley region with those of Texcoco far to the south, and the purpose was to better control the city-states (tlatoque) (Evans, 1980). This organizational approach, in conjunction with other strategies, was apparently quite successful in removing strife among the tlatoque. But given the results of our study, such a strategy would have increased the inefficiency of the region's spatial organization. This anomaly can be viewed from contrasting perspectives. The complex spatial organization imposed through Texcoco must have been quite important to that ruling body to justify such a drastic surrender of locational efficiency. From a different perspective, the system itself was apparently resilient enough, despite its substantial population density, to withstand such an increasingly inefficient spatial order. A system under extreme pressure from its population to provide subsistence and administration should have evolved in such a way that spatial efficiency was quite high. Previous anthropological research has suggested that severe ecological stress was the case with the Aztecs. Our present research findings call these conclusions into question.

THE FUTURE OF LOCATION-ALLOCATION MODELS IN ARCHAEOLOGY

With the two case studies presented, this chapter has attempted to demonstrate the utility of location-allocation models for the study of archaeologically documented settlement systems. Each case offers certain insights about the nature of the spatial organization underlying the cultural systems involved. Moreover, our results illuminate different aspects of these settlement systems, demonstrating the flexibility of such location-allocation models.

Developing and applying a multiobjective covering model to settlement remains from Ramessid Egypt seems to substantiate the importance of the mayor in the political bureaucracy of this sociocultural setting. The model also confirms the results of years of intensive field investigations: settlements important during later administrative regimes in Egypt were efficiently located to serve either as service centers or resource-based sites, while those that lost importance were often not well located. Their inefficiency, detected by the models, may have been a harbinger of their waning importance.

The results of the location-allocation applications to the northeastern Basin of Mexico may be problematic for archaeologists, for our results suggest that the seven designated Aztec political centers may have been quite inefficiently located. Selection of more efficiently located political centers could have considerably improved control over both the distribution of agricultural surplus and the population. The results of the location-allocation models question the degree to which the Late Horizon Aztecs in the northeastern Basin of Mexico were under severe ecological stress, for they were apparently able to feed their entire population and maintain administrative control of the area despite their inefficient organizational structure. Ignoring for the present the needs of the empire capital at Tenochtitlan, one wonders how many more people might have been supported if the settlement system in this portion of the basin were more efficiently organized.

Locational analysis in archaeology need not be confined to tests of the empirical validity of central place theory and other classic location theories, or to describing spatial distributions of phenomena with statistics such as those from nearest-neighbor methods or centrographic measures, which have little theoretical basis for interpretation. Further applications of location-allocation models could be fashioned to test alternative hypotheses of settlement formation and evolution.

The location-allocation modeling approaches suggested here can in no way replace the need for solid archaeological field work. Quite the opposite. Only through the skills of the field archaeologist can reasonable insights into underlying regional settlement systems be developed. Location-allocation models are merely tools available to explore potential cultural processes, and perhaps the mental constructs of past decision makers, that bear on settlement pattern development. Field archaeologists painstakingly attempt to reveal the cultural mechanisms affecting ancient societies, and in so doing, they gain at least partial insight into the thinking of people in those societies.

A symbiotic relationship between the archaeologist and the modeler must, therefore, be fashioned so that the model draws upon both empirical evidence and the best thinking available about the evolutionary and behavioral underpinnings of the societal structure being modeled. To be reasonable and interpretable, normative settlement configurations are always contingent on the skills of both the modeler and the field archaeologist.

REFERENCES

Bell, T., 1979, Location-allocation modeling in archaelogical settlement pattern research, a paper presented at the 75th annual meeting of the Association of American Geographers, Philadelphia, Pennsylvania.

Bell, T. L. and R. L. Church, 1985, Location-allocation modeling in archaeological settlement pattern research: some preliminary applications, *World Archaeology* **16**:354–371.

Butzer, K. W., 1960, Archaeology and geology in ancient Egypt, *Science* **132**:1617–1624.

Butzer, K. W., 1976, *Early Hydraulic Civilization in Egypt*, University of Chicago Press, Chicago.

Christaller, W., 1933, *Central Places in Southern Germany,* trans. C. W. Baskin, 1966, Prentice-Hall, New Jersey.

Church, R. L. and C. ReVelle, 1974, The maximal covering location problems, *Regional Science Association Papers* **32**:101–118.

Church, R. L. and C. ReVelle, 1976, Theoretical and computational links between the *p*-median, location set-covering, and maximal covering location problems, *Geographical Analysis* **8**:406–415.

Clarke, D. L., ed., 1977, *Spatial Archaeology,* Academic Press, London.

Cohon, J. L., R. L. Church and D. P. Sheer, 1979, Generating multiobjective trade-offs: an algorithm for bicriterion problems, *Water Resources Research* **15**:1001–1010.

Crumley, C. L., 1979, Three locational models: an epistemological assessment for anthropology, in *Advances in Archaeological Method and Theory,* vol. 2, M. Schiffer, ed., Academic Press, New York.

Defense Mapping Agency, 1960, *Egypt 1:100,000,* Ser. P677, Defense Mapping Agency, Washington, D.C.

Evans, S., 1980, *A settlement system analysis of the Teotihuacan region, Mexico A.D. 1350–1520.* Ph.D. dissertation, Department of Anthropology, Pennsylvania State University, University Park.

Flannery, K. V., 1972, The cultural evolution of civilizations, *Annual Review of Ecology and Systematics* **3**:399–426.

Gardiner, A. H., 1947, *Ancient Egyptian Onomastica,* 2 vols., Oxford University Press, Oxford.

Gibson, C., 1964, *The Aztecs Under Spanish Rule,* Stanford University Press, Palo Alto.

Gorenflo, L., R. L. Church, and T. L. Bell (1983), Analysis of Late Horizon settlement in the northeastern Basin of Mexico: an application of location-allocation modeling, paper presented at 38th Annual Meeting of the Southeastern Division of the Association of American Geographers, Orlando, Florida, 1983.

Gorenflo, n.d., Precolumbian travel in the Temascalapa-Teotihuacan region of the Basin of Mexico, ms. on file, Department of Geography, University of California, Santa Barbara.

Hillsman, E., 1980, *Heuristic solutions to location-allocation problems: a user's guide to Alloc IV, V and VI,* Monograph No. 7, Department of Geography, University of Iowa, Iowa City.

Johnson, G., 1972, A test of the utility of central place theory in archaeology, in *Man, Settlement and Urbanism,* P. J. Ucko, Ruth Tringham and G. W. Dimbleby Schenkman, Cambridge, Mass.; pp. 769–785.

Johnson, G., 1977, Aspects of regional analysis in archaeology, *Annual Review of Anthropology* **9603**:479–508.

Kauffman, B. E., 1981, The maximal covering location problem as a simulation of decision making in Ramessid Egypt, Master's thesis, Department of Anthropology, University of Chicago.

Kauffman, B. E., R. L. Church, and T. L. Bell, 1981, Political centralization of the Nile Valley during the Ramessid period: substantiating archaeological hypotheses with a location-allocation model, paper presented at 36th Annual Meeting of the Southeastern Division of the Association of American Geographers, Atlanta, Georgia, 1981.

Keene, A., 1979, Economic optimization models and the study of hunter-gatherer subsistence settlement systems, in *Transformation: Mathematical Approaches to Culture Change*, C. Renfrew and K. Cooke, eds., Academic Press, New York.

Keene, A., 1981, *Prehistoric Foraging in a Temperate Forest: A Linear Programming Model*, Academic Press, New York.

Kees, H., 1961, *Ancient Egypt: A Cultural Topography*, University of Chicago Press, Chicago.

Linne, S., 1948, *El Valle y la Ciudad de Mexico en 1550*, Ethnographic Museum of Sweden, Publication No. 1, Stockholm.

Marcus, J., 1973, Territorial organization of the lowland classic Maya, *Science* **180:**911–916.

O'Connor, D., 1972, The geography of settlement in ancient Egypt, in *Man, Settlement and Urbanism*, P. Ucko, R. Tringham and G. W. Dimbleby et al., eds., Duckworth, London, pp. 681–698.

Parsons, J. R., 1972, Archaeological settlement patterns, *Annual Review of Anthropology* **1:**127–150.

Paso y Troncoso, F. del, 1908, *Papales de Nueva España*, vol. 2, Madrid.

Reidhead, V., 1976, Optimization and food procurement at the prehistoric Leonard Haag site, southeast Indiana: a linear programming approach, Ph.D. diss., Department of Anthropology, Indiana University, Indiana.

Reidhead, V., 1979, Linear programming models in anthropology, *Annual Review of Anthropology* **8:**543–578.

Reidhead, V., 1980, The economy of subsistence change: a test of an optimization model, in *Modeling Change in Prehistoric Subsistence Economies*, T. Earle and A. Christensen, eds., Academic Press, New York.

Rushton, G., 1971, Postulates of central place theory and properties of central place systems, *Geographical Analysis* **3:**140–156.

Sanders, W., 1965, *The Cultural Ecology of the Teotihuacan Valley*, Department of Anthropology, Pennsylvania State University, College Station.

Sanders, W., 1976a, The agricultural history of the Basin of Mexico, in *The Valley of Mexico*, E. Wolf, ed., University of New Mexico Press, Albuquerque, pp. 101–159.

Sanders, W., 1976b, The natural environment, in *The Valley of Mexico*, E. Wolf, ed., University of New Mexico Press, Albuquerque, pp. 59–67.

Sanders, W. and R. Santley, 1980, A tale of three cities: energetics and urbanization in prehispanic central Mexico, paper prepared for Burg Wartenstein Symposium no. 86.

Sanders, W., J. R. Parsons, and R. Santley, *The Basin of Mexico*, Academic Press, New York.

Scott, A. J., 1971, *An Introduction to Spatial Allocation Analysis*, Resource Paper no. 9, Commission on College Geography, Association of American Geographers, Washington, D.C.

Smith, C. A., ed., 1976, *Regional Analysis*, vols I and II, Academic Press, New York.

Smith, M. E., 1979, The Aztec marketing system and settlement pattern in the valley of Mexico: a central place analysis, *American Antiquity* **44:**110–125.

Steponaitis, V. P., 1978, Location theory and complex chiefdoms: a Mississippian example, in *Mississippian Settlement Patterns,* B. D. Smith, ed., Academic Press, New York, pp. 417–453.

Steponaitis, V. P., 1981, Settlement hierarchies and political complexity in non-market societies: the formative period in the valley of Mexico, *American Anthropologist* **83**:320–363.

Toregas, C. and C. Revelle, 1982, Optimal location under time or distance constraints, *Regional Science Association Papers* **28**:133–143.

Watson, R., 1972, The "new archaeology" of the 1960's, *Antiquity* **46**:210–215.

4

A SPATIAL MODEL OF PARTY COMPETITION WITH ELECTORAL AND IDEOLOGICAL OBJECTIVES

V. Ginsburgh

Université Libre de Bruxelles and CORE

P. Pestieau

Université de Liège and CORE

J.-F. Thisse

Université Catholique de Louvain and CORE

The mathematical theory of elections in a representative democracy is still very much dominated by the Hotelling-Downs model (Hotelling, 1929; Downs, 1957). Accordingly, political opinions are depicted as lying on a left-right line; voters vote for the party closest to their most preferred position; and parties choose their platform with the aim of maximizing their votes. The following results are then shown to hold (see the recent survey on spatial competition by Graitson, 1982); (1) If there are two parties, both are driven toward the center of the political spectrum, more precisely to the position favored by the median voter; (2) in a three-party system, however, no stable outcome exists; (3) when there are more than three parties an equilibrium occurs in which some parties are pairwise located.

The underlying assumptions as well as the results they lead to are so questionable that various modifications of that model have been proposed. Surprisingly enough, these modifications have essentially concentrated on the structure of the policy space—for example, multidimensional platforms (Davis, Hinich, and Ordeshook, 1970)—and the behavior of voters—for example, the possibility of abstention (Davis, Hinich, and Ordeshook, 1970)—but rarely have they concentrated on the objective of the parties. In that respect, the theory has remained quite faithful to Downs, who states the fundamental hypothesis of his model

Acknowledgment: This paper was written in part during the first and third authors' visit to McMaster University, whose financial support is gratefully acknowledged. The authors thank P. Champsaur for his helpful comments.

as "parties formulate policies in order to win elections, rather than win elections in order to formulate policies" (Downs, 1957, p. 28).

For political scientists, vote maximization is undoubtedly an important objective of parties. However, it is not the only one. There is also the need to stick to, or at least not to be too unfaithful to the ideology that, more than the platform, characterizes the party's historical past and its members' opinion (Kolm, 1977; Robertson, 1976). Clearly, *ideological purity* and *vote maximization* are often incompatible, and the fundamental question is to see how such a conflict is solved in reality and how it can be formalized in theory. Downs does not ignore this conflict but in his view "the desire to keep power per se plays a larger role in the practical operation of democratic politics than the desire to implement ideological doctrines" (Downs, 1957, p. 112). This idea leads to the representation of parties as single-objective agents, interested only in getting votes and irresponsive to any departure from their ideology. Put another way, ideological mobility is assumed to be free.

We contend that in some countries and in some periods, ideology may be an important concern for parties because ideological mobility is *not* free. Therefore, party competition theory should not discard ideological mobility to favor only vote maximization. Rather, it should integrate both objectives in a bicriterion approach. Under certain assumptions of concavity, this integration can be accomplished through the maximization of a convex combination of the two objectives. For given weights, the electoral platforms are then determined at the equilibrium of the noncooperative game. In that choice process, the impact of ideology is to reduce the lure of the center and, therefore, to enlarge the set of possible outcomes. Further, in a dynamic setting, the weights should not be fixed but should be made sensitive to election outcomes. The problem is now to trace the possible paths followed by the parties over time and to determine, we hope, a long-run equilibrium. By that bicriterion approach, we hope to grasp an important aspect in the electoral process that has long been ignored either because the traditional reference, the United States, is not an ideology-prone society or because modeling conflicting objectives is not an easy task.

A few recent papers, however, address this question. In a study of U.S. Senate voting on the regulation of coal strip-mining, Kalt and Zupan (1984) compare ideology voting and voting to win. They argue that the latter objective poorly explains and predicts actual political decisions, while the former yields a broader and more realistic view of political behavior. Some authors have also challenged the dominant model by adding another constraint or objective. Samuelson (1984) assumes that candidates are restricted to positions close to their initial ones. Considering a sequence of elections, he then concludes that incumbents have an advantage. For Wittman (1983), candidates actually have two objectives: they choose a platform that is closest to their most preferred positions and that also leads to winning the election. To solve this apparent conflict, Wittman supposes that each candidate maximizes the sum of the utility received if elected and that received if not elected, weighting each by the probability of winning and losing, respectively. The existence of an equilibrium is then proved in a multidimensional-issue space.

Unlike Samuelson and Wittman, this chapter does not consider a probabilistic model. Rather it stays close to the original formulation of Hotelling and Downs but, as noted above, with parties maximizing a bicriterion function involving both ideology and winning. This approach leads to the results described in the next paragraph.

In a two-party democracy, the platforms are not necessarily situated at the center, nor are they symmetrical. Generally, they fall between the parties' ideological positions and the median voter position. The electoral outcome ultimately depends on the relative importance of the parties' objectives and their ideological positions. On the other hand, when the game is imbedded in a dynamic setting where the relative weight of ideology is increased (decreased) by the winning (losing) party, both parties can be attracted to positions equidistant from the center. Again, this long-run equilibrium will often be different from the Hotelling-Downs solution. Finally, we prove that in a multiparty system, a unique equilibrium exists and that parties are scattered all over the political spectrum.

The chapter's next section introduces the bicriterion approach in the two-party model and proves the existence and uniqueness of a Nash equilibrium. The model is illustrated by using specific examples for the objectives. The section after that provides sensitivity analysis on the relative weights, and then the following section describes an adjustment process based on the parties' reassessment of the relative weights. In the next to last section, we turn to a multiparty system and show that a unique Nash equilibrium exists. The final section presents our conclusion.

THE TWO-PARTY SYSTEM

The positions of voters and parties are defined on a segment of unit length, which represents all possible political choices. The position x_1 (x_2) of party 1 (party 2) is measured starting from the left (right) origin of the segment. The same holds for the ideological position of party 1 (party 2), which is assumed to be given and is denoted by m_1 (m_2).

Remark 1.   We leave the problem of determining m_i out of the model. Assuming that each party is made up of members, a subset of the total population of voters, m_i, may correspond to the position of the median member (Robertson, 1976).

Assumption 1.   The distribution of voters is uniform over the policy segment.

Assumption 2.   A voter casts his or her vote for the party closest to his or her most preferred position.

Assumption 3.   The ideological position m_i of party i belongs to $[0, 1/2]$. The ideological positions do not coincide; moreover the first (second) party is a left-wing (right-wing) party.

Assumption 4.   The position x_i of party i is to be chosen in $[m_, 1 - m_j]$. Party 1, say, has no incentive to locate to the left (right) of its (rival's) ideological position, since it then loses voters to its right (left) without gaining on the ideological-purity objective.

Following our bicriterion approach, each party is described by two utility functions corresponding to the ideological-purity and vote-maximization objectives.

Assumption 5.   On $[m_i, 1/2]$, ideological purity is expressed as a twice continuously differentiable, strictly decreasing and strictly concave function of the distance between the platform x_i and the ideological position m_i: $I_i (x_i - m_i)$. On $[1/2, 1 - m_j]$, the ideological factor is supposed to become overwhelming: $I_i(x_i - m_i) = -\infty$ for $x_i > 1/2$.

That I_i is decreasing with the distance from m_i is consistent with the requirement to be faithful to the party ideology. The assumption of concavity expresses that a small departure from m_i is less consequential the closer to m_i the party is located. Having I_i arbitrarily low for x_i beyond $1/2$ reflects that in the present context, the left-wing party, say, cannot conceive of choosing a position to the right of the center.

The number of votes of party i, then, is

$$\frac{1 + x_i - x_j}{2}$$

Assumption 6.   Vote maximization is represented by a twice continuously differentiable, strictly increasing and concave function of $x_i - x_j$: $V_i(x_i - x_j)$. The payoff function of party i in the political game is defined by

$$U_i(x_i, x_j) = \theta_i I_i(x_i - m_i) + (1 - \theta_i) V_i(x_i - x_j), \quad i \neq j \qquad \textbf{(4.1)}$$

where $\theta_i \epsilon[0, 1]$ and $1 - \theta_i$ are to be interpreted as the relative weights of the ideological-purity and vote-maximization objectives. As I_i and V_i are concave in x_i, the set of Pareto-optimal positions for party i corresponding to a given position x_j of party i is the set of maximizers of equation (4.1) obtained when θ_i varies from 0 to 1. For a particular value of θ_i, the ratio $\theta_i/(1 - \theta_i)$ is the marginal rate of substitution between the corresponding objectives. Spatially, we therefore notice that our bicriterion approach adds to the traditional spatial competition mechanism some elements of Weber's (1909) location theory, in which distances to some fixed positions are to be minimized.

We are interested in a noncooperative game in which party i maximizes U_i (x_i, x_j) on the set of strategies $[m_i, 1 - m_j)$. A Nash equilibrium for this game is defined as a pair $(\bar{x}_1, \bar{x}_2) \in [m_1, 1 - m_2] \times [m_2, 1 - m_1]$ such that, for $i = 1, 2$, $U_i(\bar{x}_i, \bar{x}_j) \geq U_i(x_i, \bar{x}_j)$ for any $x_i \in [m_i, 1 - m_j]$ and $i \neq j$.

Note that for $\theta_i = 0$, $U_i(x_i, x_j) = V_i(x_i - x_j)$, and we end up with the Hotelling-Downs model, for which it is known that $\bar{x}_i = \bar{x}_j = 1/2$. For $\theta_i - 1$, $U_i(x_i, x_j) = I_i(x_i = m_i)$, and it easily follows that $\bar{x}_i = m_i$ and $\bar{x}_j = m_j$. We want now to concentrate on the intermediate values of θ_i.

Assumption 7. The relative weights θ_i and $1 - \theta_i$ belong to [0, 1].

Remark 2. The above model can be reinterpreted in the context of firm-location theory. As in Hotelling's paper, it is assumed that both firms try to maximize their profit by capturing the largest number of customers. But now they also have to bear the transportation cost of an input from some supply places m_i.

Our first proposition is concerned with the existence and uniqueness of an equilibrium.

Proposition 1. If Assumptions 1–7 hold, then a unique Nash equilibrium exists.

Proof. From Assumptions 5 and 6 and Friedman's (1977) Theorem 7.4, it follows that an equilibrium exists. Furthermore,

$$\frac{\partial^2 U_i}{\partial x_i^2} + \left| \frac{\partial^2 U_i}{\partial x_i \partial x_j} \right| = \theta_i$$

$I_i'' < 0$ by Assumptions 5 and 7 and the uniqueness is guaranteed by Friedman's (1977) Theorem 7.12.

Remark 3. Assumption 5 implies that it is always optimal for party i to choose x_i in $[m_i, 1/2]$, whatever $x_j \in [m_i, 1 - m_j]$. Consequently, from now on, Assumption 4 is replaced by $x_i \in [m_i, 1/2]$.

Remark 4.   Strict concavity of I_i is needed only for uniqueness.

The actual outcome of the party competition ultimately depends on the utility functions I_i and V_i, on the ideological positions m_i, and on the relative weights θ_i. Interestingly, the platforms selected by the parties are not likely to be located at the center of the political spectrum, as predicted by the Hotelling-Downs theory. Another difference with this model is the probable differentiation of the two parties in terms of their number of votes. In most cases we may expect $\bar{x}_i$ to be different from $\bar{x}_j$ so that a winner exists. Unfortunately, at the present level of generality, it is difficult to determine who the winner is. Nevertheless, in the particular, but not too unrealistic case where both parties have the same utility functions, the following statement can be easily derived from the sensitivity analysis made in the next section. (1) if $\theta_1 = \theta_2$, then $m_i > m_j$ implies that party i wins; (2) if $m_1 = m_2$, then $\theta_i < \theta_j$ means that party i wins. In what follows, two meaningful examples are dealt with to illustrate the working of the game.

Example 1.   *The logarithmic-linear model*

Party i maximizes $\theta_i \log(1 - x_i + m_i) + (1 - \theta_i)(x_i - x_j)$ on $[m_i, 1/2]$.

If λ_i and μ_i denote the multipliers associated with the constraints $m_i \leqslant x_i$ and $x_i \leqslant 1/2$, the solution must verify

$$-\theta_i(1 - \bar{x}_i + m_i)^{-1} + (1 - \theta_i) + \lambda_i - \mu_i = 0 \qquad \text{(4.2a)}$$

$$\lambda_i(\bar{x}_i - m_i) = 0 \qquad \text{(4.2b)}$$

$$\mu_i(1/2 - \bar{x}_i) = 0 \qquad \text{(4.2c)}$$

$$\lambda_i, \mu_i \geqslant 0 \qquad \text{(4.2d)}$$

Some simple manipulations show that $\bar{x}_i = m_i$ for $\theta_i \geqslant 1/2$ and that $\bar{x}_i = 1/2$ for

$$\theta_i \leqslant \frac{2m_i + 1}{2m_i + 3}$$

For the values of θ_i in between, $\bar{x}_i$ is interior and given by

$$m_i + \frac{1 - 2\theta_i}{1 - \theta_i}$$

In all cases we note that $\bar{x}_i$ depends only on the characteristics of party i. This fact is due to the linear structure of V_i. The results are summarized in Table 4.1.

Table 4.1. Solutions in the Logarithmic-Linear Case

θ_1 \ θ_2	$\leqslant \dfrac{2m_2+1}{2m_2+3}$	$\left]\dfrac{2m_2+1}{2m_2+3}, \dfrac{1}{2}\right[$	$\geqslant \dfrac{1}{2}$
$\leqslant \dfrac{2m_1+1}{2m_1+3}$	$\bar{x}_1 = \dfrac{1}{2}, \bar{x}_2 = \dfrac{1}{2}$	$\bar{x}_1 = \dfrac{1}{2}, \bar{x}_2 = m_2 + \dfrac{1-2\theta_2}{1-\theta_2}$	$\bar{x}_1 = \dfrac{1}{2}, \bar{x}_2 = m_2$
$\left]\dfrac{2m_1+1}{2m_1+3}, \dfrac{1}{2}\right[$	$\bar{x}_1 = m_1 + \dfrac{1-2\theta_1}{1-\theta_1}, \bar{x}_2 = \dfrac{1}{2}$	$\bar{x}_1 = m_1 + \dfrac{1-2\theta_1}{1-\theta_1}$ $\bar{x}_2 = m_2 + \dfrac{1-2\theta_2}{1-\theta_2}$	$\bar{x}_1 = m_1 + \dfrac{1-2\theta_1}{1-\theta_1}, \bar{x}_2 = m_2$
$\geqslant \dfrac{1}{2}$	$\bar{x}_1 = m_1, \bar{x}_2 = \dfrac{1}{2}$	$\bar{x}_1 = m_1, \bar{x}_2 = m_2 + \dfrac{1-2\theta_2}{1-\theta_2}$	$\bar{x}_1 = m_1, \bar{x}_2 = m_2$

Remark 5.   Notice that the equilibrium solutions contained in the main diagonal boxes of Table 4.1 are obtained while ignoring the constraints $x_i \leqslant 1/2$. Consequently, they can be shown to be Nash equilibria in the entire interval $[m_i, 1 - m_j]$ when I_i is supposed to take fine values for $x_i > 1/2$. This situation does not necessarily hold for the other solutions.

Example 2.   *The logarithmic model*
Party i maximizes $\theta_i \log(1 + m_i - x_i) + (1 - \theta_i) \log(1 + x_i - x_j)$ on $[m_i, 1/2]$, and the solution obeys

$$-\theta_i(1 + m_i - \bar{x}_i)^{-1} + (1 - \theta_i)(1 + \bar{x}_i - \bar{x}_j)^{-1} + \lambda_i - \mu_j = 0 \quad \textbf{(4.3)}$$

together with $(4.2b) - (4.2d)$.

It follows from equation (4.3) that $\bar{x}_i$, when it is interior to $[m_i, 1/2]$, is a function of $\bar{x}_j$, unlike in Example 1. As a consequence, $\bar{x}_i$ depends in general on some characteristics of party j. Thus, in the case of an interior solution, we have

$$\bar{x}_i = (1 - \theta_i\theta_j)^{-1}[(1 - \theta_i)(1 + m_i) + \theta_i(1 - \theta_j)m_j - 2\theta_i\theta_j] \quad \textbf{(4.4a)}$$

while for $\bar{x}_j = m_j$ and $1/2$ we obtain, respectively,

$$\bar{x}_i = (1 - \theta_i)(1 + m_i) - \theta_i(1 - m_j) \quad \textbf{(4.4b)}$$

$$\bar{x}_i = \frac{2 - 3\theta_i}{2} - (1 - \theta_i)m_i \quad \textbf{(4.4c)}$$

Determining the regions of parameters corresponding to the different solutions does not lead to a simple partition, as in Table 4.1, and is therefore omitted here. Nevertheless, it is still true that the structure of solutions is similar to that of Example 1.

From these examples it should be clear that the classical Hotelling-Downs solution is the exception and not the rule in a model where ideological purity is of concern for the parties.

SENSITIVITY ANALYSIS

This section studies the impact on the game's solution of varying some parameters. Let $\bar{x}_i (\theta_1, \theta_2)$, $i = 1, 2$, be the equilibrium strategy of party i corresponding to the values θ_1 and θ_2. We know that $\bar{x}_i (\theta_1, \theta_2)$ is univocally determined for $\theta_j = 0$ and $\theta_i - 1$ and, by Proposition 1, that $\bar{x}_i(\theta_1, \theta_2)$ is a well-defined function for any $\theta_1, \theta_2 \in]0, 1[$. In addition, $\bar{x}_i(\theta_1, \theta_2)$ is continuous on $[0, 1]$.[2] This situation follows from the continuity and the monotonicity in $x_j, \theta_1,$

and θ_2 of the best reply functions $\bar{x}_i(x_j, \theta_1, \theta_2) = \max U_i(x_i, x_j)$, subject to x_i $\epsilon[m_i, 1/2]$. Finally, for any interior equilibrium strategy $\bar{x}_i(\theta_1, \theta_2)\epsilon]m_i, 1/2[$, we deduce from the implicit function theorem that $\bar{x}_i(\theta_1, \theta_2)$ is differentiable. A similar argument applies, *mutatis mutandis*, to the function $\bar{x}_i(m_1, m_2)$. Those properties are illustrated by the solutions depicted in Table 4.1 for the linear-logarithmic model and by equations (4.4a, b, and c) in the logarithmic model.

Proposition 2.   Let $\bar{x}_i \epsilon]m_i, 1/2[$, $i = 1, 2$, be the unique equilibrium strategy of party i for some given θ_i and m_i. Then,

$$(a) \quad \frac{\partial \bar{x}_i}{\partial \theta_i} < 0, \quad \frac{\partial \bar{x}_j}{\partial \theta_i} \leq 0 \quad \text{and} \quad \left| \frac{\partial \bar{x}_i}{\partial \theta_i} \right| > \left| \frac{\partial \bar{x}_j}{\partial \theta_i} \right|$$

$$(b) \quad \frac{\partial \bar{x}_i}{\partial m_i} > 0, \quad \frac{\partial \bar{x}_j}{\partial m_i} \geq 0 \quad \text{and} \quad \frac{\partial \bar{x}_i}{\partial m_i} > \frac{\partial \bar{x}_j}{\partial m_i}$$

Proof.   For an interior equilibrium we have

$$\frac{\partial U_i}{\partial x_i} = \theta_i I'_i + (1 - \theta_i)V'_i = 0, \quad i = 1, 2$$

Totally differentiating these conditions yields the system

$$\begin{pmatrix} \theta_i \, I''_i + (1 - \theta_i) \, V''_i & -(1 - \theta_i) \, V''_i \\ -(1 - \theta_j) \, V''_j & \theta_j \, I''_j + (1 - \theta_j) \, V''_j \end{pmatrix} \begin{pmatrix} d\bar{x}_i \\ d\bar{x}_j \end{pmatrix} = \begin{pmatrix} (V'_i - I'_i) \, d\theta_i + \theta_i \, I''_i \, dm_i \\ (V'_j - I'_j) \, d\theta_j + \theta_j \, I''_j \, dm_j \end{pmatrix}$$

The solution of this system is

$$\begin{pmatrix} d\bar{x}_i \\ d\bar{x}_j \end{pmatrix} = D^{-1} \begin{pmatrix} \theta_j \, I''_j + (1 - \theta_j) \, V''_j & (1 - \theta_i) \, V''_i \\ (1 - \theta_j) \, V''_j & \theta_i \, I''_i + (1 - \theta_i) \, V''_i \end{pmatrix}$$

$$\begin{pmatrix} (V'_i - I'_i)d\theta_i + \theta_i I''_i \, dm_i \\ (V'_j - I'_j)d\theta_j - \theta_j I''_j \, dm_j \end{pmatrix}$$

with $D = \theta_i\theta_j \, I''_i \, I''_j + \theta_i(1 - \theta_j) \, I''_i \, V''_i + \theta_j(1 - \theta_i) \, V''_i \, I''_j > 0$ by Assumptions 5 and 6.
Hence

$$\frac{\partial \bar{x}_i}{\partial \theta_i} = D^{-1}[\theta_j I''_j + (1 - \theta_j) V''_j] (V'_i - I'_i) < 0$$

$$\frac{\partial \bar{x}_j}{\partial \theta_i} = D^{-1} (1 - \theta_j) V''_j (V'_i - I'_i) \leq 0$$

and

$$\left| \frac{\partial \bar{x}_i}{\partial \theta_i} \right| > \left| \frac{\partial \bar{x}_j}{\partial \theta_j} \right|$$

since

$$I_j'' (V_i' - I_i') < 0.$$

Also

$$\frac{\partial \bar{x}_i}{\partial m_i} = D^{-1} [\theta_j I_j'' + (1 - \theta_j) V_j''] \theta_i I_i'' > 0$$

$$\frac{\partial \bar{x}_j}{\partial m_i} = D^{-1} (1 - \theta_j) \theta_i V_j'' I_i'' \geqslant 0$$

and

$$\frac{\partial \bar{x}_i}{\partial m_i} > \frac{\partial \bar{x}_j}{\partial m_i}$$

given that

$$I_j'' I_i'' > 0.$$

These results are very intuitive. (1) If the weight party i gives to ideological purity decreases or if i's ideological position moves to the center, then i moves to the center and j either sticks the previous platform ($V_j'' = 0$) or also moves to the center ($V_j'' < 0$); in general, both parties therefore react to a decrease in the role of ideology for party i by selecting more moderate platforms. (2) Party i's move is larger than party j's move, because i's reaction expresses both the direct effect of the change in θ_i (in m_i) and the interdependence effect associated with the game-theoretic framework, whereas j's reaction takes into account only the second effect.

   To illustrate, let us consider the cases depicted in Examples 1 and 2. In the linear-logarithmic model, a glance at Table 4.1 shows that when θ_i decreases, party i gradually moves from m_i to 1/2, while party j keeps its platform unchanged. A similar effect may be observed when m_i moves toward the center. In the logarithmic model, we find from (4.4a) that

$$\frac{\partial \bar{x}_i}{\partial \theta_i} = -(1 - \theta_i \theta_j)^{-2} [(1 - \theta_j)(1 + m_i - m_j) + 2\theta_j] < 0 \quad \text{and} \quad \frac{\partial \bar{x}_j}{\partial \theta_i} = \theta_j \frac{\partial \bar{x}_i}{\partial \theta_i}$$

$$\frac{\partial \bar{x}_i}{\partial m_i} = (1 - \theta_i \theta_j)^{-1} (1 - \theta_i) > 0 \quad \text{and} \quad \frac{\partial \bar{x}_j}{\partial m_i} = \theta_j \frac{\partial \bar{x}_i}{\partial m_i}$$

Here, as $V_j'' < 0$, the cross-effects on j are different from zero. With the help of Proposition 2, the sensitivity analysis can be extended to deal with the case of equilibrium strategies on the boundary. Let us start from $\theta_i = 0$ so that $\bar{x}_i(0, \theta_j) = 1/2$. By continuity, there always exists $\theta_i^1 \epsilon]0, 1[$ such that $\bar{x}_i(\theta_i, \theta_j) \epsilon]m_i, 1/2[$ for some $\theta_i > \theta_i^1$. At θ_i^1, two cases may arise. In the first one, $\bar{x}_j(\theta_j, \theta_i^1) \epsilon \{m_j, 1/2\}$. By differentiating the first-order condition relative to party i, in which $\bar{x}_j$ is replaced by its value, a routine calculation shows that $\bar{x}_i(\theta_i, \theta_j)$ decreases as θ_i increases above θ_i^1. In the second case, $x_j(\theta_j, \theta_i^1) \epsilon]m_i, 1/2[$, and Proposition 2 applies so that $\bar{x}_i(\theta_i, \theta_j)$ similarly decreases when θ_i increases. Again, by continuity, there exists a value $\theta_i^2 \epsilon]0, 1]$ such that $\bar{x}_i (\theta_i^2, \theta_j) = m_i$. From now on $\bar{x}_i (\theta_i, \theta_j) = m_i$ for any θ_i above θ_i^2. If not, this would mean that $\theta_i^3 > \theta_i^2$ may be found such that $\bar{x}_i (\theta_i^3, \theta_j) \epsilon]m_i, 1/2[$. But then, by repeating the argument developed for θ_i^1, we would have $\bar{x}_i (\theta_i^2, \theta_j) > m_i$, a contradiction. To sum-up, as the weight given to ideological purity by party i increases from 0 to 1, the party's equilibrium platform gradually moves from the central position to its ideological position.

As for the impact of the change in θ_i on the choice of the other party, it is easy to verify that party j either sticks to a certain platform (which may be interior to $[m_j, 1/2]$, as in the linear-logarithmic model) or moves away monotonically from 1/2 to reach, possibly, m_j (as in the logarithmic model).

Note, finally, that similar results can be derived when party i readjusts its ideological position from 0 to 1/2.

LONG-RUN ANALYSIS

In our bicriterion model the equilibrium platforms are not likely to attract the same number of voters. It would not be reasonable to leave the analysis at that. Indeed, one can expect the parties to react to their electoral results by reassessing the relative weights of the ideological-purity and vote-maximization objectives for the next campaign. In this way an adjustment process is generated that might converge to a certain long-run equilibrium. Formally, this process can be described by the following differential equation system: for $i = 1, 2$, and $i \neq j$,

$$\dot{\theta}_i(t) = \alpha_i \left[\bar{x}_i(t) - \bar{x}_j(t) \right] \quad \text{if} \quad \begin{cases} \bar{x}_i(t) \epsilon [m_i, \ 1/2[\quad \text{and} \quad \alpha_i(.) < 0 \\ \qquad\qquad\qquad \text{or} \\ \bar{x}_i(t) \epsilon] m_i, \ 1/2] \quad \text{and} \quad \alpha_i(.) > 0 \end{cases} \quad (4.5a)$$

and

$$\dot{\theta}_i(t) = 0 \quad \text{if} \quad \begin{cases} \bar{x}_i(t) = 1/2 \quad \text{and} \quad \alpha_i(.) < 0 \\ \qquad\qquad \text{or} \\ \bar{x}_i(t) = m_i \quad \text{and} \quad \alpha_i(.) > 0 \end{cases} \quad (4.5b)$$

where $|\alpha_i|$ is an increasing function with $\alpha_i(0) = 0$, and $[\bar{x}_1(t), \bar{x}_2(t)]$ is the unique Nash equilibrium of the game with $\theta_i = \theta_i(t)$. Let θ_1^ℓ and θ_2^ℓ be an equilibrium ($\dot{\theta}_1 = \dot{\theta}_2 = 0$) of the system (4.5a)–(4.5b). A long-run equilibrium of the political game is then defined by $x_1^\ell = \bar{x}_1(\theta_1^\ell, \theta_2^\ell)$ and $x_2^\ell = \bar{x}_1 (\theta_2^\ell, \theta_2^\ell)$.

As the parties' results are opposite it is assumed that parties' reactions are opposite in the adjustment process, so that both functions α_1 and α_2 are sign-inversing or sign-preserving.

(1) The first case (sign-inversing) corresponds to the situation in which the winning party wants to stay in power by increasing the relative weight of the vote-getting objective, while the losing party moves back to its ideological sources. In view of the analysis provided in the preceding section it is easy to see that the winner of the initial elections, say i, continuously reinforces its constituency. As a consequence, θ_i decreases and θ_j increases monotonically until the long-run equilibrium $x_i^\ell = 1/2$ and $x_j = m_j$ is reached.

(2) In the second case (sign-preserving), the winning party, concerned by party unity and pressure from militants, decides to strengthen its ideological objective, whereas the other party tries to look more attractive through a moderate platform. This situation corresponds, in our bicriterion approach, to Downs' adjustment process, in which the winning party sticks to its position, while the losing party selects a platform to win the next election.

The process works as follows. The winning party, say i, is pushed toward m_i because of the increase in θ_i and, at the same time, is pulled by the central position because of the decrease in θ_j. The losing party, here j, is subjected to similar pressures but for opposite reasons. The total impact on the constituency of i and j depends on the relative magnitude of these effects and cannot be predicted on the basis of the comparative statistics of the preceding section. Hence, proving the global stability of the system (4.5a)–(4.5b) in the general case is probably too demanding. Nevertheless, in the case of interior solutions, a long-run equilibrium, corresponding to a point on the ray $x_1 = x_2$, can be shown to be locally stable.

Proposition 3.   Assume that $\bar{x}_i(t) \epsilon] m_i, \ 1/2[$ for any t, and $i = 1, 2$. Then the system (4.5a) is asymptotically locally stable.

Proof. Let (x_1^ℓ, x_2^ℓ) be a point on the ray $x_1 = x_2$ and corresponding to the values $\theta_1^\ell, \theta_2^\ell$ of the weights. Taking the first-order part of the Taylor expansion of $\dot\theta_i(t) = \alpha_i [(\bar x_i(t) - x_i^\ell) - (\bar x_j(t) - x_j^\ell)]$ and using Proposition 2 yields

$$\begin{pmatrix} \dot\theta_1 \\ \dot\theta_2 \end{pmatrix} = D^{-1} \begin{pmatrix} \alpha_1' \, \theta_2^\ell \, I_2''(V_1' - I_1') & -\alpha_1' \, \theta_1^\ell \, I_1'' \, (V_2' - I_2') \\ -\alpha_2' \, \theta_2^\ell \, I_2'' \, (V_1' - I_1') & \alpha_2' \, \theta_1^\ell \, I_1'' \, (V_2' - I_2') \end{pmatrix} \begin{pmatrix} \theta_1 - \theta_1^\ell \\ \theta_2 - \theta_2^\ell \end{pmatrix}$$

with $D > 0$. This system has two roots. The first one is given by $\alpha_1' \, \theta_2^\ell \, I_2'' \, (V_1' - I_1') + \alpha_2' \, \theta_1^\ell \, I_1'' \, (V_2' - I_2')$, which is negative since $\alpha_i' > 0$, $V_i' > 0$, $I_i' < 0$, and $I_i'' < 0$. The second one is equal to zero (because of linear dependence), but this does not prevent local stability, as any point on the ray $x_1 = x_2$ is a possible long-run equilibrium.

When the initial constituencies are not too different, the above adjustment process therefore leads to an equal sharing of votes. This fact does not mean, however, that the long-run equilibrium is equivalent to the Hotelling-Downs solution. As any point on the bisecting ray can be a long-run equilibrium, the corresponding platforms are generally different.

The global stability of a symmetric long-run equilibrium can be established in some particular cases. As an example, let us consider the logarithmic-linear model with $m_i \in \,]0, \, 1/2[$ and $m_i \neq m_j$. Assuming $\alpha_i(\bar x_i - \bar x_j) = \bar x_i - \bar x_j$, we obtain $\dot\theta_1 + \dot\theta_2 = 0$, so that $\theta_1 + \theta_2 = k = \theta_1(0) + \theta_2(0)$. The system (4.5a)–(4.5b) then reduces to the single differential equation

$$\dot\theta_1 = g(\theta_1) = \begin{cases} 0, & \text{if } \theta_1 = 0 \text{ and } x_1(\theta_1) - x_2(k - \theta_1) < 0 \\ 0, & \text{if } \theta_1 = 1 \text{ and } x_1(\theta_1) - x_2(k - \theta_1) > 0 \\ f(\theta_1) = x_1(\theta_1) - x_2(k - \theta_1), & \text{otherwise} \end{cases} \tag{4.6}$$

where $f(\theta_1)$ is piecewise defined by $\underline\theta_i$ denoting $(2m_i + 1)/(2m_i + 3)$:

(a) $m_1 - m_2 + \dfrac{1 - 2\theta_1}{1 - \theta_1} - \dfrac{1 - 2(k - \theta_1)}{1 - (k - \theta_1)}$, for $\theta_1 \in]\underline\theta_1, \, 1/2[; \, \theta_2 \in] \, \underline\theta_2, \, 1/2[$

(b) $m_1 - m_2 - \dfrac{1 - 2(k - \theta_1)}{1 - (k - \theta_1)}$ for $\theta_1 \in [1/2, \, 1]; \, \theta_2 \in] \, \underline\theta_2, \, 1/2[$

(c) $\dfrac{1}{2} - m_2 - \dfrac{1 - 2(k - \theta_1)}{1 - (k - \theta_1)}$ for $\theta_1 \in [0, \, \underline\theta_1]; \, \theta_2 \in] \, \underline\theta_2, \, 1/2[$

(d) $m_1 - m_2 + \dfrac{1 - 2\theta_1}{1 - \theta_1}$ for $\theta_1 \in] \, \underline\theta_1, \, 1/2 \, [; \, \theta_2 \in [\, 1/2, \, 1]$

(e) $m_1 - \dfrac{1}{2} + \dfrac{1 - 2\theta_1}{1 - \theta_1}$ for $\theta_1 \in] \, \underline\theta_1, \, 1/2 \, [; \, \theta_2 \in [0, \, \underline\theta_2]$

(f) $m_1 - m_2$ for $\theta_1 \epsilon [\ 1/2, 1\ ]$; $\theta_2 \epsilon\ [1/2, 1]$

(g) 0   for $\theta_1 \epsilon [0, \underline{\theta}_1]$; $\theta_2 \epsilon [0, \underline{\theta}_2]$

(h) $\dfrac{1}{2} - m_2$ for $\theta_1 \epsilon [0, \underline{\theta}_1]$; $\theta_2 \epsilon\ [1/2, 1]$

(i) $m_1 - \dfrac{1}{2}$ for $\theta_1 \epsilon [1/2, 1]$; $\theta_2 \epsilon [0, \underline{\theta}_2]$

with $\theta_2 = k - \theta_1$. By construction of g, equation (4.6) has an equilibrium, denoted by θ_1^ℓ. When $\theta_1^\ell \epsilon [0,1]$, the corresponding long-run equilibrium does not necessarily belong to the bisecting ray. (Such is the case, for example, when $k = 3/2$ and $m_1 - m_2 > 0$, $x_1^\ell = m_1$, and $x_2^\ell = m_2$). For that reason, we assume $\theta_1^\ell \epsilon\ ]0,1[$. This assumption implies that f is defined by at least one of the pieces (a) to (e) and (g). Let us exclude, for the moment, piece (g). Then, by some simple manipulation it can be shown that at most, one equilibrium of (4.6) is associated with each possible piece. Furthermore, we can easily verify that for each possible piece, $df/d\theta_1$ is negative at the equilibrium. Given that f is continuous, there is therefore, at most, one equilibrium θ_1^ℓ in $]0,1[$. This yields a long-run equilibrium $x_1^\ell = \bar{x}_1(\theta_1^\ell)$ and $\bar{x}_2(k - \theta_1^\ell)$ such that $x_1^\ell = x_2^\ell$. In addition, as $df/d\theta_1$ is negative, θ_1^ℓ the equilibrium of (4.6), and therefore the long-run equilibrium, is *globally asymptotically stable*.

Consider now the case when piece (g) appears in the definition of f. Two cases may arise. In the first one, $k \leqslant \min\ (\underline{\theta}_1, \underline{\theta}_2)$ so that $f = 0$ everywhere. Therefore, $\theta_1(0)$ is an equilibrium, and since $\overline{\theta}_i \leqslant \underline{\theta}_i$, $\bar{x}_i\ (t) = 1/2$ for any t. In other words, the long-run equilibrium, given by $(1/2, 1/2)$, is globally asymptotically stable. In the second case, $k > \min\ (\underline{\theta}_1, \underline{\theta}_2)$. It is then easy to check that (g) is observed only if it is patched with (c) on the left or with (e) on the right, or both. By continuity of f and since $df/d\theta_1$ is negative along (c) and (e), (4.6) is globally asymptotically stable but θ_1^ℓ is not unique. Nevertheless, the long-run equilibrium is still given by $(1/2, 1/2)$ and is globally asymptotically stable.

THE MULTIPARTY SYSTEM

Ideology tends to favor the emergence of a multiparty system. First, when the two existing parties do not sufficiently weight ideological purity, some of their members (and voters) may decide to create new parties in the ideologically uncovered parts of the political spectrum (the "hinterlands"). Second, and inversely, when the two existing parties strongly favor ideological purity, space can be created between them for a new party (the "competitive fringe"). Therefore, it is worthwhile to investigate our bicriterion model in the case of several parties.

Let us assume that there are n (>2) parties. For notational convenience, the platform x_i and the ideological position m_i of party i are here measured from the left extremity of the unit segment. Assumptions 1 and 2 of the section on the two-party system are unchanged. Assumptions 3 and 4 are replaced by the following ones.

Assumption 3'. The parties are ranked in such a way that $m_1 < m_2 < \ldots < m_n$. Thus the ideological positions do not coincide, and there may be several left-wing ($m_i < 1/2$) and right-wing ($m_i > 1/2$) parties.

Assumption 4'. The position x_i of party i, $i = 2 \ldots n - 1$, is to be chosen in [max (m_{i-1}, x_{i-1}), min (m_{i+1}, x_{i+1})] and the position $x_1(x_n)$ of party 1 (party n) is in $[m_1, x_2]$ ($[x_{n-1}, m_n]$).

Party i cannot jump over the "neighboring" parties $i - 1$ and $i + 1$. Such a constraint is imposed by the concern of party i to be faithful to ideology.

Given a set $x_1 \ldots x_n$ of platforms, the number of votes of party i, $i = 2 \ldots n - 1$, is given by

$$\frac{x_i - x_{i-1}}{2} + \frac{x_{i+1} - x}{2} = \frac{x_{i+1} - x_{i-1}}{2}$$

Thus the electoral result of party i does not depend on its platform but only on the neighbors' platforms. On the other hand, the number of votes of party 1 (party n) is equal to

$$x_1 + \frac{x_2 - x_1}{2} = \frac{x_1 + x_2}{2} \left(1 - x_n + \frac{x_n - x_{n-1}}{2} = 1 - \frac{x_{n-1} + x_n}{2}\right)$$

Finally, Assumptions 5 and 6 are slightly modified in order for I_i to be a function of the absolute value $|x_i - m_i|$ and V_i a function of the number of votes of party i.

The payoff function of party i is defined by equation (4.1). For $i = 2 \ldots n - 1$, U_i is a function of x_{i-1}, x_i and x_{i+1} while $U_1(U_n)$ depends on x_1 and $x_2(x_{n-1}$ and $x_n)$. We than have a game-theoretic model with a chain effect, as in spatial-competition theory. A Nash equilibrium for this game is an n-uple of strategies $(\bar{x}_1, \ldots, \bar{x}_n)$, such that

$$U_1(\bar{x}_1, \bar{x}_2) \geqslant U_1(x_1, \bar{x}_2) \quad \text{for any } x_1 \epsilon[m_1, \bar{x}_2]$$
$$U_i(\bar{x}_{i-1}, \bar{x}_i, \bar{x}_{i+1}) \geqslant U_i(\bar{x}_{i-1}, x_i, \bar{x}_{i+1}) \quad \text{for any } x_i \, \epsilon[\max{(m_{i-1}, \bar{x}_{i-1})},$$
$$\min{(m_{i+1}, \bar{x}_{i+1})}]$$

and

$$U_n(\bar{x}_{n-1}, \bar{x}_n) \geq U_n(\bar{x}_{n-1}, x_n) \quad \text{for any } x_n \epsilon [\bar{x}_{n-1}, m_n]$$

The above denote by $\bar{x}_1(m_1, m_2)$ $(\bar{x}_n (m_{n-1}, m_n))$ the best platform of party 1 (party n) when the platform of party 2 (party $n-1$) is $m_2 (m_{n-1})$.

Proposition 4. If Assumptions 1, 2, 3', 4', and 5–7 hold, then $\bar{x}_1 = \bar{x}_1(m_1, m_2)$, $\bar{x}_2 = m_2$, ..., $\bar{x}_{n-1} = m_{n-1}$, $\bar{x}_n = \bar{x}_n (m_{n-1}, m_n)$ is the unique Nash equilibrium.

Proof. As V_i is independent of x_i for $i = 2 \ldots n - 1$, party i maximizes U_i by choosing m_i, and this choice holds for any value $x_{i-1} < m_i$ and $x_{i+1} > m_i$. Then, given $m_2(m_{n-1})$, $\bar{x}_1 (m_1, m_2)$ $(\bar{x}_n (m_{n-1}, m_n))$ is the best platform of party 1(party n).

The bicriterion approach proposed here links each party to its ideological position and prevents parties from leapfrogging. As a result, multiparty competition is stabilized and yields a set of separated platforms that cover most of the political spectrum. The spreading out of political platforms thus mirrors the spreading out of the voters' and party members' opinion. These results, again, are in sharp contrast with those derived within the Hotelling-Downs framework.

CONCLUSION

This chapter has considered a single but important modification in the Hotelling-Downs model: rather than just seeking to win elections, each party tries to choose a platform on the political scale that is as close as possible to its ideology and at the same time brings the largest number of votes. Our results can be summarized as follows. (1) In a two-party system, a unique equilibrium exists; it depends on, among other factors, the two parties' ideological positions and the weight they give to ideology; generally this equilibrium is not the median voter's position. (2) If the weight given to ideology is assumed to increase for the winner and diminish for the loser, the process will lead to positions such that both parties are equidistant from the center and thus tie. (3) In a multiparty system, a unique equilibrium exists in which all moderate parties stick to their ideological positions and the two polar parties locate in a stable position.

As compared to the current state of the theory, our findings bring positive results as to the stability of voting equilibria. The chapter also attempts to reflect more realistically what is going on in the real world of politics and policy making. It is indeed clear that a total convergence toward the center occurs very rarely. In the same way, very few democracies are or have been governed at either extreme of the political scale. In general, one witnesses governments that with alternation or permanence, follow a center-to-left or a center-to-right policy. For example, alternation occurred in postwar Germany and Britain, and there have been center-to-right governments in Italy and in France for about two decades. In the case of alternation, both parties often have modified their programs throughout their electroal histories. In the case of a long spell of center-to-right government, parties on the left often have not wanted to move further from their ideological basis.

Our model takes into account the observed alternation between parties having quite different programs. Such an alternation is precluded by the Hotelling-Downs model, in which alternation may occur but without any policy implication, as both parties have the same program. In the field of economic policy, our findings are consistent with the often observed modifications after a change of administration. In other words, the alternation between expansionary budgetary policies and restrictive monetary and fiscal policies that has occurred in several countries is consistent with the present analysis.

REFERENCES

Davis, O., M. J. Hinich, and P. C. Ordeshook, 1970, An expository development of a mathematical model of the electoral process, *American Political Science Review* **64**:426–448.

Downs, A., 1957, *An Economic Theory of Democracy,* Harper, New York.

Friedman, J. W., 1977, *Oligopoly and the Theory of Games,* North-Holland, Amsterdam.

Graitson, D., 1982, Spatial competition à la Hotelling: a selective survey, *Journal of Industrial Economics* **31**:13–25.

Hotelling, H., 1929, Stability in competition, *Economic Journal* **39**:41–57.

Kalt, J. P. and M. A. Zupan, 1984, Capture and ideology in the economic theory of politics, *American Economic Review* **74**:279–300.

Kolm, S. C., 1977, *Les élections sont-elles la démocratie?* Cerf, Paris.

Robertson, D., 1976, *A Theory of Party Competition,* Wiley, New York.

Samuelson, L., 1984, Electoral equilibria with restricted strategies, *Public Choice* **43**:307–327.

Weber, A., 1909, *Über den Standort der Industrie,* Mohr, Tübingen. Trans. C. J. Freidrich, 1929, *The Theory of the Location of Industries,* University of Chicago Press, Chicago.

Wittman, D., 1983, Candidate motivation: a synthesis of alternative theories, *American Political Science Review* **77**:142–157.

Part II

APPLICATIONS OF LOCATION-ALLOCATION MODELS

5

LOCATION-ALLOCATION AND IMPULSIVE SHOPPING: THE CASE OF GASOLINE RETAILING

Michael F. Goodchild and Valerian T. Noronha

University of Western Ontario

Location-allocation models are concerned with the selection of sites for a number of central facilities from which service can be provided to a spatially dispersed population or pattern of demand. This definition suggests that the models should be as suitable for application in the private sector, for retail site selection, as they are for such public sector facilities as fire stations and day-care centers. Yet retail applications have been relatively slow in developing (see, for example, Achabal, Gorr, and Mahajan, 1982; Ghosh and Craig, 1984; Goodchild, 1984), and there appear to be several reasons, some more valid than others. First, in the retail sector a consumer of service is free to choose the supplier, which leads to a complex pattern of allocation that may or may not be predictable using spatial-choice models. Second, objective functions are likely to be more complex in the private sector. If profit is to be maximized, it is necessary to incorporate models of each facility's operating cost, leading to a much more complex objective than the simple distance-based functions common in public facility location problems. Third, location-allocation modelers may perceive public sector decision makers and planners as more accessible and more receptive to modeling in general than their private sector counterparts.

These problems suggest that the most suitable areas for initial applications in retailing would be those where spatial behavior is relatively simple, where facilities are of fixed size and have relatively insensitive operating costs, and where sites and facilities are constantly being opened and reassessed. These features characterize low-order retail facilities such as convenience stores, gas stations, branch banks, fast-food restaurants, and supermarkets. For example, a convenience store typically has a threshold population of approximately 1,000 households, a size of up to 2,500 square feet, and a market area that corresponds closely to the simplistic model of nearest-place behavior. Goodchild (1984) has proposed two relatively simple location-allocation models for such applications, based on maximizing market share. The first, Market Share Model (MSM), assumes that demand is inelastic with respect to distance; in other words, all consumers make full use of the facilities irrespective of how far they are from them. The locations of existing stores are known and fixed, and the objective is to select sites for additional stores so as to maximize the total market share captured by the chain's new and existing stores. Goodchild's second model, Competion-Ignoring Model (CIM), assumes that the probability of a consumer using a facility is a decreasing linear function of distance. In both models travel is assumed to be to the nearest facility.

Although these models capture certain aspects of reality well, they fail to deal with a number of significant problems, particularly those having to do with departures from simple nearest-place travel behavior. For example, it is common to position convenience stores on the "home" side of the main entrance to major suburban residential developments; the aim is to capture consumers on their way home from recreational or work trips. Yet the nearest-place behavioral rule assumes that a trip to the convenience store begins and ends at the residence and is made for a single purpose.

The important distinction in this example is between a single-purpose trip made to the retail store from a known reference location, such as the home, and a multipurpose trip during which convenience shopping may be no more than an impulsive event along the predetermined route to work, school, church, or wherever. The relative importance of the two types of trips will vary across different functions. This chapter considers the example of gasoline retailing, which among retail activities is subject to perhaps the greatest proportion of impulsive behavior.

Single-purpose trips for gasoline frequently originate from the home but may also originate from a place of business or recreation or from a temporary residence. Unfortunately, although there are abundant sources of data on population patterns by place of residence, there are very few on daytime locations; so it is difficult to assess the associated distribution of demand. However, this problem would seem to be much more serious for functions such as restaurants, which are oriented toward the daytime distribution of population. We make the simplifying assumption that single-purpose trips for gasoline originate according to the nighttime, or census, distribution of population.

Multiple-purpose trip activity is another area in which few useful and comprehensive sources of information exist, and there have been few successful attempts at modeling this class of consumer spatial behavior. A truly impulsive purchase has, by definition, no impact on a predetermined trip; an impulsive gasoline purchase during the journey to work, for example, in no way affects the choice of route between the workplace and residence. We might therefore recognize an entire spectrum of degrees of impulsiveness as the gasoline purchase exerts more or less influence on the choice of route. However, if we assume that all such purchases on multipurpose trips are purely impulsive, then potential demand from this source is indicated simply by volume of traffic. If one is willing to make these assumptions, then demand for gasoline can be represented by the residential distribution of population in the case of single-purpose trips, and by the traffic counts on links in the street network in the case of impulsive buying. Both of these indicators are readily available from standard data sources.

Clearly these simplifying assumptions do not describe all types of real behavior. However, they can be implemented using existing data sources, whereas any greater accuracy would require specific market research and in the context of applied location-allocation would therefore incur a large marginal cost for a relatively small marginal benefit.

The next section describes conventional approaches to siting gasoline stations and the application of a sales forecasting model. Following that, there is a formal description of the proposed location-allocation model and an application to a particular chain operating in London, Ontario, a city of about 270,000. In the fall of 1983 there were 137 gasoline outlets within the city's official limits. Of these, the chain of interest, referred to as the client chain, controlled 31 outlets together with six undeveloped, or "green" sites.

SALES FORECASTING MODELS

In attempting to establish factors responsible for a site's success or failure, as well as its ultimate potential, conventional approaches treat each site as an independent unit. Such information underlies decisions regarding the acquisition and development of sites and the closing of unprofitable facilities. Site factors usually considered include the physical layout of facilities, visibility and access from the street, traffic conditions, the presence of neighboring facilities such as convenience stores, and the characteristics of the neighborhood (for recent reviews, see Craig, Ghosh, and McLafferty, 1984; Davies and Rogers, 1984). Goodchild (1983) has pointed out some of the methodological problems inherent in this approach, including the difficulty of establishing adequate sample sizes for model calibration, the impossibility of identifying causality, and paradoxes in the eventual use of the model to plan new outlets.

To illustrate some of these problems, a number of suitable variables were collected for each of the client chain's outlets operating in the fall of 1983:

number of service bays

number of nozzles

number of islands

hours of business (24 hours or less than 24)

presence of canopy (yes or no)

presence of car wash (yes or no)

availability of self-service (yes or no)

number of curb cuts, main and side streets

convenience store attached (yes or no)

distance to nearest competitor, same or different chain

traffic count on major street

Although price is a major factor, it has fluctuated very rapidly in recent years in the Canadian retail market, making it very difficult to construct useful representative measures.

The variables listed were used in an attempt to develop a predictive model, employing stepwise multiple regression, of total 1982 liters pumped by each outlet. Over the 31 cases, total volume varied by a factor of roughly 17, from 283,440 to 4,769,117 l. The first four predictors, accounting for roughly 86% of the variation in sales, were, in decreasing order of importance, as follows:

$$
\begin{aligned}
\text{Volume} = 706,257 &+ 1,980,000 \text{ if open 24 hours} \\
&+ \quad 885,399 \text{ if canopy present} \\
&+ \quad 773,171 \text{ if carwash present} \\
&+ \quad 602,320 \text{ if self-service} \qquad \textbf{(5.1)}
\end{aligned}
$$

Although this example clearly is highly simplified, it nevertheless illustrates many of the problems of the approach. First, strong collinearity exists between many of the variables in the data set. Consequently there is no implication that the presence of a canopy causes high sales, or that installing one would increase sales by approximately 850,000 l. Instead the variable may appear as a strong predictor because it acts as a surrogate for other, causative variables, so that the presence of a canopy is a predictor of newness of facilities, self-service, low prices, and so on. We suspect that many more than four variables are required to explain gasoline purchase behavior, but the existence of strong collinearities allows four surrogate variables to achieve a high degree of explanation. On the other hand, the individual's choice process may itself involve the use of surrogates, so that a visible feature like a canopy is perceived as an indicator of other features and may be used as a legitimate causal variable in the choice process. Although a very large number of factors potentially affect choice, such perceptual surrogates may be used as a means of reducing the dimensionality of the individual's choice space.

Second, that 86% of the variation in volume can be attributed to facilities on the site does not mean that only 14% can be attributed to all other types of factors, including location. The facilities present at each site have been determined in expectation of sales, and chains will develop and modernize a site only if evidence suggests that the location can generate enough return. We therefore expect collinearity between quality of location and quality of facilities, making it impossible to allocate cause between these two sources of variation. In the extreme, where a chain had adjusted each of its sites to match market potential, with large, high-overhead facilities in areas of high market potential and small, low-overhead facilities elsewhere, there would be perfect collinearity. In reality, historical inertia and complex leasing arrangements between the chain and the individual operator displace this ideal.

Third, in treating each facility as independent, the model pays no attention to competition and constraints on demand. The factors loosely referred to above as location and market potential are responsive to the demand for gasoline and the competition for that demand, which makes each site sensitive to the locations of nearby sites. To some extent, these effects can be captured in simple surrogates such as the distance to the nearest competitor or the number of competing outlets in the same neighborhood. However, location-allocation provides a direct method of dealing with the interactions between several outlets competing in the same marketplace.

Figure 5.1 usefully illustrates some of these points. In this figure the total 1982 sales volume of each of the 137 stations in the city has been plotted against the traffic count at the nearest intersection. In general, the points lie below an inverted V, representing maximum potential at each level of traffic density. At low density the maximum potential increases with traffic, but there is a peak at about 30,000 vehicles per day followed by a decline, due perhaps to the negative effects of very high traffic density. Each outlet lies somewhere below this inverted V; the distance below is determined by such factors as the intensity of local competition, the inadequacy of facilities, and low residential market potential.

LOCATION-ALLOCATION MODEL

We define two types of demand: residential, with associated single-purpose trips; and traffic, forming the basis for impulse purchases. Let the distribution of residential demand be represented by a series of points at locations x_i, y_i ($i = 1, \ldots, n$), with multivariate weights w_{ik} ($k = 1, \ldots, p$) associated with each point. Since the threshold for a gasoline outlet is relatively low, it is necessary to use a spatially disaggregate data base. Let the traffic demand

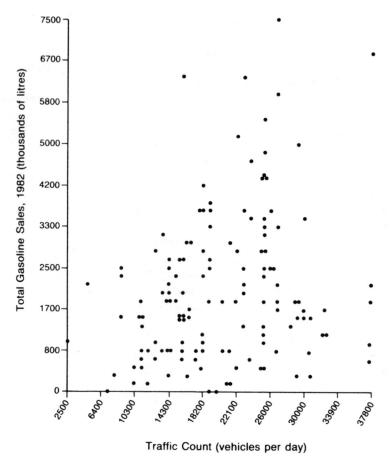

Figure 5.1.   Total 1982 sales volume against street traffic count for each of 137 gasoline outlets in London, Ontario.

be represented by a number of traffic counts t_{jm} ($m = 1, \ldots, s$) associated with each link j in the study area's street network. Multiple weights and traffic counts allow for various market segments through the use of different residential subpopulations and different types of traffic or hours of the day.

The use of traffic and network-link data suggests that it would be appropriate to structure the location-allocation model in discrete space, limiting locations to points in the network. This structure would require that the points representing residential demand become nodes of the network, in addition to the street intersections. Such a setup may well be feasible, but this study has used a continuous space approach. Goodchild (1984) gives other arguments concerning alternative representations of space in the retail context. This chapter represents traffic demand as a point located at the midpoint of each link, with coordinates u_j, v_j ($j = 1, \ldots, r$).

Assume that there is a constant probability per unit of distance driven that a driver will make an impulsive gasoline purchase. Then the potential for impulsive purchase for a single link j in the network is proportional to the traffic count multiplied by the length of the link. Thus we treat the products $t_{jm}L_j$ ($m = 1, \ldots, s$) as a set of impulse-purchase weights associated with each link, where L_j is the link length, and t_{jm} is a count of a particular type of traffic. To some extent the impulse-purchase or traffic weights $t_{jm}L_j$ are comparable to the residential market weights w_{ik}, and m and k are traffic and residential market segments respectively. By weighting each segment, we can obtain composite market weights for each sector:

$$W_i = \sum_k w_{ik}\alpha_k \qquad (5.2)$$

and

$$T_j = \sum_m t_{jm}L_j\beta_m \qquad (5.3)$$

where α_k ($k = 1, \ldots, p$) and β_m ($m = 1, \ldots, s$) are vectors of segment weights.

Let the gas stations be located at points X_a, Y_a ($a = 1, \ldots, q$); some of these stations may already exist and some may be hypothetical. The spatial behavior or allocation pattern is represented by two matrices: I_{ia} ($i = 1, \ldots, n$; $a = 1, \ldots, q$), which gives the amount of the weight W_i in residential unit i, which is assigned to station a; and J_{ja} ($j = 1, \ldots, r$; $a = 1, \ldots, q$), which similarly gives the amount of the weight T_j in traffic link j allocated to a. The allocations will of course depend on the locations of the stations, and they may also be influenced by the distributions of demand if spatial behavior affects store attraction (see Goodchild, 1978).

To maximize market share, sites must be selected to

$$\text{Max } Z = A \sum_{a \in S} \sum_i I_{ia} + B \sum_{a \in S} \sum_j J_{ja} \qquad (5.4)$$

where A represents the sales attributable to the residential market per unit of residential weight allocated, B is the corresponding sales per unit of traffic weight allocated, and S is the set of stations in the client chain.

The I and J matrices describe the spatial behavior in the system with respect to actual or hypothetical distributions of facilities. By including entries for all origin units in relation to all destinations, these matrices allow for the possibility that customers may be attracted to some station a from a distant neighborhood i or a distant traffic link j. In principle, the distances of single-purpose trips are much greater than for impulse buying, so we would expect J to be more sparse than I. However, the model in this way allows impulse buying to occur from links other than the one on which a facility is located, and thus it permits some degree of interaction between the impulse and the choice of multipurpose route. In other words, the model is not limited strictly to the pure-impulse end of the spectrum discussed earlier.

To implement the model it is necessary to make further assumptions about the I and J matrices. Predictors of consumer spatial behavior such as the gravity, Multiplicative Competitive Interactive (MCI), or spatial-interaction models might be used as the basis for I, taking a sample of observed behavior to provide data for calibration. Several studies have combined locational search with probabilistic allocation models in this way, using a variety of objective functions (Achabal, Gorr, and Mahajan, 1982; Beaumont, 1980; Goodchild and Booth, 1980; Hodgson, 1978, 1981; Mirchandani and Oudjit, 1982; O'Kelly and Storbeck, 1984). However, this approach presents numerous problems both in the stability and the generality of the model (Eagle, 1984; Ghosh, 1984; Goodchild, 1984) and in the need to know the locations and attractions of all competing stations. In practice this market is extremely dynamic and knowledge of competing locations is merely temporary. Competitors can be expected to respond to changes in the market (Ghosh and Craig, 1983, 1984; Ghosh and McLafferty, 1982) in ways that are frequently unpredictable. Finally, few clients will likely bear the cost of the data collection and modeling necessary to establish reasonable predictors for either I or J.

In these circumstances we assume that each element of I and J can be estimated by a simple function of distance, and we take linear functions in the absence of other information:

$$I_{ia} = W_i(c_r - m_r d_{ia}), \; d_{ia} < d_{ie} \quad \text{if } e \, \epsilon \, S \text{ and } e \neq a, \text{ else } I_{ia} = 0 \quad (5.5)$$

$$J_{ja} = T_j(c_t - m_t d_{ja}), \; d_{ja} < d_{je} \quad \text{if } e \, \epsilon \, S \text{ and } e \neq a, \text{ else } J_{ja} = 0 \quad (5.6)$$

where m and c represent constants, d_{ia} is the distance from the representative point of neighborhood i to facility a, and similarly d_{ja} is the distance from the center of link j. The conditions imply that nonzero weight will be assigned only to the nearest of the client's stations; the allocation to nonclient stations $e \, \epsilon \, S$ is irrelevant to the objective function. Finally, to avoid negative allocations we assume that the nearest facility will always be closer than c_r/m_r and c_t/m_t. Normally we would expect m_r to be significantly smaller than m_t, indicating a slower response to distance in the residential market.

Clearly any of these assumptions can be avoided if adequate information is available. We are concerned, therefore, with making the best possible assumptions for a given level of information about the real system; and since such information has an associated cost of data collection and analysis, we are trying to make the best possible assumptions for a given level of project expenditure.

The model simplifies under these circumstances to

$$\text{Min } Z = Am_r \sum_{a \in S} \sum_i W_i d_{ia} + Bm_t \sum_{a \in S} \sum_j T_j d_{ja} \qquad (5.7)$$

which is a p-median problem with two independent distributions of demand. Both it and the earlier version can be solved using standard algorithms for continuous-space location-allocation.

Implementation

The client chain had undergone recent expansion through a series of major mergers and takeovers. As a result, the spatial pattern of facilities in the fall of 1983 was far from optimal (Fig. 5.2), and included a number of apparent paradoxes. For example, the chain operated two similar facilities within a few hundred feet of each other; one had been acquired by takeover in the previous twelve months as the other was being built. There were large holes in the spatial coverage of the market, and the facilities present at each developed site varied from older (1930s) buildings with service bays and one full-service island to contemporary canopied multiisland self-service outlets. As noted previously, the sales for each outlet varied by a factor of 17. The client's intention was to rationalize this system by renovating or closing existing facilities and perhaps developing new sites. As an informal guideline, they felt that reducing facilities from 31 to 20 would bring their percentage of the outlets in the London market in line with their recent penetration in terms of sales volume (15–20%). Given the complex variety of contractual arrangements under which the outlets operate, maximizing market share for a given number of stations was felt to be desirable despite the lack of any clear relationship between sales volume and return. Thus, we now report on the use of the location-allocation models discussed earlier to optimize a system of 20 outlets. It is assumed that the facilities at each site are designed to exploit fully the site's potential, in other words to return maximum profitability to the chain.

Residential demand was aggregated to the 399 enumeration areas used in the 1981 census. To crudely evaluate the importance in predicting residential market of each of the many available demographic variables, we made a series of multiple regressions using the total sales of all outlets of all chains in 1982 in each census tract as the dependent variable, and demographic and socioeco-

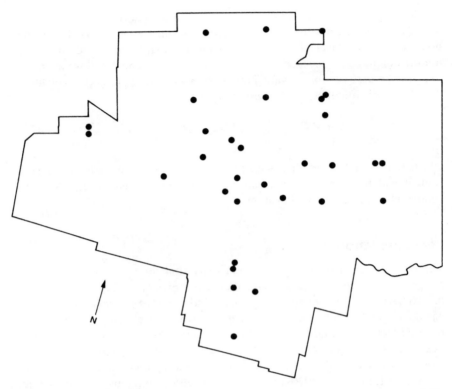

Figure 5.2.   Distribution of the client chain's operating outlets in the fall of 1983.

nomic variables aggregated to the census tract level as the independents. As a result, the following six variables were found to be about equally successful as predictors of residential demand:

w_{i1} = adults aged 20–69

w_{i2} = total population

w_{i3} = households

w_{i4} = census family households

w_{i5} = families

w_{i6} = total income

No combination achieved a substantial improvement in fit over any single variable.

Four traffic counts were available for each of the 560 links in the network of major city streets:

t_{j1} = 1981 total traffic per day

t_{j2} = 1981 light traffic per day

t_{j3} = 1991 total traffic per day

t_{j4} = 1991 light traffic per day

"Light traffic" eliminates trucks and buses. Total traffic counts varied from a high of 37,800 to a low of 8,000. Regressions at the census tract level of aggregation indicated that of the four counts, total traffic miles projected for 1991 were best able to predict the observed total sales volume in 1982. This future projection may provide better predictions either because the majority of future growth is expected in peripheral areas, where sales per vehicle mile are already higher, or because demand from outside the city limits benefits the periphery most. In either case, it is rational to base planning on 1991 projections because it will take several years to fully complete any program of rationalization.

Statistical analysis based on gross sales can give little firm basis for the assignment of relative weights to the residential and traffic markets (A and B). A more precise analysis would require extensive market research to establish correct values for the α and β weight vectors, A and B, and the elasticities m_r and m_t. The very strong correlations within the group of residential market variables ensure that solutions will be insensitive to the values in the α vector, and the case is similar for the traffic group. The solutions described below are therefore based on the use of total adults aged 20–69 (α_1 = 1, all others 0) and the 1991 total traffic forecast (β_3 = 1, all others 0). The remaining constants were defined by

$$Am_r = K\gamma \tag{5.8}$$

$$Bm_t = K(1 - \gamma) \tag{5.9}$$

ignoring K as irrelevant to the minimization of A and varying γ between 0 and 1 in steps. This procedure yields a series of solutions varying from a pure traffic orientation to purely residential.

Solutions were obtained using an enhanced version of Interactive Location-Allocation in Continuous Space (ILACS) (Goodchild, 1984). This package uses a graphics terminal to display and map elements of the market area, including residential zones, traffic links, facilities, patterns of allocation, and contours of the objective function. Distances are evaluated in straight lines but can be modified by the insertion of barriers, which force travel around their ends, and freeways, which provide faster routes. Locations can be optimized either by using the alternating (Cooper, 1964) or Tornqvist et al. (1971) algorithms or by interactively using a graphics cursor.

Four of the existing outlets were regarded as being sufficiently profitable and as having sufficiently ideal facilities that they should be retained in any solution. These four outlets were the highest-ranked in 1982 sales and were located at widely separated locations. Optimization of the location-allocation model obtained the remaining 16 locations in each solution. Starting positions were arrived at by selecting 16 of the remaining 33 sites owned by the chain so as to give a crudely uniform coverage of the city. Following optimization, these mobile locations were adjusted to nearby owned sites where doing so caused little loss of optimality and where other conditions such as site size

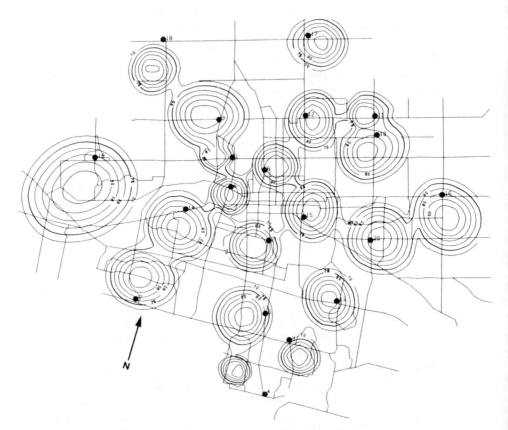

Figure 5.3. Optimal solution for 20 sites using residential market weights, adjusted to feasible locations, with contours showing loss of optimality.

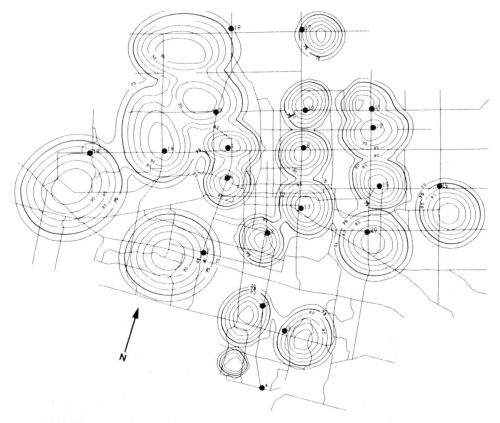

Figure 5.4. Optimal solution for 20 sites using traffic weights, adjusted to feasible locations.

and traffic conditions were favorable. In the remaining cases, locations were adjusted to links known to have favorable zoning and traffic conditions, and the client was advised to attempt to acquire a site in the immediate vicinity.

Two solutions are shown as examples. Fig. 5.3 is the residential solution ($\gamma = 1$), and Fig. 5.4 shows the traffic solution ($\gamma = 0$); both are superimposed on the network of major streets. In each case, the contours show the deterioration in optimality from adjusting a location away from its optimum, holding all other locations constant. More specifically, the height of the surface at any point is found by taking the contribution to the objective function from the demand allocated to a given optimal location and determining its ratio to the contribution if the optimum were moved to the point,

$$E_{ag} = (Am_r \sum_i W_i d_{ia} + Bm_t \sum_j T_j d_{ja})/(Am_r \sum_i W_i d_{ig} + Bm_t \sum_j T_j d_{jg}) \quad \textbf{(5.10)}$$

where g is the point at which the surface is being evaluated. The value of E_{ag} shown at point g is for the facility a with the highest value of E_{ag},

$$E_g = \underset{a}{\text{Max}}\ (E_{ag}) \qquad (5.11)$$

The contoured surface should be interpreted with caution since it is impossible to show the combined effects of moving more than one site away from its optimal location. In principle, the movement of any one site requires a reoptimization of the remaining $q - 1$ sites.

Discussion

Since a continuous space model assumes all locations in the plane to be potentially feasible, we would not expect identical solutions under different objective functions except under special circumstances. However, because the rationalization process has tended to favor a discrete set of locations, Figs. 5.3 and 5.4 show that the adjusted, suboptimal solutions share a number of common locations, 15 out of 20 to be precise. Although the two objectives are quite different, they tend to yield broadly similar solutions because of strong correlation between the underlying spatial demand patterns. Thus the optimal traffic solution is close to optimal on the residential objective, and vice versa. The strong correlation between residential distributions and traffic patterns is understandable because the former is a primary cause of the latter. It follows that although current traffic conditions may be important for short-term planning, in the long term, residential patterns may be the best indicator of future market conditions. In either case, the chain will have to progress toward the ideal 20-site solution through a series of closures, reconstruction of facilities, and new site acquisitions—a process that will take many years. It would be wise, therefore, to review and if necessary revise the present analysis at no more than five-year intervals.

The traffic and residential objectives used in this study, and the mixtures generated by the constant γ, can be placed in the framework of multiple-objective optimization (see, for example, Cohon, 1978). The observed similarity between the two extreme solutions suggests that in practice the noninferior set may contain only a small number of points, which could be readily enumerated by search. If this implication is true, it suggests that the spectrum of retail site selection, from its high traffic orientation at one end to its orientation toward activities relying on single-purpose trips from residence at the other, can be characterized by a limited number of intervals of one parameter, γ.

The main assumption made in the model is that behavior can be partitioned into two classes, one based on residential locations and the other on traffic.

Other simplifying assumptions made in the implementation may or may not be necessary, depending on the data available for the market and the willingness of the client to invest in market research. The practical issues associated with the latter, although not discussed here, are certainly not simple and deserve considerable attention. In sum, this chapter has presented no more than a first attempt at applying location-allocation to a significant practical problem.

REFERENCES

Achabal, D. D., W. L. Gorr, and V. Mahajan, 1982, MULTILOC: a multiple store location decision model, *Journal of Retailing* **58**(2):5–25.

Beaumont, J., 1980, Spatial interaction models and the location-allocation problem, *Journal of Regional Science* **20**:37–50.

Cohon, J. L., 1978, *Multiobjective Programming and Planning,* Academic Press, New York.

Cooper, L., 1964, Heuristic methods for location-allocation problems, *SIAM Review* **6**:37–52.

Craig, C. S., A. Ghosh, and S. McLafferty, 1984, Models of the retail location process: a review, *Journal of Retailing* **60**(1):5–36.

Davies, R. L. and D. S. Rogers, 1984, *Store Location and Store Assessment Research,* Wiley, New York.

Eagle, T. C., 1984, Parameter stability in disaggregate retail choice models: experimental evidence, *Journal of Retailing* **60**(1):101–123.

Ghosh, A., 1984, Parameter nonstationarity in retail choice models, *Journal of Business Research* **12**:425.

Ghosh, A. and C. S. Craig, 1983, Formulating retail location strategy in a changing environment, *Journal of Marketing* **47**:56–68.

Ghosh, A. and C. S. Craig, 1984, Facility planning in competitive environments: a location-allocation approach, *Geographical Analysis* **16**(1):39–56.

Ghosh, A. and S. L. McLafferty, 1982, Locating stores in uncertain environments: a scenario planning approach, *Journal of Retailing* **58**(4):5–22.

Goodchild, M. F., 1978, Spatial choice in location-allocation problems: the role of endogenous attraction, *Geographical Analysis* **10**:65–72.

Goodchild, M. F., 1983, Methodological problems in the calibration of sales forecasting models, *Applied Geography Conferences Papers and Proceedings* **6**:158–167.

Goodchild, M. F., 1984, ILACS: a location-allocation model for retail site selection, *Journal of Retailing* **60**(1):84–100.

Goodchild, M. F. and P. J. Booth, 1980, Location and allocation of recreation facilities: public swimming pools in London, Ontario, *Ontario Geography* **15**:35–51.

Hodgson, M. J., 1978, Towards more realistic allocation in location-allocation models: an interaction approach, *Environment and Planning, A* **10**:1273–1286.

Hodgson, M. J., 1981, A location-allocation model maximising consumers' welfare, *Regional Studies* **15**:493.

Mirchandani, P. B. and A. Oudjit, 1982, Probabilistic demands and costs in facility location problems, *Environment and Planning, A* **14**:917–932.

O'Kelly, M. E. and J. E. Storbeck, 1984, Hierarchical location models with probabilistic allocation, *Regional Studies* **18**(2):121–129.

Tornqvist, G., S. Nordbeck, B. Rystedt, and P. Gould, 1971, *Multiple Location Analysis,* Lund Studies in Geography ser. **C** (General, Mathematical and Regional Geography), no. 12, Lund, Sweden.

6

HIGH SCHOOL LOCATION DECISION MAKING IN RURAL INDIA AND LOCATION-ALLOCATION MODELS

Vinod K. Tewari and Sidheswar Jena

*Center for Human Settlements and
Environmental Studies
Indian Institute of Management, Bangalore*

Contrasting the situation in developed countries, where many schools are being closed as the number of school-age children declines, in rural India the number of school-age children continues to increase, along with the inadequacy of school facilities. Although a large-scale expansion of school facilities has taken place in the country since the launching of the first five-year-plan in 1951, in 1978 37% of rural habitations did not have a high school within a distance of 8 km (NCERT, 1978). It has also been observed that several of the schools are badly located, many of them are too small, and some are too big to be manageable (Government of India, 1966).

Following the recommendations of the Education Commission (Government of India, 1966), the government, since the fourth five-year plan, has been emphasizing the planned expansion of educational facilities to achieve economy and efficiency. The sixth five year-plan advocated greater efficiency, effectiveness, and optimization of benefits in the delivery of education services. The plan approach has emphasized systematic planning of school facilities "to achieve a larger measure of equalization of educational opportunities, both in regard to access and achievement" (Government of India, 1980, p. 353).

Although the management and financing of general education in India come under the jurisdiction of state governments, the central government plays an important role in formulating and coordinating the national policy on education.

Acknowledgment: Partial support for this work was provided by the National Science Foundation, U.S.A. (Grant No. SES 7925069). The authors gratefully acknowledge the contributions of Gerard Rushton and Michael McNulty.

Within the overall national policy framework, the state governments outline administrative procedures for making decisions on opening new schools. The decision-making process involves (1) a state government organizational structure within which the decisions are made, (2) a set of administrative procedures and locational criteria, and (3) other agencies that have an interest in the opening of schools.

In the context of the national policy for improving access to school facilities, this chapter examines the decision-making process for locating school facilities in Karnataka state in India. First, we discuss the organizational structure, procedures, and administrative criteria through which locational decisions are made and the role played by various agencies in starting new schools. Then, a case study of the decision by the state government between 1981 and 1982 to open new high schools in Bellary uses a location-allocation model to assess the extent to which the decision-making process satisfies the objectives of locational planning.

ORGANIZATIONAL STRUCTURE FOR SCHOOL LOCATION PLANNING IN KARNATAKA

At present, all the state and union territories in India have adopted a uniform system of general education known as ten–plus two–plus three, which consists of ten years of schooling, two years of preuniversity studies, and three years of college studies leading to the first university degree.

In Karnataka state, the ten years of schooling is divided into two stages: the primary stage of seven years and the secondary stage of three years (Government of Karnataka, 1978). Although a primary school teaches all seven classes, the Government of Karnataka distinguishes between a lower primary stage (up to the fourth class) and a higher primary stage (classes five through seven). The high schools, or secondary schools, impart the secondary stage. Many times, classes five through seven are attached to the high schools. Such schools are called composite high schools.

Planning for the location of schools takes place at five different area levels, with different government agencies and other organizations participating at each level (Table 6.1).

At the state level, the Department of Education, headed by the Minister of Education, and its Secretariat, headed by a Secretary, constitute the apex body of planning, coordination, administration, and implementation of educational programs. The Secretariat provides the linkage between the state's educational machinery and the Union Ministry of Education. It formulates policies, prepares the perspective plan, coordinates activities of various departments, oversees implementation of both substantial and procedural components, and reviews the implementation. Also at the state level, the office of the Commissioner of

Table 6.1. Agencies Involved in the School Location Planning Process in Karnataka

Area Level	Agency
State level	Department of Education
	Office of the Commissioner of Public Instruction
Divisional level	Office of the Joint Director of Public Instruction
District level	Office of the Deputy Director of Public Instruction
	District Development Council
	District Advisory Committee for Education
Taluk level	Office of the Assistant Educational Officer
	Taluk Development Board
Local level	Village panchayat
	Village community
	Private and voluntary organizations

Public Instruction, headed by an official drawn from the Indian Administrative Service Cadre, looks after the planning, implementation, and administration of education in the state. This office has two Directors of Public Instruction, who are in charge of primary and secondary education, respectively.

For educational administration, the state has been divided into four divisions, each under the charge of a Divisional Joint Director of Public Instruction. In addition to the overall administration of educational activities in the division, the functions of these directors include inspection of composite junior colleges, high schools, and teacher-training institutions at all levels in the division.

The divisions are further divided into educational districts, creating, altogether, 20 educational districts in Karnataka. One Deputy Director of Public Instruction is in charge of each educational district. The Deputy Director is assisted by Education Officers and Subject Inspectors and is responsible for the planning, coordination, and administration of high schools and primary schools; for the offices of the Assistant Education Officers at the taluk level (subdivision of a district); and for the education wing of the Taluk Development Board. The Deputy Director is the exofficio chairperson of the District Educational Advisory Committee and a member of the District Development Council. The committee and the council play an indirect role in the review and planning of school education in the district.

The Assistant Educational Officer, in charge of taluk level educational activities, is responsible for the administration of primary schools in each taluk and works in close cooperation with the Taluk Development Board and Block Development Officer. The Taluk Development Board is not directly involved in educational administration. However, it assumes the responsibility for maintaining primary school buildings in its area, for which it receives a government grant every year. In urban areas, the municipality also undertakes the responsibility of constructing and maintaining primary-school buildings.

At the local level, the community, village panchayats, local leaders, private registered societies, and voluntary agencies take the initiative to request that the government open schools in their areas. In return, they offer to find extra funds in the community to assist the state in financing the school expansions they request.

CRITERIA FOR LOCATING HIGH SCHOOLS

The government of Karnataka has defined a number of criteria that must be met to start a high school at a particular place. The set of criteria can be grouped into two categories: (1) location- or place-specific criteria and (2) institution- or management-specific criteria.

Location-Specific Criteria

To open a school, "the Department of Public Instruction should be satisfied that there is a need for the institution in the locality and it does not create an unhealthy competition with the existing institutions in the neighbourhood" (Govt of Mysore, 1967—before 1973, Karnataka State was called Mysore State). The department has defined specific rules and norms with respect to the concepts "need," "unhealthy competition," "maximum travel distance," and so on (Government of Mysore 1967; Government of Karnataka, 1976a, 1976b). These guidelines are as follows:

1. There should not be a school within 8 km of the proposed location. This rule is relaxed in the case of urban areas, large villages, inaccessible villages, and backward areas.
2. There must be a minimum of 50 students in class eight in the area before a school is started. This threshold requirement is relaxed in the case of inaccessible and backward areas.
3. There should not be an objection from the neighboring schools to opening a school at the proposed location. The term *neighboring school* is not clearly defined anywhere.

Management-Specific Criteria

Compared to the location-specific criteria, the management-specific criteria are more clearly defined and universally followed. These criteria are as follows:

1. The institution shall be open to all communities without any distinction as to caste, creed, or religion.
2. Premises and location: For a government high school, the public should provide 10 acres of land for the construction of a building and for laying out playgrounds (Government of Karnataka, 1973). For a nongovernment high school, the private management must provide the buildings and 5 acres of land for playgrounds (Government of Mysore, 1967).
3. Accommodation: For a nongovernment high school, a new building, with accommodation as prescribed in the rules, should be ready before the school is opened. If this is not possible, a suitable rented building may be provided for one year. If the permanent building is not ready within a year, sanction for additional classes or sections may not be given. For a government high school, the local public is expected to provide free temporary accommodation for a year, during which the government constructs the new building.
4. Financial resources: For nongovernment high schools, the management must be financially sound and must provide a stability fund of R50,000. This requirement is reduced to R10,000 in the case of high schools started by scheduled castes, scheduled tribes, or backward tribe associations (Government of Karnataka, 1977). For Government high schools, the public should donate to the government R50,000 toward the construction of the building and R10,000 toward the purchase of equipment (Government of Karnataka, 1973). For nongovernment schools, the private management must provide all necessary equipment.
5. Private management must agree to abide by both the rules and regulations laid down in the grant-in-aid code and those stipulated from time to time by the Department of Education.

DECISION MAKING IN LOCATING HIGH SCHOOL FACILITIES

The decision-making process for locating high school facilities has three components:

1. A request to open a school from a local group or private management to the authorities at the taluk or district level.
2. Recommendation of the proposal, after verification against the stated criteria, by authorities at different levels.
3. Final sanction from the Department of Education, in light of the recommendations, for opening the school at the proposed location.

Different agencies at different administrative levels perform these functions.

The decision-making process starts when the Deputy Director of Public Instruction issues a press note every year in October inviting proposals from local communities and private managements to open government and nongovernment high schools. The proposals are submitted in a prescribed form on or before October 31. The proposals for government high schools are normally prepared by the Office of the Deputy Director, either in response to requests from local communities or on the initiative of education officials of the government. The Deputy Director sends his Subject Inspectors to verify the information in the proposals, such as the population of the place; the existing high schools within 16 km and 8 km; the number of students in class seven in the feeder higher primary schools; the financial position of the management; the availability of sufficient accommodation and of local contributions in the form of land, buildings, and cash; objections from the neighboring high schools; and so on. Some of this information is collected by the Subject Inspectors themselves. Sometimes the Deputy Director himself makes field visits to ascertain this information.

Based on such detailed scrutiny, the Deputy Director sends recommendations on each proposal to the Joint Director of Public Instruction at the divisional level. The Joint Director further scrutinizes proposals and sends recommendations to the Director of Public Instruction in charge of secondary education at the state level. The Director of Public Instruction, with the help of the Deputy Director of Public Instruction (Planning), prepares a list of such proposals for the whole state, with summary information that includes the remarks of the district- and divisional-level officials. The list is placed before the Department of Education. There the Minister for Education makes the final decision, within budgetary constraints, as to the number and location of high schools, both government and nongovernment, for the state as a whole. This decision is communicated to concerned officials at the district level.

According to official procedures, proposals should be submitted through a Deputy Director; however, public and private applications are sometimes sent directly to the Joint Director or Director of Public Instruction, or to the Minister of Education. In such circumstances, the applications are referred back to the Deputy Director for recommendations before any decision takes place. Several interviewed officials at the district level felt that political and local pressure groups do greatly influence locational decisions. In the case of a rejected proposal, the concerned local group of private management can appeal to the government within a month of the decision.

The government does not officially recognize a new school until the school shows that it has met all the conditions made for its opening. This evidence may be presented several months after the opening. One consequence of failure to obtain recognition is the withdrawal of financial support from the government, without which the great majority of schools would be unable to function. Moreover, no parent would admit a child to a school that is not recognized by the government, because graduates of such schools are unable to secure admission to another school or college. Thus, unrecognized schools are forced to close.

EFFICIENCY OF RECENT DECISIONS IN BELLARY DISTRICT ON LOCATING HIGH SCHOOLS

The locational decision-making procedures just described were designed to improve the geographical accessibility to school facilities for rural populations, given resource and policy constraints. In this context, it is imperative to ask the following questions: How do the decision makers select locations in practice, and what information base do they use in the decision-making process? Are the decision makers able to find the best location satisfying the prescribed criteria? How do the procedural constraints affect the efficiency of the locational decision?

The following sections examine these questions. They analyze a case study of government decisions in 1981–1982 to locate new high schools in the rural areas of Karnataka's Bellary district.

The Study Area

The study area, Bellary district, is located in the eastern part of Karnataka state. The district is divided into eight taluks. It is enclosed by the river systems of the Tungabhadra, in the north and northwest, the Hagari, in parts of east, and the *Chinna Hagari* in some parts of south, and by an interstate boundary in the east. The district represents rural conditions in India fairly well.

In 1981, the district's population was 1,489,225 (997,065 rural and 492,160 urban). Its gross population density was 150 persons per square kilometer. The rural population density was 109 persons per square kilometer, with a little variation among the taluks.

There were 13 urban centers in the district in 1981. Most of the towns were small, except for Bellary City, which had a population of over 200,000, and Hospet City, with a population of just over 90,000. Of the 590 inhabited villages, 36% had populations below 1000, 51% were between 1000 and 2999, 10% were between 3000 and 4999, and only 3% were over 5000.

There were 821 lower primary, 460 higher primary, and 89 high schools in the district in 1981. The higher primary schools also have lower primary sections, so a total of 1,281 schools were educating at the lower primary level. In all, 322 higher primary schools and 48 high schools were located in the rural areas of the district. Of the 322 rural higher primary schools, only 186 were complete higher primary schools that had classes one to seven. The western half of the district had a higher proportion of high schools and higher primary schools. In

Table 6.2. High Schools in Bellary District, 1951–1981

Year	Total Number of Schools	Schools in Rural Areas[a]	Year	Total Number of Schools	Schools in Rural Areas
1951	15	0	1971	69	37
1961	23	4	1972	72	39
1962	26	5	1973	72	39
1963	37	14	1974	72	39
1964	38	14	1975	72	38
1965	43	17	1976	75	39
1966	53[b]	26	1977	75	39
1967	58	28	1978	77	40
1968	65[c]	34	1979	79	41
1969	65	34	1980	82	44
1970	67	34	1981	89	48

[a] Rural areas are as defined in 1981 census.
[b] One school opened in 1966 was closed down in 1973.
[c] One school opened in 1968 was closed down in 1975 and again opened in 1981.

fact, of schools in rural areas, 56% of higher primary schools, 67% of complete higher primary schools, and 65% of high schools were found in the western four taluks of the district, which together had 49% of the rural population (Government of Karnataka, 1981).

In 1951, there were only 15 high schools in the district, and of these, 13 were located in places classified as urban areas. The number increased to 23 in 1961, 77 in 1971, and as noted, 89 in 1981 (Table 6.2). In 1980, the 82 high schools were in 56 locations, of which 12 were urban locations and 44 were villages. A large number of schools (48) were opened during the eight-year period from 1961 to 1968–33 in rural areas and 15 in urban areas. Schools were opened almost every year, although the total number of schools opened during the decade 1961–1971 was higher than during the decade 1971–1981. In 1981, 7 high schools were opened—4 in rural areas and 3 in urban areas. The four schools in rural areas were sanctioned from 13 proposals submitted to the government by public and private managements. The next section examines the case history of these proposals to find out how decision makers approached locational decisions in practice.

Anatomy of Recent High School Location Decisions

Of the 13 proposals to open high schools in rural areas for the academic year 1981–1982, 6 were for government high schools and 7 were for nongovernment high schools. The proposals were received from 12 villages, as shown in Fig. 6.1. One village submitted two proposals—one for a government high school and the other for a nongovernment one. Since this study's main purpose is to examine the decision-making process without questioning any particular decision, the names of the villages have been deliberately omitted.

The proposals for government high schools were initiated at various levels of government, under different circumstances, following requests from local public and political leaders. The proposals for nongovernment high schools were submitted to the Deputy Director of Public Instruction by the respective managements.

The Deputy Director and his Subject Inspectors carried out field visits to the proposed locations to obtain information required in the pro forma proposals for government high schools, and collect additional information. The information in the proposals was related to the need for a school at the location and to the

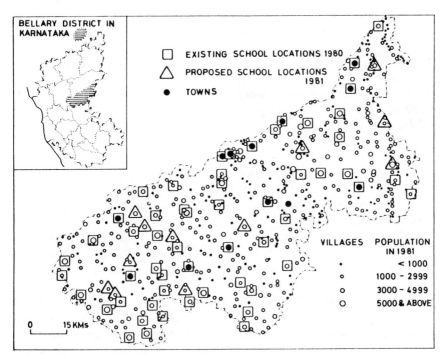

Figure 6.1.   Existing and proposed high school locations in Bellary district, India.

fulfilment of management-specific criteria, and it concerned the following variables:

1. Language of instruction.
2. Type of school—boys only, girls only, or for both.
3. Population of the village.
4. Distance in kilometers to the nearest existing high school.
5. Objections from neighboring high schools, if any.
6. Number of feeder higher primary schools.
7. The feeder higher primary schools' distance in kilometers from the proposed location.
8. Enrollment in class seven in the higher primary school at the proposed location.
9. Total enrollment in class seven in all the feeder higher primary schools.
10. Expected enrollment in class eight, if the high school is opened at the proposed location.
11. Total population of the villages where the feeder schools are.
12. Distribution of children at the proposed location, by age groups.
13. Distribution of the number of persons, by various languages.
14. Scheduled caste and scheduled tribe populations at the proposed location.
15. Distance in kilometers from the proposed location to the nearest metaled road.
16. Distance in kilometers from the proposed location to the nearest railway station.
17. Number of existing high schools within 8 km of the proposed location.
18. Distance in kilometers to existing high schools within 16 km.
19. Distance in kilometers from the proposed location to the nearest existing high school.

There were differences in the information required for government and non-government high schools. The information provided in the 13 proposals is presented in Table 6.3 for the government high schools and in Table 6.4 for the nongovernment high schools.

Scrutiny of this data clearly shows that most of the information was inconsistent and inaccurate, and that both types of proposals contained information gaps. Also, in the case of proposals for nongovernment high schools, certain information provided by management was different from that given by the district officials. It was found that such inconsistencies and gaps were not accidental; the same thing was happening in other districts and had also occurred in previous years. This problem was observed by the Director of Public Instruction in a letter of October 10, 1980, addressed to all the deputy and joint directors in the state; referring to the opening of high schools in 1981 he wrote, "During the previous years, it was noticed that the Deputy Directors of Public Instruction

**Table 6.3. Information Contained in the Proposals for
Government High Schools, 1981**

Variable Number	Proposal				
	G_1	G_2	G_3	G_4	G_5
1	Kannada	Kannada	Kannada	Kannada	Kannada
2	Boys & girls	Boys & girls	Boys & girls	Boys & girls	Boys & girls
3	5029	3893	6492	2779	3462
4	11	4	15	8	16
5	No	Yes	No	No	No
6	2	2	4	3	4
7	0,3	0,4	5,5,8,16	0,12,13	0,8
8	32	40	ND	19	2
9	62	61	101	ND	53
10	62	ND	101	ND	53
11	ND	ND	21,285	2779	5784
12 age group:					
6–11	257	ND	705	230	388
11–14	76	ND	301	64	251
14–17	22	ND	265	2	210
17–18	ND	ND	259	ND	274
13 language:					
Kannada	6000	ND	8500	4500	5000
Telugu	3000	ND	630	400	ND
Others	1000	ND	ND	200	ND
14	690	61	891	510	350
15	14	5	15	0	0
16	14	40	15	40	30

Notes: See text for description of the variables.
ND—no data was provided in the proposal.
Information on proposal G_6 was submitted directly to the Department of Education and could not be obtained from that office.
Source: The proposals for government high schools, Office of the Deputy Director of Public Instruction, Bellary, and Office of the Commissioner of Public Instruction, Bangalore, 1981.

had sent incomplete and vague reports, for example the columns 15 to 18 in the proforma were left blank in most cases and in remaining cases incomplete information was furnished. . . . which caused considerable delay in settling the cases.''

Table 6.4. Information Contained in the Proposals for Nongovernment High Schools, 1981

Variable Number	Proposal						
	N_1	N_2	N_3	N_4	N_5	N_6	N_7
Provided by the management							
1	Kannada	Kannada	Kannada	Kannada	Kannada	Kannada	Kannada
2	Boys & girls	Boys & girls	Boys & girls	Boys & girls	Boys & girls	Boys & girls	Boys & girls
3	10,000	7555	3500	3000	3500	3500	2500
4	11	6	11	8	12	8	10
6	2	7	3	4	4	4	3
7	0,3	0,1,5	ND	ND	ND	0,1,2,3	ND
8	31	18	36	25	38	25	36
9	61	92	85	85	103	80	78
11	11,000	23,280	7500	10,500	9000	6000	5000
17	0	1	0	0	0	0	0
18	11	6,10, 10,16	11,12, 30,15	8,8,8	12,18, 12	8	10
Provided by the district officials							
5	No	No	Yes	No	No	Yes	Yes
6	2	7	3	4	4	4	3
8	31	18	36	25	29	15	36
9	61	92	85	85	61	44	78
19	>8	>8	5	8	12	>8	<8

Notes: See text for description of the variables.
ND—no data was provided in the proposal.
Source: The proposals for nongovernment high schools, Office of the Deputy Director of Public Instruction, Bellary, and Office of the Commissioner of Public Instruction, Bangalore, 1981.

As stated earlier, the management-specific criteria for government high schools were availability of R50,000 for building, R10,000 for equipment, 10 acres of land, and a rent-free temporary building; and for the nongovernment high schools the criteria were R50,000 as a stability fund, 5 acres of land, accommodation for the school, furniture, and other facilities. The information on these criteria was provided by the public for government schools and by management for nongovernment schools. In their field verifications, the district officials had ascertained whether these conditions could be fulfilled by the local people or the management if each school was sanctioned. According to the officials, the requisite financial support and other facilities were available for proposals G_1, G_3, N_1, and N_2.

Recommendations on the Proposals

Of the 12 proposals submitted to the district office, the Deputy Director determined a need for 7 schools. To justify his approval of the need for a school, the Deputy Director, in letters addressed to the Joint Director in February 1981, mentioned the following two factors: (1) the distance to the nearest existing high school was not less than 8 km, and (2) no neighboring schools objected that the proposed school would affect their student strength. However, one proposal (G_3) that had fulfilled the two conditions was denied approval because it contained the following remarks from the Subject Inspector: ". . . there is no higher primary school with Kannada as a medium of instruction in the village and therefore, there is no need for a high school" (translated from Kannada). The village was predominantly Telugu speaking but had asked for a Kannada high school. According to the Subject Inspector's report, four feeder Kannada higher primary schools in the catchment area had a total enrollment of 101 in the terminal class. It is not clear why this aspect was not considered in determining the need for a school.

The Deputy Director found that considerations regarding management-specific criteria were satisfied in only three out of the seven cases—G_1, N_1, and N_2—

Table 6.5. Conclusions and Recommendations for the 12 Proposals

Proposal	Need	Fulfilment of the Criteria	Recommendation for Sanction	Sanctioned by the Government
G_1	Yes	Yes	Yes	Yes
G_2	No	No	No	No
G_3	No	Yes	No	No
G_4	Yes	No	No	No
G_5	Yes	No	No	No
$G_6{}^a$				Yes
N_1	Yes	Yes	Yes	No
N_2	Yes	Yes	Yes	No
N_3	No	No	No	No
N_4	Yes	No	No	Yes
N_5	Yes	No	No	No
N_6	No	No	No	No
N_7	No	No	No	Yes

^{a} Proposal not available.
Source: The proposals for government and nongovernment high schools, Office of the Deputy Director of Public Instruction, Bellary, and Office of the Commissioner of Public Instruction, Bangalore, 1981.

and therefore these three were recommended for the sanction of a school (Table 6.5). Based on the Deputy Director's recommendation, the Joint Director of Public Instruction recommended the same three proposals for opening new high schools in the district during the academic year 1981–1982.

Decisions of the Government

The Joint Director of Public Instruction submitted the proposals, with detailed information in the prescribed form, and their accompanying recommendations to the Commissioner of Public Instruction. The proposals were further scrutinized at the commissioner's office, and a consolidated statement on them was submitted to the government for its sanction. This statement included information on name of the management, place name, language of instruction, population at the school location, neighboring high schools within 16 km, number of students in class 7 of the feeder higher primary schools, recommendations of the deputy and joint directors, objections from neighboring schools, if any, and proportion of scheduled caste or scheduled tribe population in the village. At the government level, the Department of Education—with input from its own director, commissioner, and secretary—accorded sanction to four proposals G_1, G_6, N_4, and N_7. Thus, of the three schools recommended by the officials at the district, division, and commissioner's levels, only one was sanctioned. Schools sanctioned on proposals N_4 and N_7 did not satisfy all the criteria, and no information was available on proposal G_6 detailing why the school was sanctioned. Thus, sanctioning proposals G_6, N_4, and N_7 was not in accordance with the government's prescribed procedure and criteria.

The decision-making process took nearly eight months, starting from October 10, 1980, when the government issued its press note inviting proposals, and ending in June 1981, when the official order sanctioning the schools in Bellary district was issued. The anatomy of this decision-making process shows that the final decisions were neither founded on a proper and relevant information base nor subject to any explicit objective function, although a large amount of data was collected during the process and a number of administrators participated in the decision making. However, the criteria specified in the procedure and the data required do indicate the desire of the service providers to improve the geographic accessibility and efficiency of the service system. The decisions were based on an evaluation of the few formal proposals that were submitted. Further, the facts gathered to describe proposed locations were not always accurate.

The system of determining new locations is clearly based on the assumption that inviting people to come forward and propose a place will give a "choice set" from which a smaller subset of alternatives can be selected. This system can therefore be evaluated from two points of view: the degree to which it generates a good choice set and the degree to which the process of selection from the choice set is efficient.

The Analysis System

To find the efficiency of the locational decisions, the improvement in accessibility of school facilities yielded by the proposed and sanctioned school locations was computed and compared with that of the same number of optimal locations. The method used to determine the optimal locations was location-allocation modeling (Rushton and Kohler, 1973; Rushton, 1979; Hodgart, 1978; Bach, 1980; Church and ReVelle, 1974, 1976). Location-allocation heuristics, based on the vertex substitution algorithm (Tietz and Bart, 1968), were used to find optimal locations with respect to the following commonly used accessibility criteria: (1) to minimize the average traveled distance of the population to the nearest school facility, and (2) to maximize the population covered within a maximum service distance of 8 km.

The first optimality criterion of minimizing average distance to a system of facilities is equivalent to maximizing access of the population to the facility system. As access is closely related to the utilization of facilities, in this case high schools, minimizing average distance could be an important consideration in determining facility locations. Another consideration could be to locate facilities in such a way that as many users as possible are covered within a specified maximum service distance from their closest facility. In this case, the second optimality criterion of maximizing population covered within a maximum specified distance of 8 km becomes more relevant. The choice of a maximum service distance of 8 km was based on the government objective of providing the rural population with a high school within 8 km.

The index of locational inefficiency for a set of locations added to the existing system was defined as $|Z - \bar{Z}| / Z$, where Z is the improvement in the objective functions gained by adding a set of locations to the existing system, and $\bar{Z}$ is the improvement gained by adding the same number of optimal locations.

The proposed, recommended, and sanctioned locations and corresponding sets of optimal locations were also compared on other measures: maximum distance of a user from the nearest facility in the total system, average size of villages proposed for new schools, average size of population in the total catchment area, and average size of population and average number of villages in the catchment area within 8 km of school locations.

The computations of accessibility measures and optimal locations were carried out using the computerized location-allocation analysis system ALLOC VII, earlier developed by Rushton and Kohler (1973) and Hillsman (1980) and further improved as a part of this study. The analysis system was sufficiently flexible that suitable data editing made it possible to find the optimal locations of a large number of facilities for a number of objective functions (Hillsman, 1984).

The analysis relied on a computer-based geocoded information system developed for the district. The information system contained the locations of 600 villages and towns in 1971, their population in 1981, and the inter-place distances. (The geocoded information system was developed in 1980, before the availability of the 1981 census. Therefore the information system was based on the locations of 600 villages and towns given in the 1971 census.) The villages and towns were identified on a 1:25,000-scale map that was created using Government of Karnataka Survey and Settlement Maps, the Survey of India's topographical sheets, and road maps from the Karnataka Road Transport Corporation. Metaled roads, unmetaled roads, cart tracks, and foot paths were identified as networks connecting all the villages and towns. Each link on the network was identified and measured and this data was used to compute a 600-x-600 distance matrix using a shortest-path algorithm (Ostresh, 1973).

The demand for high school educational facilities at a village at any given time is the number of children who, after completing class seven of a higher primary school, are willing to pursue studies in a high school. However, in view of the national educational policy of providing universal education to all children up to age 14, the number of children in the high school age group of 13 to 15 years would be a more appropriate measure of potential demand. In the absence of this information at the village level and on the assumption that the age structure of village populations did not vary significantly across the district, the total 1981 population of each village and town was used as the demand variable.

Locational Efficiency of Proposed Locations

Using the analysis system described, the 12 available proposals were first examined to measure their inefficiency compared with 12 optimal locations. Using the existing high school service system in the district as fixed locations, two sets of optimal locations were determined based on the two objective functions, minimizing average distance and maximizing coverage of population within 8 km of the school locations. There were 82 high schools in Bellary district during the academic year 1980–1981; they were located in 44 villages and 12 towns comprising a system of 56 high school locations. The 12 optimal locations were selected from 452 feasible locations, which were defined as the total number of villages minus the 56 villages with existing schools and the 82 villages with a population of less than 500. For the 56 school locations existing in 1980–1981, the average distance of the rural population in the district to the nearest high school was calculated as 6.8 km. Some people in the district were more than 26 km away from the nearest high school; only 63% of the rural population and 55% of village locations were within 8 km of a high school.

Compared to adding the 12 proposed locations to the existing system, the addition of 12 optimal locations resulted in a greater improvement in various systemwide accessibility measures (Table 6.6). While the proposed locations had no effect on the maximum distance to the nearest school, the optimal locations reduced it from 27 km to 17 km, thus improving the accessibility of those who were most disadvantaged.

The locational inefficiency of the proposed locations was 52% for the criterion of minimizing average distance and 62% for the criterion of maximizing coverage. Improvements in population coverage within 8 km using the two optimal sets of locations, based on the two criteria respectively, were 43% and 62% (Tables 6.7 and 6.8). The two sets of optimal locations were also superior to the proposed locations on other measures of accessibility characteristics, such as the average population in the catchment area and the average population and number of villages in the catchment within 8 km of the school location. (All the measures of accessibility in the analysis refer to the rural population.) The average size of the proposed villages for new schools was larger than the average size of

Table 6.6. Systemwide Accessibility Characteristics, 1980–1981, Proposed and Optimal

Accessibility Characteristics	56 Existing Locations	56 Existing plus 12 Proposed	56 Existing plus 12 Optimal		% Improvement of Optimal over Proposed	
			$0-1^a$	$0-2^b$	$0-1^a$	$0-2^b$
Average distance traveled (kilometers)	6.8	5.6	5.0	5.2	10.7	7.1
Maximum distance traveled (kilometers)	26.9	26.9	17.1	17.1	57.3	57.3
Percent rural population in villages with schools	18.8	22.8	22.2	20.9	−2.6	−8.3
Percent rural population within 8 km of a school	63.3	73.6	78.8	80.1	7.1	8.8
Percent villages within 8 km of a school	55.1	64.8	71.6	73.3	10.5	13.1

[a] 0–1 was based on the objective function of minimizing the average distance.

[b] 0–2 was based on maximizing coverage of population within 8 km.

Table 6.7. Improvement in Accessibility Owing to Proposed and Optimal Locations (Minimizing Average Distance)

Accessibility Characteristics	12 Proposed	12 Optimal	% Improvement of Optimal over Proposed
Systemwide improvement over existing system			
Person-kilometers saved	1,141,974	1,732,256	51.7
Increase in population covered within 8 km	103,740	147,948	42.6
Increase in number of villages covered within 8 km	56	95	69.6
Locational characteristics of new locations			
Average population of village at school location	3,327	2,817	−15.3
Average population per catchment area	13,586	16,231	19.5
Average population per catchment area (within 8 km)	9,264	13,896	50.0
Average number of villages per catchment area (within 8 km)	5.3	8.6	62.3

the villages in the 12 optimal locations. The two sets of 12 locations—proposed and optimal—had two locations in common for the criterion of minimizing average distance and one for the criterion of maximizing coverage (Fig. 6.2). Whereas 7 of the 12 proposals came from the western half of the district, which was relatively better off in terms of school facilities, and 6 from the

Table 6.8. Improvement in Accessibility Owing to Proposed and Optimal Locations (Maximizing Coverage Within 8 km)

Accessibility Characteristics	12 Proposed	12 Optimal	% Improvement of Optimal over Proposed
Systemwide improvement over existing system			
Person-kilometers saved	1,141,974	1,546,113	35.4
Increase in population covered within 8 km	103,740	168,444	62.4
Increase in number of villages covered within 8 km	56	105	87.5
Locational characteristics of new locations			
Average population at school location	3,327	1,840	−44.7
Average population per catchment area	13,586	16,540	21.7
Average population per catchment area (within 8 km)	9,264	14,726	59.8
Average number of villages per catchment area (within 8 km)	5.3	9.0	69.8

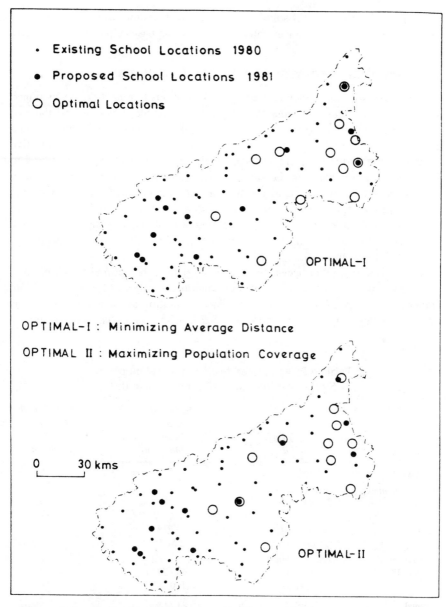

Figure 6.2. Optimal solutions corresponding to the 12 proposed high school locations.

eastern half, in the optimal solution 11 of the 12 locations were from the eastern half; in fact, 7 were concentrated in the extreme eastern part of the district.

Locational Efficiency of Sanctioned Locations

Four optimal solutions, each consisting of 4 locations to reflect the 4 sanctioned locations, were obtained using the two sets of feasible locations and the two objective functions described earlier. When the 4 optimal locations were obtained using the 12 proposed locations as the feasible location set, based on the objective function of minimizing average distance, the optimal locations were found to be 98% more efficient than the sanctioned locations. On the other hand, when the choice set was enlarged to all the feasible 452 village locations, with the same objective function, the optimal locations turned out to be 115% more efficient (Table 6.9). For the objective function of maximizing the coverage of population withint 8 km, the two sets of optimal locations were, respectively, 100% and 150% more efficient.

On measures of other accessibility characteristics, the optimal locations were also 100–150% better than those of the sanctioned locations (Table 6.10). More importantly, whereas the minimum distance separating the sanctioned locations and an existing neighboring school was 5.5 km, the same distance for optimal

Table 6.9. Accessibility Characteristics of the Four Sanctioned and Optimal Locations

Locations	Person-km Saved	Increased Coverage of Population (Within 8 km)	Increased Coverage of Villages (Within 8 km)
Sanctioned	428,581	32,821	20
Optimal: minimizing average distance			
Optimal 1[a]	848,263	65,677	40
	(97.9*)	(100.1)	(100.0)
Optimal 2[b]	923,116	73,748	39
	(115.4*)	(124.7)	(95.0)
Optimal: maximizing coverage within 8 km			
Optimal 1[a]	848,263	65,677	40
	(97.9)	(100.1*)	(100.0)
Optimal 2[b]	885,857	81,868	49
	(106.7)	(149.4*)	(145.0)

* Figures in parentheses are per cent improvement over sanctioned locations. Those marked * provide a measure of locational inefficiency of the sanctioned locations compared to the optimal ones.
[a] Optimal 1: candidates—12 proposed locations of 13 proposals.
[b] Optimal 2: candidates—452 feasible locations.

**Table 6.10. Locational Characteristics of the
Four Sanctioned and Optimal Locations**

Locations	Average Population of Village at Proposed Location	Population per Catchment Area	Population per Catchment Area (Within 8 km)	Number of Villages per Catchment Area (Within 8 km)	Minimum Distance from Nearest Existing School (km)
Sanctioned	3,280	17,938	10,165	6.8	5.5
Optimal: minimizing average distance					
Optimal 1	4,027	28,272	16,419	10.5	10.0
Optimal 2	3,529	29,512	18,437	10.5	10.0
Optimal: maximizing coverage within 8 km					
Optimal 1	4,027	28,272	16,419	10.5	10.0
Optimal 2	1,130	25,390	20,467	12.3	12.1

Note: Refer to Table 6.9 for interpretation of optimal 1 and 2.

locations was 10 km or more. Thus, while the sanctioned locations violated the criterion that no school should be opened within 8 km of an existing school, the optimal locations gave a very good solution in this regard.

The assignment of villages to their closest school locations clearly showed that in the western half of the district most villages were served by a high school within 8 km, but in the eastern half most villages were beyond 8 km from the closest school. Yet it was the better-served part of the district in which 3 of the 4 new high schools were located (Fig. 6.3). The inefficiency of these 3 locations compared to the optimal ones was further emphasized when the sanctioned locations and the four sets of optimal locations were plotted along with their assignments on the district map. The four optimal locations in all the solutions were selected from the poorly served north-eastern part of the district (Fig. 6.4).

CONCLUSION

In India, where the population has been increasing at a rate of about 25% per decade and a sizable proportion of those living in rural areas have never had adequate access to school facilities, a large number of new schools are started every year by both government and private organizations. For example, in Karnataka state, which had a population of 37 million in 1981, about 900 new high schools were opened during the years 1983 and 1984. State and national planning documents have often emphasized the need to plan school locations to avoid inefficient use of scarce resources and to bring rural children

○   EXISTING SCHOOL LOCATIONS IN 1980

●   SANCTIONED SCHOOL LOCATIONS IN 1981

VILLAGES ASSIGNED TO THE NEAREST
SCHOOL LOCATIONS

———————   Within 8 Kilometers

- - - - - - -   Beyond 8 Kilometers

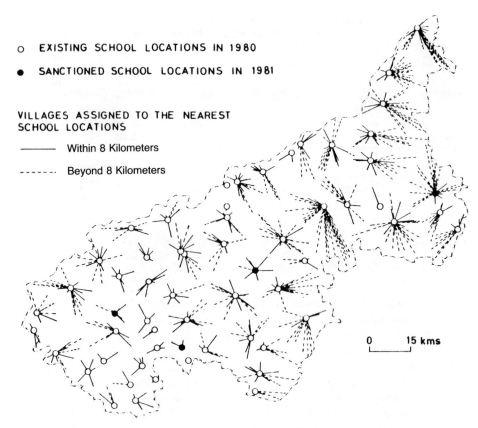

0   15 kms

Figure 6.3.   High school locations (existing and sanctioned) and their catchment areas.

within easy access of schools (Government of India, 1966, 1980; Government of Karnataka, 1980). Procedures have been devised by those involved in the planning process to provide school facilities, according to prescribed criteria, in areas that are poorly served. But the decision makers following these procedures are not able to take into account the total situation in which the facilities are planned. Therefore, they often come up with solutions that are not the best among available alternatives.

The analysis here of the decision-making process for locating high schools in rural areas of Karnataka revealed that (1) the criteria prescribed for making locational decisions did not specify a clear-cut accessibility measure; (2) the absence of a proper information system, the complexity of the situation, the magnitude of the problem, and the subjective bias of various interest groups

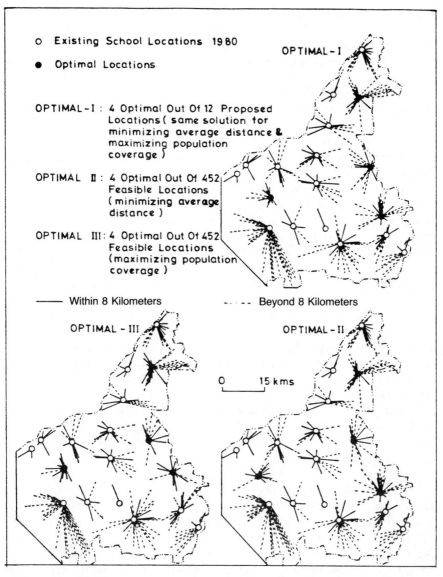

Figure 6.4. Optimal solutions corresponding to the 4 sanctioned high school locations (showing north-eastern part of the district, which contained all the optimal locations, is shown here).

resulted in an incomplete and inaccurate information base for decision making; and (3) the final decisions were inconsistent with the planning objectives and prescribed criteria.

Although the locational criteria did not reflect the plan objective, the criterion of minimum distance of 8 km between a proposed location and the nearest existing location did reflect the planners' intention to improve the spatial distribution of facilities and their coverage of the population. However, the criteria did not specify any way to measure the systemwide efficiency of a selected location compared with those not selected.

The management criteria, considered necessary in view of the limited resources, imposed the precondition that a proposal must furnish the evidence satisfying the criteria before any location could be recommended by the district and divisional offices for a new high school. Such restrictions are likely to lead to a lopsided spatial distribution of facilities favoring the relatively advanced areas where local initiative and contributions are more easily forthcoming, and this bias defeats the very purpose of the planning exercise in this regard. Six out of the seven proposals submitted for nongovernment high schools, for example, came from the relatively advanced western part of the district.

The average size of the villages proposed for new schools was found to be larger than the average size of villages in an equal number of optimal locations determined using the location-allocation model. This evidence supports the conclusion that the currently used decision process, by placing the burden on individual places to propose new facilities for themselves, does not help identify the places suited to serve the clusters of villages commonly found in rural areas where population densities are low. One would expect that the larger villages, with better local resources and leadership, would propose themselves for the location of new high schools, and the evidence suggests that this situation occurred in the study area—to the detriment of providing accessible education to scattered populations in backward areas.

Furthermore, three out of the four sanctioned locations, selected by limiting the decision making to the proposals submitted, were in that part of the district already better served by existing locations. On the other hand, the four optimal locations for new schools, selected out of all eligible locations using the objective function of maximizing coverage, were villages of smaller average size than the sanctioned locations, and they would cover a larger average population and a larger number of villages in their catchment areas. Moreover, the optimal solution selected locations from the areas that were poorly served.

The findings of this study also showed that the decision makers made decisions that were inconsistent with their own prescribed criteria. It is suggested that political pressures influence the highest levels of decision making, but no evidence could be obtained to implicate political influence as a possible reason for such inconsistent outcomes.

The results of applying location-allocation models in this study illustrate the potential usefulness of such models in public facility location planning and decision making. It is often argued that "sophisticated" mathematical models are not useful in a developing country like India, where decision makers have neither the necessary training nor the access to computing facilities required for the usage of such models. Nevertheless, this attitude is slowly changing with the growing realization that in the future, electronic computers and information technology will have to play a major role in the planning process. At the same time, while making a strong case for decentralised district planning, a working group of the planning commission observed; "since the resources available for social services are so limited, it seems desirable that an attempt should be made to evolve some principles for the proper location of all types of social service facilities" (Government of India, 1984). This problem context is amenable to location-allocation models having a sound theoretical base, a well-documented computer software, and demonstrated applications to various service facility location problems. Such models can be very helpful to decision makers in their search for alternative locational arrangements that help fulfill defined objectives.

REFERENCES

Bach, L., 1980, Locational models for systems of private and public facilities based on concepts of accessibility and access opportunity, *Environment and Planning, A* **12**:301–320.

Church, R. and C. Revelle, 1974, The maximal covering Location Problem, *Regional Science Association Papers* **32**:101–118.

Church, R. and C. Revelle, 1976, Theoretical and computational links between the *p*-median, location set-covering, and the maximal covering location problem, *Geographical Analysis* **8**:406–415.

Government of India, 1966, *Education and National Development: Report of the Education Commission, 1964–66*, Ministry of Education, New Delhi.

Government of India, 1980, *Sixth Five Year Plan: 1980–85*, Planning Commission, New Delhi.

Government of India, 1984, *Report of the Working Group on District Planning*, vol. 2, Planning Commission, New Delhi.

Government of Karnataka, 1973, Government Order no. ED. M. H. S. 73, dated 13–7–73, Bangalore.

Government of Karnataka, 1976a, Proforma for Application for Opening New Non-Government Secondary Schools in Karnataka State, Department of Education, Bangalore.

Government of Karnataka, 1976b, Proforma for Opening New Government Secondary Schools in Karnataka State, Department of Education, Bangalore.

Government of Karnataka, 1977, Government Order no. ED.155.PGC.74, dated 19–8–1977, Bangalore.

Government of Karnataka, 1978, *Draft Five Year Plan, 1978–83*, Planning Department, Bangalore.

Government of Karnataka, 1980, *Karnataka: Draft Sixth Five Year Plan, 1980–85*, Planning Department, Bangalore.

Government of Karnataka, 1981, *Educational Statistics: Bellary District, 1980*, Office of the Commissioner of Public Instruction, Bangalore.

Government of Mysore, 1967, *Uniform Grant-In-Aid Code for Non-Government Secondary Schools*, Bangalore.

Hillsman, E. L., 1980, *Heuristic Solutions to Location-Allocation Problems: A User's Guide to ALLOC IV, V and VI, Monograph Number 7, Department of Geography, University of Iowa, Iowa City*.

Hillsman, E. L., 1984, The p-median structure as a unified linear model for location-allocation analysis, *Environment and Planning, A* **16**:305–318.

Hodgart, R. L., 1978, Optimizing access to public services: A review of problems, models and methods of locating central facilities, *Progress in Geography* **2**:17–48.

NCERT, 1978, *Report of the Fourth Educational Survey*, National Council of Educational Research and Training, New Delhi.

Ostresh, L. M., 1973, SPA: A shortest path algorithm, in *Computer Programs for Location-Allocation Problems*, G. Rushton, M. F. Goodchild, and L. M. Ostresh, eds., Monograph 6, Department of Geography, University of Iowa, Iowa City, pp. 141–162.

Rushton, G., 1979, *Optimal Location of Facilities*, COMpress, Wentworth, N.H.

Rushton, G. and J. A. Kohler, 1973, Heuristic solutions to multi-facility location problems on a graph, in *Computer Programs for Location Allocation Problems*, G. Rushton, M. F. Goodchild, and L. M. Ostresh, eds., Monograph 6, Department of Geography, University of Iowa, Iowa City, pp. 163–187.

Teitz, M. B. and P. Bart, 1968, Heuristic methods for estimating the generalized vertex median of a weighted graph, *Operations Research* **16**:955–961.

7

HIERARCHICAL LOCATION ANALYSIS USING COVERING OBJECTIVES

Richard L. Church

University of California, Santa Barbara

David J. Eaton

University of Texas at Austin

Over the past twenty years, location-allocation analysts have developed a variety of models termed *hierarchical* in that they incorporate multiple levels of either goals or facilities. Several investigators have used *hierarchy* to mean incorporation of a set of goals organized by priority to determine the best facility configuration for one service. Examples of this genre include Daskin and Stern (1981) and Church (1974). A more common use of *hierarchy* refers to the problem of coordinating multiple levels of facilities.

Formulations to locate facilities for multiple levels of service can be classified by (1) the geographical relationship between demand and services, (2) the nature of the relationship among service levels, and (3) the measure of system effectiveness. A review of the literature indicates that these models are difficult to classify (see Table 7.1). It may be useful to describe and contrast the papers cited in Table 7.1 as a prelude to developing hierarchical referral formulations using covering objectives.

All hierarchical service delivery systems involve a set of sites operated in a coordinated manner. Systems differ in the relationship between upper and lower service levels (see Fig. 7.1).

A school system is a set of parallel facilities because a client goes to one appropriate service—primary, junior high, or high school. Multilevel warehousing is an example of sequential service because higher-level (regional) warehouses ship to lower-level (local) warehouses when the latter call for additional shipments. A two-tiered emergency medical service system involving basic and advanced life support can be a referral service because a critical emergency can

Table 7.1. Previous Formulations of Hierarchical Service Location

Problem Structure	Objectives	Reference
I. Planar relation of facilities to demands	Minimize transportation and facility costs Maximize utilization	Dokmeci (1973) Dokmeci (1979)
	Maximize per capita net social benefits	Schultz (1970)
II. Network relation of facilities to demands		
A. Nonreferral models	Minimize respond time to demand Minimize transport time from demand	Berlin, ReVelle, and Elzinga (1976)
	Maximize accessibility Minimize user costs Maximize utilization and utility	Calvo and Marks (1973) Tien, El-Tell, and Simons (1983)
	Minimize number of facilities to cover demand Minimize average total time	Banerji and Fisher (1974)
	Minimize maximum distance Minimize average distance	Fisher and Rushton (1979)
	Minimize transportation and facility costs	Marks (1969) Marks and Liebman (1971) Scott (1971) Walker, Aquilina, and Shur (1974) Vasan (1975) Whitlatch (1977) Ravindran and Hanline (1980) Osleeb, Rattick, and Lewis (1981)
	Minimize travel time	Harvey, Hung, and Brown (1974) Tien and El-Tell (1984)
	Minimize costs to assure closest service facility	Ruth (1979)
	Maximize demand coverage	Charnes and Storbeck (1980)
	Maximize demand coverage	Moore and ReVelle (1981) Moore and ReVelle (1982)
B. Referral models	Minimize total travel	Narula, Ogbu, and Samuelson (1975) Narula and Ogbu (1979) Narula and Ogbu (1985)
	Minimize transportation and facility costs	Law (1976)

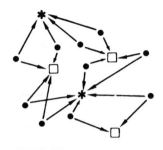

Parallel Service

Example: a school system

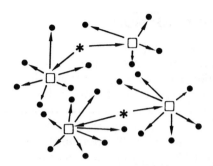

Sequential Service

Example: multilevel warehousing

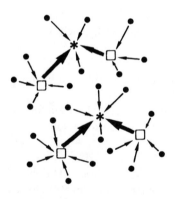

Referral Service

Example:
a two-tiered emergency
medical service system

Figure 7.1. Hierarchical service delivery systems. Points represent sources of demand; Squares represent lower service level; stars represent higher service level; arrows indicate contact patterns but not necessarily movement.

be treated initially by a basic life support vehicle and referred to an advanced life support vehicle. Hierarchical location models have been developed to minimize weighted distance, maximize coverage, minimize cost, maximize utilization, or achieve a combination of these or other objectives. Some of the problems have been defined in a network and others on the plane.

PLANAR MODELS

Dokmeci (1973) defined a planar, hierarchical facility location problem and developed a heuristic procedure for solving it. This problem minimizes transportation and facility costs for an n-level hierarchy where demand is fixed and the size and number of each facility on each level is determined. Dokmeci (1979) suggested a multiobjective version of this problem that maximizes utilization as a function of distance.

Schultz (1970) determined a nested central place geometry that maximizes net social benefits for a hierarchical medical system of health posts, general hospitals, and medical centers. The objective was to maximize per capita net social benefits of the population served, assuming homogeneously distributed demand and that each facility meets demand within its service area.

NETWORK NONREFERRAL MODELS

Scott (1971) defined a sequential facility transportation problem containing a hierarchical system of facilities. A facility at level R receives (dispatches) goods only from level R-1 and dispatches (receives) goods only to level $R+1$, with no permitted circumventing of hierarchy levels. Scott structured this sequential facility model as a mathematical programming problem and utilized a heuristic approach to solve the model.

The hierarchical model is similar to sequential network facility location formulations that generate the best plan of facilities and transportation for solid-waste planning (see Marks and Liebman, 1971; Vasan, 1974; Walker, Aquilina, and Schur, 1974). These models handle solid-waste flows from their origin, to intermediate processing points, and on to final disposal facilities. The sizes, treatment technologies, and transfer functions of the intermediate levels between origins and the final disposal sites, along with the siting of the final disposal sites, represent the possible alternatives. The pure transfer problem formulated by Marks (1969) is similar to Scott's model. Other examples have been developed for coal washing and blending (Ravindran and Hanline, 1980), port expansion (Osleeb, Rattick, and Lewis, 1981), and coal-gasification facility location (Whitlatch, 1977).

Harvey, Hung, and Brown (1974) applied the p-median problem in Sierra Leone to determine the number and optimal positions of intermediate order centers in a central place hierarchy. Although they used the p-median model as a one-level problem, they did recognize the interactions between lower and higher levels.

Berlin, ReVelle, and Elzinga (1976) discuss a "dual facility" hospital and ambulance location problem to minimize (1) ambulance response time from bases to demand areas and (2) transport distances to hospitals from demand areas. This problem is "pseudohierarchical"; although two facility levels are defined, the formulation can be decomposed into independent hospital and ambulance location problems and solved optimally.

Calvo and Marks (1973) formulated a multiobjective model to locate health care facilities within a hierarchical format: the model maximizes accessibility, minimizes user costs, maximizes demand or utilization, and maximizes utility. It is based on assumptions that (1) users go to the closest appropriate level; (2) there is no referral to higher levels; and (3) at any level, each facility offers all lower-level services.

Tien, El-Tell, and Simmons (1983) have described several problems associated with the Calvo and Marks formulation and have suggested an improved successively exclusive service formulation. They also introduce a new feature whereby a demand cannot assign to a place more than once even if additional service levels may be available at that point.

Banerji and Fisher (1974) have used both median and covering concepts in a hierarchical location-planning problem in India. The hierarchy was nested; higher-level settlements also perform services available at lower levels. The location set covering problem was used to determine the minimum number of necessary facilities for a given level. The p-median problem was used to shift locations to maximize accessibility. The solution procedure first solved the highest level; then, after incorporating the highest-level facilities as fixed, the procedures located the next lower level. Fixing that level, the process proceeded downward in this fashion to the lowest level. The p-median problems were solved by the Teitz and Bart heuristic (Teitz and Bart, 1968).

Fisher and Rushton (1979) analyzed hierarchical locations in India in terms of the maximum distance and the average distance of any user from the closest facility. They compared heuristic and existing spatial efficiencies for several services. The heuristic solution procedure involved using the Tietz and Bart procedure in one of three modes: top-down hierarchy (same as Banerji and Fisher, 1974, bottom-up hierarchy (opposite of top down), and middle-level hierarchy (start at the middle and then solve up and down).

Ruth (1979) modeled a hierarchy of regional hospitals. Hospitals can be either located or fixed, and service quantities or levels are allocated via a mixed-integer programming model. Costs are minimized by assigning each patient to the closest hospital having capacity at the needed level of service. Standards of service are introduced in the constraint set.

Charnes and Storbeck (1980) have described a two-tiered model of an emergency medical service system. The model locates basic life support (BLS) and advanced life support (ALS) vehicles. The authors defined their problem within a goal programming context and used the covering of demand by vehicles as a goal. The model attempts to ensure that if all services cannot cover all calls, then at least the BLS vehicle should cover demand.

Moore and ReVelle (1982) developed a nested, parallel hierarchical covering model for locating medical services. Clinic and hospital service levels are located using a maximal covering problem framework. The services are nested; hospitals provide clinic services as well as hospital services. The authors considered no referral services, in terms of patients traveling from clinics to hospitals, and no interaction between hospitals and clinics, except that hospitals themselves provide clinic services. The model was solved by linear programming and was applied to a region of Honduras (Moore and ReVelle, 1981).

NETWORK REFERRAL MODELS

Law (1976) developed a hierarchical location-allocation model and applied it to providing health facilities in a ten-county area of Texas. This model minimizes transportation and facility costs to provide services for all levels of the hierarchy. Demand for a service level located at another site generates interfacility flow. The model was solved by a branch-and-bound algorithm using special network procedures in resolving each node.

Narula, Ogbu, and Samuelsson (1975) modeled a nested hierarchical health facility network that locates a fixed number of facilities at each level while minimizing patients' total travel. Referrals are based on a proportion of patients treated at a given level. Narula and Ogbu (1979) introduced additional heuristics to this hierarchical median problem, including substitutions from the bottom-up (forward) and top-down (backward) and add, drop, and greedy substitutions. Narula and Ogbu (1985) used lagrangian relaxation with subgradient optimization to solve a two-level median problem. The relaxation that was used often resulted in sizable duality gaps.

Tien and El-Tell (1984) developed a quasi-hierarchical model and applied it to a 31-village area of Jordan. The quasi-hierarchy consisted of village clinics and regional clinics. Regional clinics have doctors who make daily visits to the village clinics. The model sites both village and regional clinics and allocates physician time among village clinics. The formulation minimizes the weighted distance of assigning villages to clinics and village clinics to regional clinics. Tien and El-Tell solved their formulation with the IBM MPSX linear programming software.

This review has described a diverse set of successful location-allocation models for hierarchical service delivery. Our work in Colombia and the Dominican Republic has indicated that these existing formulations do not adequately address problems that arise in rural health settings where (1) there is a hierarchy of services, (2) there are different types of demand or dissimilar services, (3) the goal of accessibility is paramount for each service, and (4) referral occurs among the hierarchy levels.

HIERARCHICAL REFERRAL SYSTEMS AND ACCESSIBILITY

Health services in rural areas involve a coordinated set of interacting providers serving different demands. People seek services directly, and patients or professionals move up and down among facilities. Consider the example of the rural health service paradigm of the Colombian Ministry of Public Health (Bennett, 1978). The idealized system consists of four service levels. Health outposts in villages offer first aid and advice. Hospitals in small towns provide outpatient treatment, limited inpatient care, first aid, and advice. Regional hospitals in larger towns have a laboratory and inpatient care wards and provide all the services of local hospitals. Medical centers in cities provide a full range of medical and hospital support services. Each level provides a certain type of service. Patients reach the upper levels primarily through referral, although each level can provide all services of the levels below. Health professionals move from higher to lower levels for health center visits, supervision, material supply, and training purposes. Such medical service hierarchies have also been described by Engle (1968), Wells and Klees (1980), Sorkin (1976), and King (1966).

A wide range of investigators not involved with location-allocation analysis have observed that accessibility is or should be the goal of such rural health systems. For example, King (1966) has stated that 70% of the outpatients of health centers in Kenya live within 10 mi of a center. Similar results have been measured in India (King, 1966). In Uganda, the average number of outpatient visits was found to decline by 50% for every (1) 2 mi that people live from a hospital, (2) 1.5 mi from a dispensary, and (3) 1 mi from an aid post (King, 1966; Sorkin, 1976). It was found that 71% of the patients who visited aid posts lived within 5 mi of a facility (Gershenberg and Haskell, 1972). In Tanzania, 91% of all inpatients at the national hospital in Dar es Salaam lived within 9 mi of the hospital. Sorkin has stated that the catchment area of rural health centers in Tanzania seems to be limited to an effective maximum of 10 mi (1976). Thus, accessibility is a surrogate for utilization and for net benefits; the closer people are to a health service, the more likely they are to use and benefit from it. There are also upper bounds on travel that appear to limit accessibility.

USE OF COVERING MODELS

Two early location formulations explicitly consider facility accessibility to sources of demand within upper bounds on time or distance: the set and maximal covering models. Toregas et al. (1971) defined the set covering problem as: Find the minimum number of facilities and their locations such that each point of demand has a facility within s time units. Due to its clarity and strong relation to stated need, the set covering approach has been widely applied in planning public facility locations. ReVelle et al. (1977) found that the set covering model had been requested for use in nearly 100 cities, and Chaiken (1978) noted that it had been used in 54 communities to locate fire stations.

The maximal covering location problem, defined by Church and ReVelle (1974), can be stated as allocate p facilities to positions on the network so that the maximum fraction of demand will find service within a stated time or distance standard. The maximal covering approach can find the location of sites for facilities to cover as much of the demand as possible. As ReVelle et al. (1977) have noted, this statement is responsive to the regulations associated with the Emergency Medical Service Act of 1973. An application of maximal covering to ambulance deployment in Austin, Texas, was designated as a runner-up for the 1984 Management Science Achievement Award (Eaton et al., 1985).

Both the set covering and maximal covering methods locate facilities for one service level. Moore and ReVelle (1981, 1982) have extended covering models to nonreferral hierarchical systems. The problem is to locate a fixed number of two facilities (clinics and hospitals) to (1) cover as many people as possible by clinic services and (2) assure that hospital services are as accessible as possible. Moore and ReVelle formulated their model (hereafter designated H1) using these two objectives, as follows:

$$\text{H1:} \quad \text{Max } Z = \left[\sum_i a_i y_i \, , \quad \sum_i b_i z_i \right] \tag{7.1}$$

subject to

$$\sum_{j \in N_i} c_j + \sum_{j \in V_i} h_j \geq y_i \qquad \text{for each } i \tag{7.2}$$

$$\sum_{j \in U_i} h_j \geq z_i \qquad \text{for each } i \tag{7.3}$$

$$\sum_j c_j = P_c \tag{7.4}$$

$$\sum_j h_j = P_h \tag{7.5}$$

$$c_j, h_j, y_i \text{ are zero-one variables} \tag{7.6}$$

where

I = the set of demand points i,
J = the set of potential facility sites j,
a_i = the population at i needing clinic service,
b_i = demand for hospital service at i,
y_i = 1 if demand i is covered by clinic service, 0 otherwise,
z_i = 1 if demand i is covered by a hospital, 0 otherwise,
P_c = the number of clinics to be located,
P_h = the number of hospitals to be located (hospitals can also provide clinic service),
U_i = $(j|$ a hospital at site j can serve demand i with hospital services), the set of hospital sites that can serve demand i,
N_i = $(j|$ a clinic at site j can serve i with clinic services), the set of clinic sites that can serve demand i,
V_i = $(j|$ a hospital at site j can serve i with clinic services), the set of hospital sites that can provide clinic services to demand i,
c_j = 1 if a clinic is allocated to site j, 0 otherwise
h_j = 1 if a hospital is allocated to site j, 0 otherwise.

The problem is to maximize both coverage objectives simultaneously while locating P_h hospitals and P_c clinics. The first term of the objective deals with clinic service coverage and the second term deals with hospital coverage. Equation (7.2) defines whether clinic service covers demand i. The coverage variable y_i is constrained to be zero unless a hospital (providing clinic service) or a clinic is placed at a site that can serve demand node i. A similar type of constraint defines the value of the z_i variable; coverage means that a hospital is available to serve area i (equation 7.3). Equations (7.4) and (7.5) constrain the number of clinics and hospitals to P_c and P_h, respectively.

This model is called a nonreferral hierarchy because the second level (hospital) provides service directly to the demand population; both levels of service are delivered directly. The model would decompose into two separate location problems (a hospital location model and a clinic location model) if hospitals did not also provide clinic service. The single-objective version of the Moore-ReVelle model deals with maximizing the population that is covered by both hospital and clinic services (Moore and ReVelle, 1982).

The Moore and ReVelle (1982) and Charnes and Storbeck (1980) models are related in that they consider both levels of service directed to the demand population. Although the models deal with the location of different types of service, strong similarities exist. First, both models constrain the total number of facilities (or units) located at each level of service. Second, both models seek to cover as much as possible in terms of both levels of service. The major distinction is that the Charnes and Storbeck model assigns a benefit to covering secondary-level demand when primary-level coverage is not provided.

THE HIERARCHICAL REFERRAL PROBLEM

Moore and ReVelle's nonreferral formulation provides services (hospitals and clinics) directly to people; both levels serve the demand population. What if a hospital not only treats the demand population but also serves clinics? This happens when professional services and supplies are delivered directly from hospitals to clinics and only indirectly to patients.

For example, in Colombia, a well-defined support relationship exists between hospitals and rural clinics (Penn, 1977; Hamon, 1979). Each rural clinic is supplied by its closest hospital with health education material, periodic visits from physicians and dentists, and drugs and supplies. Hospitals and clinics are linked in the supportive relationship. Hospitals also serve a client population directly. In addition, rural patients enter the system at health clinics and can proceed to a higher-level hospital upon referral, usually after an examination. Hospitals should be accessible to the clinics, as referred patients would travel from clinics to hospitals. There are thus two directions of referral—bottom up (patient referral) and top down (professional travel to clinics).

Referral systems can be classified as bottom up and top down depending on the relative importance of (1) patient referral from clinics to hospitals for care or (2) health professional referral from hospitals for supervision and supply of clinics. Both assume that the primary system level (the clinic) provides first aid and advice. Primary demand can be represented by the size of the resident population served or by some other measure of demand. A planner might wish to limit the maximum time or travel distance from the patient's residence to the health center so that people should not have to travel too far for service.

In bottom-up referral, the secondary level (local hospital) serves both people referred to it by clinics and persons who come directly to it for primary-level services. A planner might wish to limit the maximum time or distance referral patients travel from a clinic to a hospital. An upper limit on travel is particularly important for life threatening events, such as birth complications, accidents, poisonings, and the like. The number of persons referred is a good surrogate for the volume of secondary-level demand.

In top-down referral, the hospital level serves people directly. In addition, a hospital sends health professionals to clinics. These ''itinerant'' doctors, nurses, or dentists supervise local health workers, supply the clinics, and provide visiting professional services. If professionals travel to and from clinics in a day, the period spent on the road reduces the time available for providing services or supervision. If a hospital is far away from a clinic (above some upper time limit), it may not be cost effective to send out a health professional. Each clinic represents a service demand for the top-down system.

In the case of either referral problem, a planner would like to deploy clinics and local hospitals to function as an integrated rural health system. This system should be designed so that the fewest possible clinics or hospitals serve as large a fraction of the primary demand and secondary referrals as possible. Both problems of hierarchical referral service can be defined as follows: in a two-level service system with referral, simultaneously find the set of P_1 level-one sites and P_2 level-two sites that can provide services to the largest primary and referral demands.

These problems are formulated below. The section after that describes applications to locating health facilities in rural areas in Colombia and the Dominican Republic. Although these problems are defined in this paper for two levels, they can be extended to a system with three or more levels by adding additional constraints to represent the pattern of the hierarchical referral service for each level.

A Top-Down Referral Formulation

The top-down referral hierarchical covering problem (called R1) includes two objectives, maximizing the number of people that are covered by clinics and minimizing the number of clinics that are not covered by a hospital:

$$\text{R1:} \quad \text{Max } Z = \left(\sum_i a_i y_i \ , \quad -\sum_j q_j \right) \tag{7.7}$$

subject to

$$\sum_{j \in N_i} c_j + \sum_{j \in V_i} h_j \geq y_i \qquad \text{for each } i \tag{7.8}$$

$$\sum_{k \in M_j} h_k - c_j + q_j \geq 0 \qquad \text{for each } j \tag{7.9}$$

$$\sum_j h_j = P_h \tag{7.10}$$

$$\sum_j c_j = P_c \tag{7.11}$$

$$c_j, h_j, y_i, q_j \text{ are zero-one variables} \tag{7.12}$$

where

$q_j = 1$ if a clinic allocated to site j is not covered by a hospital, 0 otherwise

$M_j = (k|$ a hospital at site k can serve a clinic at site $j)$ the set of hospital sites that can serve a clinic sited at j (M_j must include at least $k = j$, as by definition a hospital can accept referrals for hospital service from its outpatient clinic).

and where all other notation is as previously defined in the Moore and ReVelle model (equations 7.1–7.6).

The second term of the objective vector is written with a negative sign, as maximization of the negative of a term is equivalent to the minimization of the term. The sum of the q_j's equals the number of clinics that are not covered by hospitals. The first constraint defines whether demand point i is covered by either clinics or hospitals. A clinic or a hospital can cover demand if that facility is accessible within some maximum distance or time standard. If a clinic or a hospital is placed at a site that can cover area i (i.e., a member of the N_i or V_i sets, respectively), the sum of all c_j and h_j will be at least one; the value of y_i can equal one, indicating that area i is covered.

The second constraint determines whether a clinic j is covered by any hospital. A hospital can cover a clinic if patients' travel time does not exceed a specified maximal service time. If any c_j equals one, either a hospital covers this clinic or q_j must equal one to satisfy the constraint. If q_j equals one, by definition clinic j is not covered by a hospital.

The third and fourth constraints define the total number of hospitals and clinics (P_h and P_c respectively). This approach does not include explicit cost variables in site selection. Equation (7.10) implicitly treats as equal the fixed and variable costs of all hospitals. Equation (7.11) performs the same role for clinics. ReVelle and Swain (1970) have shown that when site costs and the costs of expansion or development are equivalent between facilities, a constraint on the number of sites is equivalent to a budget constraint. In addition, Toregas (1971) has explicitly optimized the number of sites used to provide service at a constrained level. He showed that if site-development costs are homogeneous, then the results are equivalent to minimizing costs of facility development subject to the same constraints on the level of service provision. In the R1 case, the costs of facilities are treated as homogeneous by level. If an investigator prefers to use a budget constraint instead of constraints (7.10) and (7.11), it could be inserted in formulation R1.

Both objectives assure accessibility of health services. The first provides clinic service to the largest possible number of people. The second ensures that hospitals are as accessible as possible to clinics making referrals. Even when cases are somewhat serious most patients enter the system at the clinic level and then travel to a hospital. If the hospital is inaccessible from a clinic, the referral patients' level of service is degraded. The second objective allows for travel from hospitals to clinics; note that clinic coverage by hospitals is different from covering the number of people referred by clinics to hospitals.

The R1 formulation measures benefits in an either/or fashion, determining whether demand is or is not covered. If an investigator prefers to weight benefits as a function of distance, then it would be possible to incorporate elements of weighted-benefit maximal covering location models (Church and Roberts, 1983).

A Bottom-Up Referral Formulation

A bottom-up multiobjective formulation, labeled R2, can find facilities that maximize hospital accessibility for the persons referred from clinic services. The change in objectives from R1 to R2 reflects an emphasis on referral accessibility versus efficient clinic coverage. Recall that a top-down model (clinic coverage) asserts that hospitals supply clinics and send professional staff for periodic visits. It reflects a priority that staff avoid spending more than a minimum portion of the day traveling to a clinic so that maximum time is left for the purpose of the visit—serving patients at the clinic. That objective is complementary with explicit weighting of referral-patient demand. It is straightforward to formulate a third hierarchical model that includes both of these secondary-level service objectives. Further, referral and nonreferral hierarchy objectives can be combined to obtain an even more comprehensive model, as follows:

$$\text{R2:}\quad \text{Max } Z = \left[\sum_i a_i y_i \ , \ \sum_i \left(\sum_{j \in N_i} \alpha_i a_i x_{ij} \right) \right] \tag{7.13}$$

subject to

$$\sum_{j \in N_i} c_j + \sum_{j \in V_i} h_j \geq y_i \qquad \text{for each } i \tag{7.14}$$

$$x_{ij} \leq c_j + h_j \qquad \text{for each } i \text{ and } j \in N_i \tag{7.15}$$

$$x_{ij} \leq h_j \qquad \begin{array}{l} \text{for each } i \text{ and } j \in V_i \\ \text{where } j \notin N_i \end{array} \tag{7.16}$$

$$x_{ij} \leq \sum_{k \in M_j} h_k \qquad \text{for each } i \text{ and } j \in N_i \tag{7.17}$$

$$\sum_{j \in N_i} x_{ij} \leq 1 \qquad \text{for each } i \tag{7.18}$$

$$\sum_j c_j = P_c \tag{7.19}$$

$$\sum_j h_j = P_h \tag{7.20}$$

$$c_j, h_j, x_{ij}, y_i \text{ are zero-one variables} \tag{7.21}$$

where

$x_{ij} = 1$ if demand i is covered by clinic service at j and j is covered by hospital service, 0 otherwise,

$\alpha_i =$ referral rate at i—the frequency at which patients from i are referred to a hospital from the clinic where they entered the medical service system.

and where all other notation is as previously defined.

The bottom-up formulation differs from the top-down case in that constraints (7.15)–(7.18) define whether a demand covered by clinic service is backed up by accessible hospital service. The x_{ij} variable is constrained to be zero (in equation 7.15) unless a hospital or clinic is placed at j where j is an element of N_i. Constraint (7.16) forces x_{ij} to be zero unless a hospital with clinic services is located at site j. Constraint (7.16) is added for each site j not in the set N_i but in the set V_i. Any x_{ij} will be zero unless a hospital covers site j (constraint 7.17). Together, constraints (7.15)–(7.17) keep x_{ij} equal to zero (no coverage of demand by clinics and clinics by hospitals) unless j represents a hospital site or has a clinic that is covered by a hospital. Constraint (7.17) allows, at most, only one x_{ij} variable for a given i to be equal to one, which ensures that a given demand population is counted only once in clinic coverage with back-up hospital referral service.

Both R1 and R2 are multiobjective, zero-one integer, linear programming problems. Each can be solved by linear programming and branch-and-bound procedures, given a weighted function of the two objectives.

APPLICATIONS OF REFERRAL HIERARCHICAL COVERING MODELS

Formulations R1 and R2 have been applied to analyze rural health facilities in Colombia. They also could be used in the Dominican Republic to select sites for itinerant permanent physicians serving health outposts.

The Colombian application relates to Zarzal, a region of approximately 1232 square kilometers of mostly mountainous terrain in the Valle del Cauca. Approximately 78,000 persons live in the area in 79 towns and villages. These people are served by 1 regional and 4 local hospitals, 6 ambulances, and 24 rural clinics (Ministerio de Salud, 1977) (see Fig. 7.2). Both bottom-up and top-down referral patterns operate in Zarzal (Bennett et al., 1978). Patients are referred from health centers to hospitals for outpatient and limited inpatient care. Physicians, nurses, and dentists travel from hospitals to clinics.

In earlier research, the authors used a maximal covering approach to determine the best pattern of health clinics in the Zarzal region (Eaton et al., (1982). As a result of that study, the Zarzal and Valle planners incorporated maximal covering results into their planning process for analyzing the extension of health services (Ministerio de Salud, 1978b).

The Valle planners established a maximum travel distance of 15 km (or a 2 hour walk) for health clinic coverage (Bennett, 1978). Patient ascents and descents

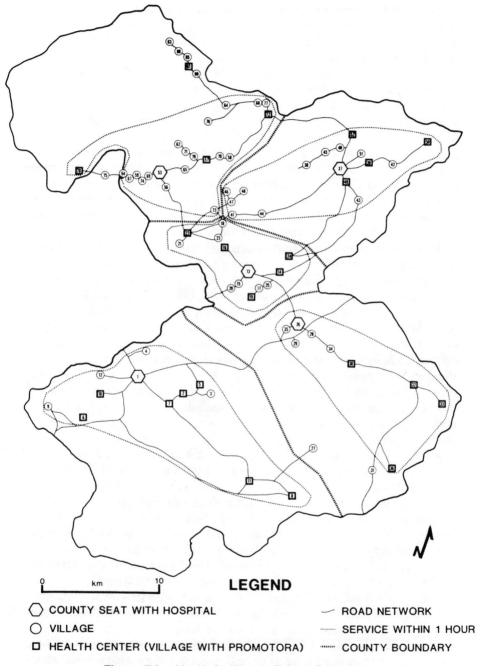

Figure 7.2. Health facilities in Zarzal, Colombia.

in the mountains were limited to 30 m up and 100 m down from home to a clinic. Planners intended these constraints to limit the effort a patient would have to exert to travel to a health center. The same limits were used to determine whether clinics could be served by a hospital.

Valle planners established 1 hour by jeep or ambulance as an upper bound on travel time to a hospital from a clinic: any patient should reach a hospital via auto from a clinic in 1 hour or less. Doctors, dentists, and nurses visiting a health clinic for a day and returning to their base hospitals should travel no more than 2 hours out of the working day.

The hierarchical referral models R1 and R2 were applied to the Zarzal region using the upper bounds on service distance, elevation differences, and times. We estimated referral rates for each village based on the planning survey; these ranged from .0628 to .0943 (Ministerio de Salud, 1978a; Valencia, Bustamante, and Patino, 1978). Results of the computer runs for several different combinations of hospitals and clinics are presented in Tables 7.2 and 7.3.

To solve these multiobjective models, we constructed a single combined objective as the sum of the two objectives, each multiplied by a weight. By systematically varying the weights and solving a series of problems, the noninferior trade-off curve can be generated (Zadeh, 1963); this approach has been called the weighting method (Cohon, 1978). Both models were solved by relaxing the integer requirements and using the MPSX linear programming system on an IBM model 370/3033 at the University of Tennessee. If any fractional solutions occurred, a branch-and-bound algorithm would have been used to resolve them. However, all optimal linear programming solutions generated were integer solutions. A typical R1 problem required 8 seconds per solution and the comparable time for solving an R2 problem was 20 seconds. The difference in solution times for R1 and R2 results because R2 is a larger problem in terms of variables and constraints.

Tables 7.2 and 7.3 present the results generated by the top-down and bottom-up models, respectively. Although the top-down formulation (R1) optimizes only population covered by clinic services and clinic coverage by hospitals, Table 7.2 also presents results for three other coverage criteria: population referred to hospitals, rural population (for clinic service), and referred rural population. Table 7.3, showing the results of the bottom-up referral formulation, includes these five criteria for comparison (although R2 maximizes only coverage for total population and referred population). Also for comparison, we generated combinations of hospitals (3 or 5 sites) and clinics (15, 20, and 24).

The solutions from the R1 and R2 models are quite similar; indeed, some configurations selected by the models are identical. Two reasons for this result are (1) the 1-hour travel-time constraint for hospital coverage of clinics or clinic referral population is not restrictive in such a small region, and (2) the number of facilities (within the indicated range) is not too binding. Results of the two models would not be so similar for a smaller number of facilities or lower upper bounds on travel time.

Tables 7.2 and 7.3 also include statistics associated with the current system of 5 hospitals and 24 health outposts. Note that a shift from the current system to 3 hospitals and 15 clinics would not lead to great reductions in coverage— the two solutions are comparable. These results suggest that two hospitals are of limited value for delivering referral services. In Zarzal, as in many areas of developing nations, the quality of health care in hospitals is limited by the scarcity of funds for equipment, supplies, and personnel. If two hospitals could be eliminated and their budgets shifted to the three remaining hospitals (which are located optimally in terms of both the R1 and R2 models), there would be no reduction in access to care. Such a concentration of health resources could improve the quality of health care with no change in the regional health-sector budget.

Health planners in Zarzal have already accepted the use of systems analysis (maximal covering) as an adjunct to common sense in devising health service facility extensions. The referral hierarchy covering approach can more comprehensively convert a vision of interacting rural health services into a deployment plan for operational facilities.

The idea that it is possible to reduce the expense of health service, redirect scarce health-sector funds, and ultimately improve health care is not unique to Zarzal. The top-down referral formulation (R1) has been discussed by the Dominican Republic's Ministry of Public Health and Welfare for use in assessing the number and deployment of physicians in rural areas in Health Region V, an area on the eastern tip of the island of Hispaniola, east of and including San Pedro de Macoris.

One common problem in developing nations is how to induce physicians to practice in rural areas. In the Dominican Republic, as in other countries, one strategy that works poorly is to require all physicians to spend one year after medical school in national service. Assigning some of these doctors to rural clinics does assure a level of medical attention, but few doctors remain in such posts after the required year.

In the Dominican Republic, many such inexperienced "national service year" physicians are assigned to work with health promoters in rural clinics. Reducing the number of physician posts, increasing salaries, and providing other incentives would be means to recruit a more permanent rural physician corps while retaining a stable health-sector budget. Doctors who live for a number of years in an area are more likely to integrate in the community and provide better care because they know their patients. For health care access to remain at the current level, the doctors would need to travel regularly from their base of operations to health clinics where doctors served under the old system. Thus the proposal is a top-down referral system with patients traveling to rural clinics for advice

Table 7.2. Coverage Patterns of the Top-Down Referral Formulation (R1)

Objectives and Other Criteria	Objective Weights		Coverage Results (% of Maximum Coverage or Number of Clinics)				
			Three Hospitals		Five Hospitals		
	#1	#2	15 Clinics	20 Clinics	20 Clinics	24 Clinics	Current 24[a]
Objective #1: Total Zarzal population covered (maximum = 77,871)	1	0	0.957	0.971	0.976	0.983	(0.953)
	1	100	0.954	0.969	0.974	0.981	
	1	5,000	0.947	0.961	0.965	0.972	
Objective #2: Clinics covered by hospital (maximum = 24)	1	0	13	16	16	19	(23)
	1	100	14	19	19	23	
	1	5,000	15	20	20	24	
Total referred patients covered (maximum = 6,093)	1	0	0.976	0.955	0.959	0.965	(0.947)
	1	100	0.950	0.963	0.967	0.974	
	1	5,000	0.952	0.965	0.970	0.976	
Rural population covered (maximum = 25,007)	1	0	0.866	0.911	0.925	0.949	(0.852)
	1	100	0.858	0.903	0.918	0.941	
	1	5,000	0.834	0.878	0.891	0.912	
Referred rural population covered (maximum = 1,918)	1	0	0.830	0.857	0.871	0.888	(0.832)
	1	110	0.840	0.882	0.896	0.918	
	1	5,000	0.848	0.889	0.904	0.922	

[a] Refers to the current deployment of health facilities (see Fig. 7.2).

Table 7.3. Coverage Patterns of the Bottom-Up Referral Formulation (R2)

Objectives and Other Criteria	Objective Weights		Coverage Results (% of Maximum Coverage or Number of Clinics)				
			Three Hospitals		Five Hospitals		
	#1	#2	15 Clinics	20 Clinics	20 Clinics	24 Clinics	Current 24[a]
Objective #1: Total population covered (maximum = 77,871)	1	0	0.957	0.971	0.976	0.983	(0.953)
	1	13	0.953	0.969	0.973	0.981	
	0	1	0.946	0.960	0.965	0.972	
Clinics covered by hospitals (maximum = 24)	1	0	12	16	16	21	(23)
	1	13	14	19	19	23	
	0	1	15	20	20	24	
Objective #2: Total referred patients covered (maximum = 6,093)	1	0	0.943	0.954	0.959	0.973	(0.947)
	1	13	0.948	0.963	0.968	0.974	
	0	1	0.953	0.965	0.970	0.976	
Rural population covered (maximum = 25,007)	1	0	0.866	0.911	0.925	0.949	(0.852)
	1	13	0.853	0.903	0.917	0.941	
	0	1	0.833	0.876	0.891	0.912	
Referred rural population covered (maximum = 1,918)	1	0	0.819	0.852	0.871	0.913	(0.832)
	1	13	0.836	0.882	0.900	0.923	
	0	1	0.848	0.889	0.904	0.922	

[a] Refers to the current deployment of health facilities (see Fig. 7.2).

from the health promoters; patients traveling to the rural clinics having physicians for medical care and for advice from health promoters; and doctors traveling from their clinics to other clinics to provide medical care, supervise health promoters, and deliver supplies.

SUMMARY

This chapter has reviewed concepts of hierarchies and described existing relevant formulations. It has presented two referral hierarchy covering formulations appropriate for rural health or emergency medical service systems. The two models have been applied to rural health service planning in Zarzal, Colombia, and the San Pedro de Macoris area of the Dominican Republic. Numerical results from the Colombia case are presented and comparisons are made between the existing health system and other options for health service delivery.

REFERENCES

Banerji, S. and H. B. Fisher, 1974, Hierarchical location analysis for integrated area planning in rural India, *Regional Science Association Papers* **33**:177–194.

Bennett, V., 1978, The use of location analysis for siting health promoters in rural Colombia, M.S. thesis, University of Texas, Austin.

Bennett, V., A. Rodriguez, R. Bustamante, C. Valencia, and M. Madrinan, 1978, Applicacion de tecnicas de cobertura y analisis socioculturales en la comunidad escuela, Planning Report, Ministerio de Salud, Cali, Colombia.

Berlin, G. N., C. ReVelle, and J. Elzinga, 1976, Determining ambulance-hospital locations for on-scene and hospital services, *Environment and Planning, A* **8**:553–561.

Calvo, A. B. and D. H. Marks, 1973, Location of health care facilities: an analytical approach, *Socio-Economic Planning Sciences* **7**:407–422.

Chaiken, J., 1978, Transfer of emergency deployment models to operating agencies, *Management Science* **24**:719–731.

Charnes, A. and J. Storbeck, 1980, A goal programming model for siting multilevel EMS systems, *Socio-Economic Planning Sciences* **14**:155–161.

Church, R. L., 1974, Synthesis of a class of public facility location models, Ph.D. diss., Johns Hopkins University, Baltimore.

Church, R. L. and C. S. ReVelle, 1974, The maximal covering location problem, *Regional Science Association Papers* **32**:101–118.

Church, R. L. and K. L. Roberts, 1983, Generalized coverage models and public facility location, *Regional Science Association Papers* **53**:117–135.

Cohon, J., 1978, *Multiobjective Programming and Planning*, Academic Press, New York.

Daskin, M. and E. Stern, 1981, A hierarchical objective set covering model for EMS vehicle deployment, *Transportation Science* **15(2)**:137–152.

Dokmeci, V. F., 1973, An optimization model for a hierarchical spatial system, *Journal of Regional Science* **13**:439–451.

Dokmeci, V. F., 1979, Multiobjective model for regional planning of health facilities, *Environment and Planning, A* **11**:517–525.

Eaton, D. J., R. L. Church, V. L. Bennett, B. L. Hamon, and L. G. Lopez, 1982, On deployment of health resources in rural Valle Del Cauca, Colombia, in *Planning and Development Processes in the Third World,* W. Cook, ed., Elsevier, Amsterdam, pp. 331–359.

Eaton, D. J. and M. Daskin, 1981, A multiple model approach to planning emergency medical service vehicle deployments, in *Systems Science in Health Care,* Pergammon, Montreal, pp. 951–959.

Eaton, D. J., M. S. Daskin, D. Simmons, B. Bulloch, and G. Jansma, 1985, Determining emergency medical service vehicle deployment in Austin, Texas, *Interfaces* **15(1)**:96–108.

Engel, A, 1968, *Perspectives in Health Planning,* Athlone, London.

Fisher, H. B. and G. Rushton, 1979, Spatial efficiency of service locations and the regional development process, *Regional Science Association Papers* **42**:83–97.

Gershenberg, I. and M. A. Haskell, 1972, The distribution of medical services in Uganda, *Social Science and Medicine* **6**:365–369.

Hamon, B., 1979, Health resource allocation in rural Colombia, Master's thesis, University of Texas, Austin.

Harvey, M. E., M. S. Hung, and J. R. Brown, 1974, The application of a *p*-median algorithm to the identification of nodal hierarchies and growth centers, *Economic Geography* **50**:187–202.

King, M., 1966, *Medical Care in Developing Countries,* Oxford University Press, Nairobi.

Law, J., 1976, Optimal location-allocation of multi-level hierarchical facilities, Ph.D. diss., University of Texas, Austin.

Marks, D. H., 1969, Mathematical analysis of solid waste collection, Ph.D. diss., John Hopkins University, Baltimore.

Marks, D. H., and J. C. Liebman, 1971, Location models: solid waste collection example, *Urban Planning and Development Division of ASCE Journal* **97**:15–30.

Ministerio de Salud, 1977, Censo y diagnostico sanitario de la unidad regional de salud de Zarzal, Servicio Seccional de Salud del Departamento de Valle del Cauca and Universidad del Valle, Cali, Colombia.

Ministerio de Salud, 1978*a*, Modelo simplificado para censo y diagnostico sanitario rural y urbano para las unidades regionales de salud, Servicio Seccional de Salud del Departamento de Valle del Cauca and Universidad del Valle, Cali, Colombia.

Ministerio de Salud, 1978*b*, Projecto de convenio de integracion, sector salud-sector educacion en la comunidad-escuela, Servicio Seccional de Salud del Departamento de Valle del Cauca and Universidad del Valle, Cali, Colombia.

Moore, G. C. and C. S. ReVelle, 1981, The hierarchical service location problem, mimeo, Johns Hopkins University, Baltimore.

Moore, G. C. and C. S. ReVelle, 1982, The hierarchical service location problem, *Management Science* **28**:775–780.

Narula, S. C. and U. I. Ogbu, 1979, An hierarchical location-allocation problem, *Omega* **7**:137–143.

Narula, S. C. and U. I. Ogbu, 1985, Lagrangian relaxation and decomposition in an uncapacitated 2-hierarchical location-allocation problem, *Computers and Operations Research* **12(2):**169–180.

Narula, S. C., U. I. Ogbu, and H. M. Samuelsson, 1975, Location of Health Facilities in Developing Countries, presented at the ORSA/TIMS meeting, Chicago, working paper no. **221,** School of Management, State University of New York, Buffalo.

Osleeb, J. P., S. J. Rattick, and G. K. Lewis, 1981, A model to analyze coal handling in New England ports, *Applied Geography Conferences Proceedings* **4:**122–140.

Penn, R., 1977, Un punto clave en el sistema de salud Colombiano: La Promotora, Fulbright Commission, Bogota.

Ravindran, A. and D. L. Hanline, 1980, Optimal location of coal blending plants by mixed-integer programming, *American Institute of Industrial Engineers Transactions* **12:**179–185.

ReVelle, C. S., D. Bigman, D. Schilling, J. Cohon, and R. Church, 1977, Facility location: a review of context-free and EMS models, *Health Services Research* **12**(Summer): 129–145.

ReVelle, C. and R. Swain, 1970, Central facilities location, *Geographical Analysis* **2:**30–42.

Ruth, R. J., 1979, A mixed integer programming model for regional planning of a hospital inpatient service, Ph.D. diss., University of Michigan, Ann Arbor.

Schultz, G. P., 1970, The logic of health care facility planning, *Socio-Economic Planning Sciences* **4:**383–393.

Scott, A. J., 1971, Operational analysis of nodal hierarchies in network systems, *Operational Research Quarterly* **22:**25–37.

Sorkin, A. L., 1976, *Health Economics in Developing Countries,* Lexington Books, Lexington, Mass.

Teitz, M. and P. Bart, 1968, Heuristic methods for estimating the generalized vertex median of a weighted graph, *Operations Research* **16:**955–961.

Tien, J. M. and K. El-Tell, 1984, A quasihierarchical location-allocation model for primary health care planning, *IEEE Sys. Man and Cybernetics Transactions* **SMC-14(3):**373–380.

Tien, J. M., K. El-Tell, and G. R. Simons, 1983, Improved formulations to the hierarchical health facility location-allocation problem, *IEEE Trans. Sys. Man and Cybernetics Transactions* **SMC-13(6):**1128–1132.

Toregas, C., 1971, Location under maximum travel time constraints, Ph.D. diss., Cornell University, Ithaca, New York.

Toregas, C., R. Swain, C. ReVelle, and L. Bergman, 1971, The location of emergency service facilities, *Operations Research* **19:**1363–1373.

Valencia, L. G., R. Bustamante, and D. Patino, 1978, Summary of the rural sanitary diagnosis of the department of Valle del Cauca, paper presented at the joint meetings of The Operations Research Society of America and the Institute of Management Sciences, New York.

Vasan, K. S., 1975, Optimization models for regional public systems, Ph.D. diss., University of California, Berkeley.

Walker, W., M. Aquilina, and D. Schur, 1974, Development and use of a fixed charge programming model for regional solid waste planning, paper presented at the joint meetings of The Operations Research Society of America and the Institute of Management Sciences, San Juan.

Wells, S. and S. Klees, 1980, *Health Economics and Development* Praeger, New York.

Whitlatch, E. E., 1977, Coal gasification and water resources development, *Water Resources Planning and Management Division of ASCE Journal* **WR-2:**299–314.

Zadeh, L. A., 1963, Optimality and non-scalar-valued performance criteria, *IEEE Transactions on Automatic Control* **AC-8:**59–60.

8

SPATIAL DISTRIBUTION DESIGN FOR FIRE FIGHTING UNITS

Pitu B. Mirchandani

Rensselaer Polytechnic Institute

John M. Reilly

Capital District Transportation Authority

Over the past two decades, decisions concerning the deployment and allocation of fire fighting resources have been subject to considerable quantitative analysis. These analyses have come at an appropriate time—when local governments have much interest in improving their productivity and the quality of the services delivered.

One of the problems in the design of a fire fighting system is to determine the optimal spatial distribution of fire fighting units. This chapter deals with that problem. We first briefly discuss the operations of fire departments and survey existing methods for determining optimal locations for fire fighting units. These methods leave out consideration of the value of response times for units arriving after the first one. The succeeding sections develop a model that considers the first and second units' response times. In the computation of the expected response time and its distribution, the model includes some practical constraints and an explicit consideration of the sizes of fire districts. Some experience with the model is then discussed.

FIRE DEPARTMENT OPERATIONS AND LOCATIONAL ISSUES

Up to 20% of local government expenditures are spent on public fire protection. The mission of an urban fire department can be categorized in two broad areas: fire prevention and fire suppression. Fire prevention activities generally represent a small proportion of a department's operating budget and include both fire inspection and public education programs. The goal of fire prevention activity is both to reduce the occurrence of fires and to mitigate the effects of those that do occur.

Fire suppression activity constitutes the bulk of the fire department outlay. The basic organizational unit for fire suppression is a company that includes a specialized piece of equipment and a complement of between two and six fire fighters. Most companies are engine companies that apply water or chemical agents to a fire to extinguish it. Ladder companies, which represent about one-quarter to one-third of all companies, are responsible for evacuating occupants from fires and for reaching fires at high levels.

In theory, the strategic planning of a fire department is relatively simple. Most of the time, fire companies are not engaged in fire fighting but are maintaining a state of preparedness in case a fire occurs. When a fire is detected, information concerning its location is transmitted by various means to a central dispatcher. Depending on the location and type of fire, a predetermined number and type of fire fighting units are assigned to the fire. The preferred strategy is to send one or more of the closest units to the fire. However, if one or several of the closest units are unavailable, the next closest unit or units are assigned. Once the initial complement of fire fighters and equipment arrive at the scene of the fire, the chief officer at the scene can request additional companies. Occasionally, during periods of several small fires in a certain section of the city or a major fire at a single point, the fire fighting resources of an area may be depleted. In these cases, available units from other sections of the city may be called on to cover empty firehouses in an area of heavy demand. This action is referred to as redeployment.

Generally, a constant number of units are on duty regardless of the time of day, despite considerable variation in the hourly and daily rate at which fires occur. This practice holds because fire damage is related to response time and, in all but the largest cities, the number of units required to maintain an acceptable level of response time does not vary with the rate of fire incidence.

When deciding on the level of expenditures for public fire control activity, fire departments rely on a grading schedule issued by the Insurance Services Office (1974), an organization sponsored by the property and casualty insurance industry. The Insurance Services Office (ISO) acts as a rating organization and provides the industry with the statistical data required to establish fire insurance rates based on expectations of fire loss. Staff engineers from the ISO periodically assess the fire protection capability of a city. Each municipality is rated on a scale of 1 to 10; Class 1 has the highest rating. The factors included in the assessment are the quality of the water supply, the quality of the fire department, the communications system, and the extent to which adequate building codes are enforced and fire prevention activities conducted.

The grading schedule suggests a number of standards for the location of fire fighting units. These standards are all distance based; for example, in central business districts of medium-size cities, the first arriving engine company must be within 1 mi and the second within 2 mi, and there must be five engines altogether within 3½ miles. Similar standards exist for ladder companies.

It is generally recognized that response time is a better measure of fire department performance than response distance (Schaenman et al., 1977), since response time is a key factor in minimizing property damage and casualty loss. It is also more directly related to the perceived quality of service. Thus, response time standards might better assure comparable service quality to all properties regardless of speed.

LITERATURE REVIEW

As with many other public service problems, there are conflicting schools of thought on the location of emergency service facilities. Hirsch (1973) suggests three criteria for the allocation of public services among geographic areas: *input equalization, output equalization,* and *efficiency maximization.* Although the mission of the fire service is well understood, it is difficult to define its inputs and outputs. If one considers fire fighting units as inputs, the small number and indivisibility of such units preclude an accurate assessment of whether or not each sector of a city is being provided with equal service inputs.

The fire department's ultimate output is avoidance of property and casualty fire loss. There is a lack of understanding of how the fire service inputs (fire equipment and fire fighters) relate to this output specification. Travel time to fires is often used as a proxy measure of the performance of the fire system. The output equalization and efficiency maximization allocation criteria can be operationalized using this measure.

The criterion of output *equalization* would locate fire stations in such a way that travel time to fires in each area of the city would be equalized. On the other hand, the criteria for *efficiency* maximization would locate units to minimize total travel time. The difference between the two can be illustrated by considering a small zone with relatively few fires that has surrounding zones with large numbers of fires. Travel-time equalization would ensure that travel time to this zone would be comparable to that in all other zones. The minimization of total travel time may provide no such safeguard, an unacceptable response time to that zone could arise in an attempt to maximize the efficiency of the entire fire defense system.

Output equalization could be looked on as minimizing the maximum travel time to zones subject to a constraint on the number of facilities. Maximization of efficiency could be looked on as minimizing the total travel time, also subject to a constraint on the number of facilities. These and other criteria are used in the distribution design of fire fighting units.

Static Models

Two types of analytical models for the location of fire stations have been advanced: static models, which assume that all units are always available for dispatch to a fire, and dynamic models, which allow the possibility that some units may be unavailable if they are busy at other fires.

Static models should be sufficient to represent the actual fire fighting situation in all but the largest cities. In a study performed in Alexandria, Virginia, Nilson, Swartz, and Westfall (1974) observed that engine companies were busy only 2.6% of the time and ladder companies only 1.6% of the time. Although these figures include temporal and spatial variations, the busiest engine company was unavailable to fight fires 4.6% of the time and the busiest ladder company was unavailable 2.0% of the time.

The static models include both evaluation and optimization models. The evaluation models compute performance measures of alternate siting plans. The two best known of these models are the Rand Firehouse Site Evaluation Model (Dormont, Hausner and Walker, 1975) and the Public Technology Site Evaluation Program (Public Technology, Inc., 1974). These models provide sets of performance measures including response-time and work-load distributions for alternate siting plans. The optimization models, on the other hand, find the optimal location of sites according to some performance measure or measures. These models can be placed into two categories: set covering formulations and travel time (or damage) minimization formulations.

The set covering formulations seek to minimize the number of stations required to ensure that no area is more than a prespecified distance or time from a fire station. To solve this type of problem, Toregas et al. (1971) have used a binary-integer-program formulation that can be stated as

$$\text{Min } Z = \sum_{j=1}^{n} y_j \qquad (8.1)$$

subject to

$$\sum_{j=1}^{n} a_{ij} y_j \geq 1 \qquad i = 1, 2, \ldots, m \qquad (8.2)$$

$$y_j \epsilon(0, 1) \qquad j = 1, 2, \ldots, n \qquad (8.3)$$

where

m is the number of zones,
n is the number of available sites,

$a_{ij} = 1$ if zone i can be served by a unit at site j without violating constraints,
   0 otherwise,
$y_j = 1$ if a fire unit is placed at site j, 0 otherwise.

Observe that the coefficients a_{ij} can take care of travel time standards or any other constraints that the decision maker might impose. That is, if an established facility at site j is not allowed to be the primary server for zone i—for whatever reasons, be it "too far," "will result in too much additional workload for site j station," or "politically infeasible"—then a_{ij} will be made equal to zero. Furthermore, y_j can capture the situations (1) where a station is already established at site j (here y_j is fixed at 1) or (2) where a station is prohibited from being established (here y_j is fixed at 0).

The set covering formulation of equations 8.1–8.3 does not take into account the situation where stations can be built on any of the available sites but preference is to be given to existing stations, provided their use does not increase the value of the objective function Z. Hendrick et al. (1974) have modified the formulation to accommodate this practical requirement. Their formulation is a "hierarchical" linear integer program of the following form:

$$\text{Min } Z = \sum_{j=1}^{r} y_j + \sum_{j=r+1}^{n} (1 + e)y_j \qquad \text{(8.4)}$$

subject to

$$\sum_{j=1}^{n} a_{ij}y_j \geqslant 1 \qquad i = 1, 2, \ldots, m$$

$$y_j \epsilon (0, 1) \qquad j = 1, 2, \ldots, n$$

where e is a positive constant less than $1/n$.

The sites having existing stations are indexed from 1 to r, while potential sites are indexed from $r + 1$ to n. Hendrick et al. proved that the formulation will establish the same number of facilities as the original formulation, although the actual sites in the new formulation might be different. This hierarchical formulation provides a tie-breaking criterion slanted toward the selection of existing station sites over potential sites.

The advantages of set covering formulations include that they require less data and that the solutions specify not only the location of the optimal set of stations but of the total number of units as well. In the *minisum* formulations, which we discuss shortly, the number of units to be located (p) is usually exogenous to the model. There is also the following *maximal covering* problem formulation, expressed by Schilling et al. (1980), where the number of units to be located is exogenous: locate p facilities such that the largest number of people have a facility within a specified maximum distance. Schilling et al. formulated this problem for siting fire stations in Baltimore.

The second major category of emergency facility location models includes the class of minisum problems. Generally, models of this form locate a number of facilities so as to maximize some time- or distance-related utility function.

The dominant historical use of minisum location models is in private sector distribution design—for problems such as "where should we locate p warehouses." In the classical p-median problem we minimize the sum of transportation costs associated with the p facilities:

$$\text{Min } Z = \sum_{j=1}^{n} \sum_{i=1}^{m} d_{ij} x_{ij} \tag{8.5}$$

subject to

$$\sum_{j=1}^{n} y_j = p \tag{8.6}$$

$$\sum_{j=1}^{n} x_{ij} = 1 \qquad i = 1, 2, \ldots, m \tag{8.7}$$

$$y_j \geq x_{ij} \qquad i = 1, 2, \ldots, m; j = 1, 2, \ldots, n \tag{8.8}$$

$$y_j, x_{ij} \in (0, 1) \qquad i = 1, 2, \ldots, m; j = 1, 2, \ldots, n \tag{8.9}$$

where

$x_{ij} = 1$ if facility at site j serves zone i, 0 otherwise,
d_{ij} is the "cost" (e.g., travel time, disutility) of serving zone i from site j.

Constraint (8.6) restricts the number of facilities to p; constraint (8.7) assures that zone i is served; and constraint (8.8) ensures that zone i can be served only by established facilities (i.e., if $x_{ij} = 1$ then y_j must also be equal to 1).

Observe that the number of facilities p is given exogenously, which may be considered a simplification. Ideally, we would like to be able to make explicit trade-offs between the number of facilities and the "cost" of fires (damage, property loss, casualties, etc.). Objective (8.5), which basically sums the d_{ij} that are active, provides a measure of this cost. By varying p and noting the resultant minimum cost, we can conduct trade-offs.

Another trade-off that may be useful is between the cost of establishing facilities and the cost of fires. This trade-off may be easily accomplished by adding another cost component in the objective function and removing the restriction of p facilities. The problem formulation then becomes

$$\text{Min } Z = \sum_{j=1}^{n} \sum_{i=1}^{m} d_{ij} x_{ij} + \sum_{j=1}^{m} f_j y_j \qquad (8.10)$$

subject to (8.7)–(8.9), where f_j is the fixed cost of establishing a facility at site j. The above formulation is referred to in private sector distribution-design literature as the *uncapacitated facility location* model.

In her pioneering work as a researcher with the Joint Fire Research Organization (U.K.), Hogg (1971*b*) used a median approach to the problem of locating fire stations. Hogg's methodology was a linear program that suggested which p of n available sites ($p < n$) should be filled by a fire unit. The number of units to be located (p) would be determined by policy makers. The nominal objective of the optimization was to minimize the extent of fire damage for a given number of fire units. However, recognizing the lack of available data, Hogg used a proxy criterion of minimizing the total travel time of all units to all fires for a prescribed total number of units.

The solution was not subject to a constraint on maximum travel time to any sector of the service area, and it was theoretically possible for an area to be further away from a fire station than a fire official would deem reasonable and a resident of the area would deem equitable. Furthermore, the model assumed that fire damage was linearly related to travel time and that all building types exhibited a similar relationship between fire damage and fire-unit response time. Hogg suggested that a knowledge of these relationships would not only improve the result in terms of its prescribed intention of minimizing fire loss for a given level of expenditure, but would also provide valuable insights into the appropriate number of fire units to be maintained. She suggested that "each additional station should pay for its upkeep in terms of the resultant savings in life and monetary loss from fires" (Hogg, 1971*b*, p. 276). Hogg's paper is significant in that it considered average travel time as a siting criterion, rather than treating it as a coverage criterion as suggested by fire insurance rating organizations and set covering models.

Another family of location models uses a *minimax* strategy to locate facilities or units. Here the objective is not to minimize the total cost but to minimize the maximum cost that any demand point may incur. This objective may be considered one of Hirsch's (1973) "output equalization" criteria. Formally, the problem can be stated as

$$\text{Min } Z$$

subject to

$$Z \geq d_{ij} x_{ij} \qquad i = 1, 2, \ldots, m; j = 1, 2, \ldots, n \qquad \textbf{(8.11)}$$

and (8.6)–(8.9). This formulation is also called the *p-center* problem.

Rider (1975) has proposed a two-stage model referred to as a *parametric allocation model*. Instead of selecting a set of p locations out of n possible sites ($p < n$), the problem is reduced to an allocation-location problem. The first stage (allocation) involves an assignment of the total number of fire fighting units to different homogeneous regions of the city, each having a given demand rate. Rider used relationships developed by Kolesar (1975) that relate density of fire fighting units (units per square mile) with average travel time to allocate the units to the different regions. Once the allocation task is complete, the actual location of units within regions is determined in the second stage, using other techniques such as set covering.

Dynamic Models

Situations may arise where a closest unit may not be available to attend a given fire because it is busy answering another fire. In this case the next closest unit (if available) may be assigned to the unattended fire. Dynamic models can allow for such situations. These models are especially useful in regions with high demand rates, with respect to established facilities, where the probability that some facilities may run out of available units is significant.

When the probability that units will be unavailable is high, the strategic decisions made are not limited to the location of fire fighting units. Additionally, the decision maker must address issues such as how many and which units to dispatch. Static models assume that each fire is attended by the closest unit or units. However, if there is a high probability that units will not be available, a policy of sending the closest unit may not be best for some performance criteria.

Between 1969 and 1975, the New York City Rand Institute undertook considerable research in the analysis of fire departments. Rand was assigned to develop analytical tools to improve the delivery of fire protection service through more effective use of fire department resources (Walker, Chaiken, and Ignall, 1979). Besides their work on fire station placement, Rand researchers conducted analyses concerning decisions on how many units to send to each alarm, which units to send, and where to relocate available units during times of inordinately high demand for fire units in some sections of the city. In addition to the parametric allocation model discussed previously, Rand researchers developed several other models relevant to fire department operations analysis. The most significant are the hypercube queueing model and the fire operations stimulation model.

The hypercube queueing model (Larson, 1974, 1975), an evaluative model, estimates the probability that each unit will be busy under varying dispatch policies, given the fire arrival (generation) rates and service time distributions for all zones in the city. The model is based on the representation of the queueing processes that takes place when the preferred unit for an incident is unavailable. It is useful for locating units and districting the region into response areas for fire units—to minimize response time as well as to more or less equalize work load.

The fire operations simulation model (Carter, Ignall, and Walker, 1975) is a computer representation of a fire department's units and the incidents to which they respond. As such, this model provides insights on dispatch policies, the configuration of response areas, and the number of units to send to alarms. In contrast with analytical models, such as the hypercube queueing model, simulation models can evaluate rather complicated deployment strategies that do not require the restrictive assumptions frequently needed in analytical models. On the other hand, simulation models require considerable computer resources.

The simulation model developed by Carter, Ignall, and Walker consists of an incident generator, a simulation program, and a postsimulation analysis program. The incident generator takes a probabilistic description of fire characteristics from historical data and converts it to a set of incidents, to each of which it assigns a location, a type, and a level of information available to the dispatcher receiving the alarm. The simulation program gives each incident a time of occurrence, a time of receipt of the alarm, a time of notification to responding units, a time of arrival at the scene, a time of releasing units from the scene, and an arrival time when units return to the firehouse. The program essentially keeps track of performance characteristics such as response-time distribution and the state of the system (number of units available) over time.

The postsimulation analysis program performs statistical analyses of simulation runs and compares statistics with other simulation runs. The simulation model can be used to evaluate a variety of policies by examining response-time statistics, work load distribution among fire units, and other performance measures.

Walker, Chaiken, and Ignall (1979) give a detailed review of the Rand Institute's work on fire operations analysis. Interested readers may also want to review ambulance location models, since the delivery of ambulance service has some similar characteristics to that of fire protection, in that quality of service is highly related to response time. Also, the analysis for ambulance locations is similar to that for fire station placement in that it considers the trade-off between equitable and efficient provision of service and the balance of workload among service units. Savas (1969), Gordon and Zelin (1970), Volz (1971), Fitzsimmons (1973), and Daberkow and King (1977) are some relevant ambulance location studies.

ADDITIONAL MODELING CONSIDERATIONS

Set covering models—for example, those that use a distance standard that the closest unit be within, say, 1 mi—implicitly assume that distances between 0 and 1 mi are equally desirable, whereas distances above 1 mi are highly undesirable to the extent that they are not allowed in the feasible solution set. On the other hand the minisum models discussed assume that a distance of 2 mi. is twice as costly as a distance of 1 mi.—that is, that a linear relationship exists between cost and distance.

If, in fact, the relationship between cost and distance is nonlinear, then the minisum formulation must be modified to take that into account. This modification can be simply accomplished, for example in the p-median formulation, by changing the given linear objective function (8.5) to

$$\text{Min } Z = \sum_{j=1}^{n} \sum_{i=1}^{m} g(d_{ij})x_{ij} \tag{8.12}$$

where $g(ij)$ is the cost of zone i being a distance d_{ij} from site j. Observe that the implicit set covering assumption illustrated in the first paragraph of this section can be incorporated by $g(d_{ij})$ as follows:

$$g(d_{ij}) = \begin{cases} 1 & \text{if } d_{ij} \leq 1 \text{ mi.} \\ \infty & \text{if } d_{ij} > 1 \text{ mi.} \end{cases} \tag{8.13}$$

Of course, constraint (8.6) must be dropped in this formulation, since the solution of the problem will include the number units to be located.

Allowing a nonlinear function to map distance to cost in locational decision problems is not new; Handler and Mirchandani (1979) review several such models. However, no systematic approach to determine this function for locating fire fighting units has been tried. The next section discusses what this function could be and how it can be obtained.

Another consideration often missing in fire station placement models is the value of the second and later arriving units. Consider the following question: Is it better to have the first and second closest units both at distances of 1 mi, or is it better to have the closest unit at a distance of ½ mi and the second closest at a distance of 2 mi? The answer depends, of course, on how we trade-off the response times of the first and second closest units. In the section after the next one we discuss how to accommodate this trade-off in a minisum model.

Finally, in the section following that, we consider additional constraints that may be imposed by the physical aspects of the fire protection system or by the decision makers, and we illustrate how these constraints may also be incorporated in a minisum model.

The Objective Function

To obtain a relationship between the response time of fire units and the property or casualty damage is quite difficult for several reasons. First, the data collection practices in fire protection services generally do not include response time as a variable of interest. Second, the extent of fire damage is quite random even for fires in comparable structures with similar fire-unit response times. Third, considerable observer error is possible among individuals making the estimate of fire damage (Schaenman et al., 1977). Nevertheless, Hogg (1971a), Corman, Ignall, Rider, and Stevenson (1976), Ignall, Rider, and Urbach (1978), and Halpern, Isherwood, and Wand (1979), among others, have attempted to model damage–response-time relationships empirically using statistical and other methods.

In addition to the technical difficulty in assessing the damage–response-time relationship, problems exist owing to uncertainty about how decision makers value the levels of achievement of various performance measures. For example, if the fire department is intended to minimize both casualty and property damage, even if it were known how location decisions influence the level of achievement of these performance measures, subjective assessments of the relative value of the attributes associated with these measures would have to be made. Thus, complicating the locational decision problem are the difficulty in objectively measuring the performance of a fire fighting system faced with multiple attributes, and the uncertainties in the exact future location and number of fires and response times to them.

An alternative to obtaining an empirically derived cost function is to use *utility analysis*. This method uses the experience of fire fighting professionals, by assessing their utility functions, to make the trade-offs of the various attributes. Utility functions are ideal instruments for making decisions under uncertainty.

The basic concept of decision-making under conditions of uncertainty can be summarized as follows. At the time a decision-maker must choose among a set of alternatives, he is faced with some uncertainty in the consequences of his action. These consequences may be multi-dimensional. A probability density function of the consequences of each alternative can represent this uncertainty. The decision-maker should select the alternative associated with the most desirable probability distribution of the consequences.

Let us temporarily assume that the arrival time of the first unit fully determines all adverse consequences of fires. Each siting plan yields a probability density function for the first unit's response time. Rather than minimizing the expected value of this response time, utility theory suggests that the decision maker minimize some function of the distribution of response times, called the expected utility, which is defined as

$$E[u(t)] = \int_0^\infty f(t)u(t)dt$$

where $f(t)$ is the probability density function of the response time t to a random fire, and $u(t)$ is the utility of response time t.

The utility of response time need not be a linear function of response time. For example, a fire chief might prefer a 3-min expected response time with a low variance to a shorter expected response time with a high variance. This preference can be revealed by the single-attribute utility function $u(t)$.

Through a series of structured interviews with a decision maker, it is usually possible to assess the decision maker's utility function. Keeney and Raiffa (1976) describe these assessment procedures in detail. In fact, Keeney (1973) has used these procedures to assess New York City Fire Department officials' utility functions for response times.

If we partition the given region into m zones, the expected utility can be represented as

$$E[u(t)] = \sum_{i=1}^{m} p_i \left[\int_0^\infty f_i(t)u(t)dt \right] \qquad \textbf{(8.14)}$$

where p_i is the proportion of fires in zone i, and $f_i(t)$ is the probability density function of the response time t to a random fire in zone i.

Note that $f_i(t)$ depends on the location of the closest unit to zone i. Thus a suitable locational criterion for this scenario is to maximize $E[u(t)]$ by optimally placing the required p units among the available n sites, subject to given constraints.

Multiple Arriving Units

Although the above utility function is adequate for the response time of a single unit, the operating environment of a typical fire department is such that when a call for service is received, a preassigned number of units are dispatched from prespecified locations. The value of dispatching more than one unit can be included in the analyses, since utility functions can be obtained for multiple attributes. In this case, since it is likely that the response time of the second arriving unit will have some effect on the damage caused by a fire, we need a multidimensional utility function $u(t_1, t_2)$ where t_1 is the first-unit response time and t_2 the second-unit response time.

To simplify the discussion, we will assume that the two attributes (first-unit response time and second-unit response time) are mutually utility independent. This assumption means that for a fixed level of one attribute, the relative importance of varying levels of the other attribute is not dependent on the level of the first attribute.

The expected utility resulting from a given set of locations can be expressed as

$$E[u(t_1, t_2)] = \sum_{i=1}^{m} p_i \left[\int_0^\infty \int_0^\infty f_i(t_1, t_2) u(t_1, t_2) dt_1 dt_2 \right] \tag{8.15}$$

where $f_i(t_1, t_2)$ is the joint probability density function of the first- and second-unit response times to a random fire in zone i, and $u(t_1, t_2)$ is the bivariate utility function of the first- and second-unit response times.

Alternately, this can be represented as

$$E[u(t_1, t_2)] = \int_0^\infty \int_0^\infty f(t_1, t_2) u(t_1, t_2) dt_1 dt_2 \tag{8.16}$$

where $f(t_1, t_2)$ is the joint probability density function of the first- and second-unit response times to a random fire in the entire service area.

The Taylor series expansion of $u(t_1, t_2)$ around $\bar{t}_1, \bar{t}_2$ is

$$u(t_1, t_2) = u(\bar{t}_1, \bar{t}_2) + u_{10}(\bar{t}_1, \bar{t}_2)(t_1 - \bar{t}_1) + u_{01}(\bar{t}_1, \bar{t}_2)(t_2 - \bar{t}_2)$$

$$+ \frac{1}{2} u_{20}(\bar{t}_1, \bar{t}_2)(t_1 - \bar{t}_1)^2 + \frac{1}{2} u_{02}(\bar{t}_1, \bar{t}_2)(t_2 - \bar{t}_2)^2$$

$$+ u_{11}(\bar{t}_1, \bar{t}_2)(t_1 - \bar{t}_1)(t_2 - \bar{t}_2) + \ldots \tag{8.17}$$

where $u_{ij}(t_1, t_2)$ are the partial derivatives corresponding to i and j:

$$u_{ij}(t_1, t_2) = \left(\frac{\partial}{\partial t_1}\right)^i \left(\frac{\partial}{\partial t_2}\right)^j u(t_1, t_2)$$

If $\bar{t}_1$ is the mean of t_1, and $\bar{t}_2$ is the mean of t_2, then $E(t_1 - \bar{t}_1) = 0$ and $E(t_2 - \bar{t}_2) = 0$. In that case the expected value of $u(t_1, t_2)$ is approximated by

$$E[u(t_1, t_2)] = u(\bar{t}_1, \bar{t}_2) + \frac{1}{2} E[u_{20}(\bar{t}_1, \bar{t}_2)(t_1 - \bar{t}_1)^2]$$

$$+ \frac{1}{2} E[u_{02}(\bar{t}_1, \bar{t}_2)(t_2 - \bar{t}_2)^2]$$

$$+ E[u_{11}(\bar{t}_1, \bar{t}_2)(t_1 - \bar{t}_1)(t_2 - \bar{t}_2)] \tag{8.18}$$

or, equivalently, by

$$E[u(t_1, t_2)] = u(\bar{t}_1, \bar{t}_2) + \frac{1}{2} u_{20}(\bar{t}_1, \bar{t}_2)\sigma_1^2 + \frac{1}{2} u_{02}(\bar{t}_1, \bar{t}_2)\sigma_2^2$$

$$+ u_{11}(\bar{t}_1, \bar{t}_2) \, \text{cov}(t_1, t_2) \qquad\qquad (8.19)$$

where

$$\sigma_1^2 = \text{the variance of } t_1,$$
$$\sigma_2^2 = \text{the variance of } t_2,$$
$$\text{cov}(t_1, t_2) = \text{the covariance of } t_1 \text{ and } t_2.$$

Expression (8.19) gives the expected utility as a function of the means, variances, and covariances of the response times of the two first arriving units. For each siting plan we can easily compute these means, variances and covariances (see Appendix 8.A at the end of the chapter). Hence one may use (8.19) to determine the siting plan that maximizes expected utility.

The Constraints

In the minisum model we already have constraints (8.7)–(8.9) that result from physical considerations. Furthermore, if some site j has to be filled, we fix the corresponding y_j to be 1; whereas if we wish to prohibit a unit from being located there, we fix y_j to be 0. In a p-median-type formulation, for scenarios where the number of units to be located is prespecified to be equal to p, the additional constraint (8.6) is required. On the other hand, if we impose a fixed cost f_j of locating a unit at any site j and let the number of units depend on the minimization of the total cost of placement of the units, then we need to add an additional term to the objective function:

$$\sum_{j=1}^{n} f_j y_j$$

We add this term instead of constaint (8.6), as we did for (8.10) in the uncapacitated facility location problem. From this point onward, for illustrative purposes, we restrict our discussions to only the first scenario—where the number of units to be located is prespecified to be a given number p—that is, to the p-median-type minisum formulations.

Either decision makers' dictums or practice considerations may require that at least c of the existing facilities, located at, say, set C of sites, must remain in the new design. We then need the additional constraint

$$\sum_{y_j \in C} y_j \geq c \qquad\qquad (8.20)$$

Recall that unless we have a nonlinear function of the form of (8.13) in the objective (8.12), we can end up with an optimal solution where for some zones, the closest unit is too far in terms of travel cost. To prevent this problem, we can include the following constraints, which limit how far the closest unit can be from each zone:

$$x_{ij}d_{ij} \leq d_i^* \qquad i = 1, 2, \ldots, m; \quad j = 1, 2, \ldots, n \qquad (8.21)$$

where d_i^* is the maximum allowable traveling cost (e.g., travel time, distance, disutility) for fires from zone i. The d_i^* must be specified by the decision maker.

Including the consideration for the first and second arriving units requires additional constraints to (1) prevent the same unit from being assigned as the first and second unit, (2) assure that exactly two units are assigned to each zone, and (3) limit how far from each zone the second arriving unit can be located. (The constraints given below can easily be extended to the units arriving third, fourth, and so on.)

Let us introduce the additional notation ' (prime) to indicate the corresponding variables associated with the second arriving unit:

$x'_{ij} = 1$ if site j contains the second arriving unit to zone i, 0 otherwise,
$d_i'^* = $ the maximum allowable traveling cost for the second arriving unit to zone i (of course, $d_i'^* \geq d_i^*$).

To prevent the same unit from being assigned as the first and second arriving unit requires

$$x_{ij} + x'_{ij} \leq 1 \qquad i = 1, 2, \ldots, m; \quad j = 1, 2, \ldots, n \qquad (8.22)$$

To assure that each zone i also has a second arriving unit assigned to it requires

$$\sum_{j=1}^{n} x'_{ij} = 1 \qquad i = 1, 2, \ldots, m \qquad (8.23)$$

To make sure that the second unit arriving at zone i has a traveling cost of less than (or equal to) $d_i'^*$ requires

$$x'_{ij}d_{ij} \leq d_i'^* \qquad i = 1, 2, \ldots, m; \quad j = 1, 2, \ldots, n \qquad (8.24)$$

Of course, as in the single-unit model, we must ensure that if site j is assigned to zone i for the second arriving unit, then site j must have an established facility. That is,

$$y_j \geq x'_{ij} \qquad i = 1, 2, \ldots, m; \quad j = 1, 2, \ldots, n \qquad (8.25)$$

Besides the inclusion of constraints (8.22)–(8.25), the consideration of the second arriving unit adds an interesting dimension to the practical problem. The single-unit model suggests that two-unit stations would be suboptimal; a transfer of a second unit to a zone without an established facility would decrease the expected travel time or, in general, increase the expected utility. However, including in the objective function the travel cost of the second arriving unit could result in two units being located at the same physical site, that is, in a two-unit station. (Observe that cities do maintain multiple-unit stations, and the above consideration may be one of the reasons for it.) To accommodate this situation in the mathematical model, two site indices must be allocated to any physical site where two or more units can be located.

THE MODEL AND ITS SOLUTION

In the section just completed, we discussed additional considerations that must be included to make a fire station placement model more applicable in actual scenarios. The discussion took the framework of an optimization problem that asked for the optimal locations of p fire fighting units using a minisum criterion. Similar considerations apply for other optimization criteria and models.

Evaluative studies, which require the detailed evaluation of a small number of siting alternatives, must also include consideration of the performance measures of the system, the second units' response times, and the constraints imposed on the system. Perhaps, though, these considerations need not be exactly as represented in our optimization model, because decision variables are treated differently in evaluation models than in optimization models. The solution of an optimization model prescribes the decision variables, whereas evaluative studies analyze each possible set of decision-variable values (corresponding to each alternative) in turn, using an evaluation model. The performance of the system is optimized in the first case, and in the second case the alternative with the best performance is selected from the given alternatives.

An optimization model may be solved either *exactly,* often by applying algorithms based on mathematical concepts, or *approximately,* often by using heuristics based on common sense and simple mathematical approximations. When the optimization model is not standard (e.g., when it is not a standard linear programming model, a dynamic programming model, etc.), or an exact algorithm is not available, the investigator resorts to a heuristic method to solve the model. The optimization model that we present in the next section has a complicated nonlinear objective function and a large number of constraints, and an exact algorithm for its solution is not available. We thus use a *site-substitution* heuristic, originally developed by Teitz and Bart (1968) for the classic p-median problem, to obtain a good, if not optimal, solution. The heuristic is discussed in the section following the next one.

When the number of feasible solutions in an optimization model is not large and a heuristic *systematically* evaluates some, if not all, feasible solutions, then our distinction between an optimization model and an evaluation model becomes fuzzy. This condition is precisely why the developed model may also be used in an evaluative study. In general, the choice of a model depends on the number of candidate alternatives. If the decision faced by a fire chief is to decide which of two sites is better for a new fire unit, then perhaps a detailed simulation model (e.g., Carter, Ignall, and Walker, 1975) is more appropriate. If the decision involves a small number of alternatives (say, three to ten), then perhaps a descriptive model, such as the Rand firehouse site evaluation model (Dormont, Hausner, and Walker, 1975), may be more appropriate. The model and approach proposed here is probably more useful when the number of alternatives is larger, primarily because this model requires less computational effort and considers additional system aspects (as discussed in the foregoing section). Of course, when the number of alternatives has been narrowed down to two or three, detailed impact studies are necessary before the final choice is made.

An Optimization Model

In a p-median problem framework, the model can be stated verbally as follows: Optimize *given objective function*, subject to *exactly p facilities are located*, and *other constraints*. From the earlier discussion of modeling, the problem can be mathematically formulated as

$$\text{Min } Z = E[u(t_1, t_2)]$$

$$= u(\bar{t}_1, \bar{t}_2) + \frac{1}{2} u_{20}(\bar{t}_1, \bar{t}_2)\sigma_1^2 + \frac{1}{2} u_{02}(\bar{t}_1, \bar{t}_2)\sigma_2^2$$

$$+ u_{11}(\bar{t}_1, \bar{t}_2) \text{ cov}(t_1, t_2) \tag{8.26}$$

subject to

$$\sum_{j=1}^{n} y_j = p \tag{8.27}$$

$$\sum_{j=1}^{n} x_{ij} = 1 \qquad i = 1, 2, \ldots, m \qquad\qquad (8.28)$$

$$\sum_{j=1}^{n} x'_{ij} = 1 \qquad i = 1, 2, \ldots, m \qquad\qquad (8.29)$$

$$x_{ij} + x'_{ij} \leq 1 \qquad i = 1, 2, \ldots, m; \quad j = 1, 2, \ldots, n \qquad (8.30)$$

$$y_j \geq x_{ij} \qquad i = 1, 2, \ldots, m; \quad j = 1, 2, \ldots, n \qquad (8.31)$$

$$y_j \geq x'_{ij} \qquad i = 1, 2, \ldots, m; \quad j = 1, 2, \ldots, n \qquad (8.32)$$

$$x_{ij}t_{ij} \leq t_i^* \qquad i = 1, 2, \ldots, m; \quad j = 1, 2, \ldots, n \qquad (8.33)$$

$$x'_{ij}t_{ij} \leq t_i'^* \qquad i = 1, 2, \ldots, m; \quad j = 1, 2, \ldots, n \qquad (8.34)$$

$$\sum_{y_j \in C} y_j \geq c \qquad\qquad (8.35)$$

$$y_j, x_{ij}, x'_{ij} \in (0, 1) \qquad i = 1, 2, \ldots, m; \quad j = 1, 2, \ldots, n \qquad (8.36)$$

where

t_{ij} = the expected travel time from zone i fires to site j,
t_i^* = the maximum allowable expected travel time for the first unit arriving at zone i,
t_i' = the maximum allowable expected travel time for the second unit arriving at zone i.

For each set of decision-variable values that are feasible for the constraint set (8.27)–(8.36), we have a value of Z. (See Appendix 8.A for how this value can be computed.) The set of values that minimizes Z gives the optimal decisions.

A Solution Procedure

Three methods of solving (8.26)–(8.36) are exhaustive enumeration, mathematical programming, and (3) heuristic approaches.

Exhaustive enumeration requires the calculation of Z for each feasible combination of p out of n sites. The number of combinations of p out of n sites is $\binom{n}{p}$ although probably much fewer are feasible for the constraint set (8.27)–(8.36). Nevertheless, this number could be quite large if n and p are not small. Hence, exhaustive enumeration may require excessive computation.

Mathematical programming approaches are not readily available for problems of the type formulated here. For one thing, the objective function is nonlinear in a complicated fashion (see Appendix 8.A) with respect to the decision variables. For another, the number of constraints and decision variables is large even for moderate-sized problems. Excluding the integrality requirements (8.36), the number of constraints is $5mn + 2m + 2$ and the number of decision variables is $2mn + n$, both quite large numbers even for modest numbers of sites (m) and zones (n).

There are probably several heuristic approaches to solving the problem. One that is easily applicable to our model is Teitz and Bart's (1968) site-substitution approach. Essentially, their method starts with a set of seed sites and substitutes, one at a time, sites within the current set for sites outside the current set. If a site substitution reduces the value of the objective function, then the new site is included and the old site is eliminated from the current set. This process continues until no substitution yields a decrease in the value of the objective function.

Applied to our problem, Teitz and Bart's procedure can be stated verbally as

1. Select an initial set of p sites, say set P.
2. Compute the expected utility for that set P.
3. Select some site v_b not in P.
4. For each site in P, substitute v_b and compute the expected utility of the new set of p sites.
5. Find the site v_k in P such that when it is replaced by v_b, the reduction in the value of the objective function is greater than for any other of the p sites.
6. If a site meeting the criteria in step 5 is found, substitute v_b for v_k unless this substitution violates any constraints. If no such site exists, retain all sites in P and continue.
7. Select another site not contained in the current set and not previously tried as a candidate for entry. Repeat steps 4–6.
8. When all sites not in P have been tried, define the resulting set of existing sites as P' and repeat steps 2–7 using P' instead of P. Complete repetition of this series is called a cycle.
9. When a cycle of steps 2–8 results in no substitution, terminate the procedure.

Further details of the procedure are available in Reilly (1983).

Note that the final set of sites selected depends on the initial seed sites. We suggest that the procedure be executed several times, each time using a new set of seed sites, and that the best final set be selected from all the executions. Each set of seed sites may be randomly generated or may be selected by a decision maker, an expert, or someone who knows the system well.

Some Further Remarks on the Model

Implicit in our partitioning of the region into zones is an assumption that the zones are homogeneous in terms of types of fire, distribution of fires within the zones, and so on, and that fires can therefore be aggregated into incidence rates p_i. However, using additional indices, the model may easily be extended to finer classification of fire incidents. For example, p_{ik} could denote the proportion of fires of type k in zone i. The zones themselves could be defined so that fires of each type are homogeneous within them. Only for convenience and presentation purposes have we used a single index in our model development. In the application discussed in the next section, we include fire types as well as fire locations in the placement decision.

MODEL APPLICATION

The model was used to design the spatial distribution of fire fighting units in Albany, New York, a city with a population of 102,000. The Albany Fire Department operates ten engine companies and four ladder companies. Fig. 8.1 shows the geographic distribution of these units.

The Data

Fire incidence records for 1978 and 1979 were used to get an idea of the spatial and temporal distribution of fires in the city. For each structural fire (false alarms and fires outside the buildings were not included), the following data were collected: (1) date, (2) day of week, (3) hour, and (4) alarm-box number (a proxy for location). Additionally, all fires were grouped into one of three types:

Type 1: low risk—fires in one- or two-story residences.

Type 2: property risk—fires in commercial establishments.

Type 3: high risk—fires in multiple-occupancy dwellings, offices, and places of public assembly.

The data showed that the mean number of low-risk fires per day was 1.14; for property-risk fires it was 0.19 and for high-risk fires 0.92.

Statistical analyses of the data indicated that there was little significant difference between the spatial and temporal patterns for fires in 1978 and 1979 (see Table 8.1). It appears that the patterns of time and space distribution between

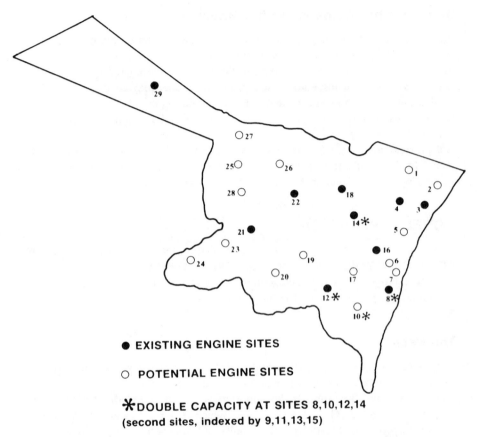

Figure 8.1. Location of existing and candidate engine sites in Albany.

the two years are similar. The exceptions (where the critical probabilities are very low) are as follows:

1. The month-of-year distribution for type-2 and 3 fires appear to be different between the two years. Later analysis showed that this characteristic exhibits a high degree of variability that cannot be explained as a seasonal phenomenon.
2. The time-of-day distribution of type-1 fires appears to be different between the two years. In a further analysis of the data, fires were classified into three time groups. Doing so greatly increased the level of significance, which suggests that distribution of fires between the three time groups remained essentially the same but that there was some shifting within time groups between the two years.

**Table 8.1. Test for Internal Consistency
Between Years (1978 and 1979 Data)**

Characteristic	Type-1 Fires	Type-2 Fires	Type-3 Fires
Month	.13	.03	.002
Hour	.03	.77	.707
Zone	.36	.30	.56
Day	.79	.77	.40
Time group	.84	.13	.75

Each entry is the level of significance at which there is no difference
between 1978 and 1979 for each fire type and characteristic.

The data also revealed that weekend days have the highest frequency of
noncommercial fires and the lowest frequency of commercial fires. This finding
could mean that fires are less likely to occur in unoccupied structures. However,
the difference between weekdays and weekend days was not statistically significant.

After further statistical analysis was performed on the data we concluded
that 1978 and 1979 data were representative of the spatial and temporal distribution of fires in Albany.

Candidate Sites

We chose a number of candidate sites distributed throughout the city. These
included all existing sites, three sites abandoned by the city in the past few
years, and several other sites. All the candidate sites are shown in Fig. 8.1,
and four of them (sites 8, 10, 12, and 14) were assumed to have a capacity of
two units.

Partitioning the Region into Zones

The region consisted of 400 alarm boxes, which were grouped into 38 zones
to simplify computation. The groupings resulted in relatively large zones in
the less densely populated areas of the city; in the most built-up areas they
resulted in zones of a few blocks. In the larger zones it was assumed that fires
of each type were uniformly distributed throughout the zone. In the small zones
it was assumed that the fires originated at the zone centroids.

Interzonal Travel Times

A matrix of rectilinear distances between pairs of zones was determined from
a map. Lacking actual travel time data, we used Kolesar's (1975) travel time–
distance model:

$$t(x) = \begin{cases} 2\left(\dfrac{x}{a}\right)^{1/2} & \text{for } x \leqslant 2d_c \\[2em] \dfrac{s_c}{a} + \dfrac{x}{s_c} & \text{for } x > 2d_c, \end{cases} \tag{8.37}$$

where

$t(x)$ = the travel time for distance x,
a = the vehicle acceleration,
s_c = the vehicle cruising speed,
d_c = the distance required to achieve cruising speed.

The model parameters do not vary much for cities of comparable size (U.S. Department of Housing and Urban Development, 1977). Based on data for cities comparable in size to Albany, the following travel time–distance relationships were assumed:

$$t(x) = \begin{cases} 2.1x^{1/2} & \text{for } x \leqslant 0.38 \text{ mi} \\ 0.65 + 1.7x & \text{for } x > 0.38 \text{ mi} \end{cases} \tag{8.38}$$

The Objective Function

Through a series of structured interviews with an official of the Albany Fire Department, we developed a utility function of the response times for the first two engines and ladders arriving at a fire. First, of course, single attribute utility functions were assessed. Then six utility functions were determined, three for arriving engines (one for each fire type) and three for arriving ladders.

To illustrate, consider the utility function for the first-engine response time to low-risk (type-1) fires. The fire department official revealed that he was "constantly risk averse." This concept is best illustrated by an example. For any 50:50 lottery of two possible first-unit response times, say 2 min and 4 min, a decision maker has an attribute value, referred to as that decision maker's *certainty equivalent* (CE) of the lottery, for which he or she is indifferent between that value for certain and the given lottery. If the CE of the lottery is greater than the expected value of the lottery (in this case 3 min), then the decision maker is said to be risk averse. If risk premium (the difference between the expected value and CE) remains constant as the consequences (i.e., the two values of the lottery) are changed by equal amounts, then the decision maker is said to be constantly risk averse. A constantly risk-averse utility function has the following form

$$u(t) = a - be^{ct} \qquad\qquad (8.39)$$

During the interview, the official indicated that he would be indifferent between a 50:50 lottery of 1-min and 5-min response times and a response time of 3.75 min with certainty. This implies

$$u(3.75) = 0.5u(1) + 0.5u(5) \qquad\qquad (8.40)$$

The utility functions were assessed over a consequence space of 0–10 min. We assigned the following arbitrary values for the utilities of the two extreme outcomes:

$$u(0) \ = 1 \qquad\qquad (8.41)$$

$$u(10) = 0 \qquad\qquad (8.42)$$

Equations (8.39)–(8.42) result in the following utility function for the first-engine response time to low-risk fires:

$$u^{E11}(t) = 1.016 - 0.016e^{0.415t} \qquad\qquad (8.43)$$

where $u^{Ek\ell}(t)$ is the utility of response time t of the kth engine to type-ℓ fire, $k = 1, 2, \ell = 1, 2, 3$.

Likewise, utility functions for ladder response times were obtained. We will use the following notation for the various utility functions in this application:

$u^{Lk\ell}(t)$ = the utility of the response time t of the kth ladder to a type-l fire, $k = 1, 2, l = 1, 2, 3$,

$u^{E\ell}(t_1, t_2)$ = the utility of response times t_1 and t_2 of the first and second engines to a type-l fire;

$u^{L\ell}(t_1, t_2)$ = the utility of response times t_1 and t_2 of the first and second ladders to a type-l fire.

Table 8.2 shows the various conditional utility functions that we assessed.

These utility functions were then scaled and combined to reflect the utility for response times for each type of fire. Table 8.3 gives the composite utility functions. (The details are given in Reilly, 1983).

Executing the Solution Procedure

The solution procedure was executed several times using the data and the utility function discussed earlier in this section. The runs were intended to evaluate the current set of locations and to provide some insights into how the model

Table 8.2. Conditional Utility Functions of Response Time

Type 1 (low-risk fires)		Type 2 (property-risk fires)		Type 3 (high-risk fires)	
$u^{E11}(t)$	$1.016 - .016e^{.415t}$	$u^{E12}(t)$	$1.016 - .016e^{.415t}$	$u^{E13}(t)$	$1.016 - .016e^{.415t}$
$u^{E21}(t)$	$1.079 - .079e^{.262t}$	$u^{E22}(t)$	$1.016 - .016e^{.415t}$	$u^{E23}(t)$	$1.016 - .016e^{.415t}$
$u^{L11}(t)$	$1.0 \ + .017t - .0117t^2$	$u^{L12}(t)$	$1.016 - .016e^{.415t}$	$u^{L13}(t)$	$1.016 - .016e^{.415t}$
				$u^{L23}(t)$	$1.0 \ + .017t - .017t^2$

Table 8.3. Engine and Ladder Response-Time Utility Functions

$$u^{E1}(t_1, t_2) = .971u^{E11}(t_1) + .763u^{E21}(t_2) - .734u^{E11}(t_1)u^{E21}(t_2)$$
$$u^{E2}(t_1, t_2) = -.556 + 1.55[.994u^{E12}(t_1) + .986u^{E22}(t_2) - .980u^{E12}(t_1)u^{E22}(t_2)]$$
$$u^{E3}(t_1, t_2) = -.0144 + [.971u^{E13}(t_1) + .763u^{E23}(t_2) - .734u^{E13}(t_1)u^{E23}(t_2)]$$
$$u^{L1}(t_1, t_2) = u^{L11}(t_1)$$
$$u^{L2}(t_1, t_2) = -2.9609 + 3.908u^{L12}(t_1)$$
$$u^{L3}(t_1, t_2) = -7.254 + 8.356[.940u^{L13}(t_1) + .763u^{L23}(t_2) - .286u^{L13}(t_1)u^{L23}(t_2)]$$

Note: The single-attribute utility functions $u^{E11}(t_1)$ $u^{E21}(t_2)$, etc., are given in Table 8.2.

could be used to make actual decisions about fire-unit placement decisions in Albany.

The first execution was an assessment of the current location set, following which three unconstrained optimization runs were executed to determine the effect of varying the objective function. Each of these three cases located a total of ten units (the current number). The three objective functions were (1) minimize the mean first-unit travel time, (2) minimize the mean of the sum of the first- and second-unit travel times, and (3) maximize the expected utility of the first- and second-unit travel times. Table 8.4 shows several performance measures for these three runs, as well as for the existing set of locations. Observe that the units are located at slightly different site sets for each optimization criterion. Based on these runs, the decision maker can select a set that appears to have the appropriate levels of performances.

The model was then executed additional times with some constraints imposed on the location of units (see Table 8.5). In each of these cases, the objective was to maximize expected utility. The first execution (run 5) was to locate nine units constrained by the requirement that all nine of them be among the existing set of stations. That is, we wished to determine which unit should be removed in case the city did not have sufficient resources to maintain all ten of the present units. The analysis suggested that the unit at site 14 should be removed. The next run (run 6) was to locate eight units also subject to the constraint that all eight be located at the existing sites. Here we were finding which two sites to close should the city elect to do so. The model suggested

Table 8.4. Analysis of Initial Runs

	Run 1	Run 2	Run 3	Run 4
Objective	Evaluate existing sites	Minimize mean of 1st-unit times	Minimize mean of 1st- & 2nd-unit times	Maximize expected utility
Sites selected	3 4 9 12 14 16 18 21 22 29	3 4 5 9 12 16 18 21 22 29	4 5 8 9 12 14 18 21 22 29	3 4 7 9 12 16 18 21 22 29
Performance measures				
Expected utility	.98617	.98621	.98611	.98624
First-unit response time				
Mean	1.57	1.56	1.59	1.57
Standard deviation	1.39	1.39	1.42	1.39
Second-unit response time				
Mean	2.23	2.22	2.14	2.19
Standard deviation	1.54	1.54	1.57	1.55

Table 8.5. Analysis of Additional Runs

	Run 5	Run 6	Run 7	Run 8	Run 9
Objective: maximize expected utility subject to constraints	Locate 9 engine sites from existing 10	Locate 8 engine sites from existing 10	Locate 11 engine sites; use existing 10	Locate 10 engine sites; use all existing fire-houses	Locate 10 engine sites; use 9 of existing 10
Sites selected	3 4 9 12 16 18 21 22 29	3 4 9 12 18 21 22 29	3 4 9 12 14 16 18 21 22 25 29	3 4 7 9 12 16 18 21 22 29	3 4 9 12 16 18 21 22 25 29
Performance measures					
Expected utility	.98601	.98572	.98683	.98624	.98661
First-unit response time					
Mean	1.60	1.66	1.52	1.57	1.54
Standard deviation	1.39	1.38	1.33	1.39	1.32
Second-unit response time					
Mean	2.29	2.43	2.17	2.19	2.23
Standard deviation	1.52	1.49	1.42	1.55	1.41

the units at sites 14 and 16 be removed. Conversely, were the city to add a unit to the existing set, its best location would be at site 25 (see run 7).

In addition to the engine locations, Albany has two firehouses where only ladder units are located. We executed the program (run 8) to locate ten engine companies subject to the constraint of using only existing firehouses (ten engine and two ladder houses). This analysis suggested that the engine unit at site 14 should be moved to the ladder house at site 7.

Finally, we executed a run to provide some insight into where a single new engine might be placed (run 9). Here we located ten units subject to the constraint that nine of the ten existing sites remain. The recommended decision was to close the facility at site 14 and open one at site 25.

None of these analyses placed any constraint on maximum response time. However, the maximum first-unit response time was under 3 min in all cases. For some runs of the program, the maximum response-time constraint could have been introduced to limit the number of acceptable solutions. For example, for certain neighborhoods, a constraint could be introduced to prohibit the first- (or second-) unit response time in any proposed solution from exceeding the current first- (or second-) unit response time. In such cases, the maximum travel time constraint can be looked on not as a technical constraint imposed by a fire professional's judgement but as a policy constraint imposed through a political process.

Note that the values of the expected utilities are quite close to one another. However, this characteristic does not suggest that the decision makers are nearly indifferent among the alternate siting plans. It is an issue of scaling the utility function. Since any positive linear transformation of the function retains the preference structure of the decision maker, we could have shown larger numeric differences among the expected utilities without affecting the actual results of the analyses.

CONCLUSIONS AND FURTHER REMARKS

We reviewed the state-of-the-art models for locating fire fighting units. Although only a few applications of the models have been documented, it appears that a number of them have been used for decision making within fire protection services. Nevertheless, certain important considerations are missing in these models.

This chapter developed a minisum location model that includes some of these considerations. The model permits substantial flexibility in the form of the objective function. For its application to the Albany Fire Department, we used a complex nonlinear objective function that depended on the type of fire and the first- and second-unit response-time distributions. That the model can include consideration of the second-unit response time is especially important, since it appears that at least in the judgement of fire officials, the response time of the second arriving unit does have significant effect on the fire department's ability to control the spread of fire.

The benefit of the ability to impose several additional constraints on the optimization model considerably improves the model's usefulness, not only in long-range strategic planning but also in making relatively small incremental changes in the spatial design of the fire fighting units. The major benefit of this model is in finding good solutions to fairly well-constrained problems. Even the results from a well-constrained and well-formulated model, though, should be looked upon more as a guide to decision making than as a substitute for judicious consideration of a range of alternatives that includes both fire fighting and non–fire fighting factors.

Future work can take several directions. Studies relating fire-unit response time to fire damage have been limited to date. Improving the development of such a relationship would have two distinct advantages. First, it would improve the quality of decision making, since locational decisions could then consider the fire department's actual performance in averting property and casualty loss, rather than relying on a utility value that acts as a proxy for such measurements of performance. Second, such a relationship between response time and damage would provide some insights into the number of units to be operated. As the model is currently constructed, the number of units to be located is an exogenous variable supplied by the user. This formation is a cost-effectiveness model that seeks to maximize an objective, subject to a limitation on the resources available to achieve that objective. When using a utility function, it is difficult to relate an increase in expected utility with the cost of resources necessary to achieve that increase.

In applying the model, one of the more difficult tasks was the assessment of the utility functions. A useful enhancement to this work would be to design procedures, possibly computer-assisted, for such assessment. These procedures would also be useful for the spatial design of other public services.

The existing model depends on a utility function based on the preferences of a single, experienced individual. It would be useful to develop techniques that permit the assessment of a composite utility function of a small group of experienced fire fighting professionals. Another alternative would be to perform a sensitivity analysis to determine the range of utility-function parameters over which the solution as proposed by the model remains the best solution.

APPENDIX 8. A: COMPUTING TRAVEL TIME MEANS, VARIANCES, AND COVARIANCES

We stated in the body of the chapter that for a given set of locations of the units, the expected utility of the set could be approximately computed as a function of (1) the systemwide means of the travel times of the first and second arriving units, $\bar{t}_1$ and $\bar{t}_2$, (2) systemwide variances of these travel times, $\text{var}(t_1) \equiv \sigma_1^2$ and $\text{var}(t_2) \equiv \sigma_2^2$, and (3) their covariance, $\text{cov}(t_1, t_2)$. We briefly show here how these values can be computed when fires originate continuously and uniformly over a region rather than at discrete points. (The discrete case is straightforward.)

We assume that the region is partitioned into square zones. This approximation will not result in errors that are too large as long as the actual zones are "fairly convex and compact" (see Larson and Odoni, 1981). Further, the analysis presented here, based on geometric probability, can easily be extended to other zones of regular shape such as rectangular or circular zones. We must, of course, approximate any irregular zone to a shape that is amenable to the analysis.

Travel from the outside of a zone to the boundary of the zone follows the existing road network. Travel within a zone takes place in directions parallel to the zone boundaries (i.e., via a rectilinear metric). Thus, the travel time means, variances, and covariances will depend on (1) the frequencies of fires among zones, (2) the spatial distribution of fires within each zone, and (3) the variation in travel speeds to and within each zone.

To make the cases manageable when considered by an algorithm, we assume that each service unit can enter a zone only at either a zone corner or an edge midpoint. We also assume that zones are homogeneous in the sense that fires are uniformly distributed within each zone and the travel speed is constant[1] within the zone. (These assumptions are not necessary for computing the means, variances, and covariances; however, numerical integration may be required in a subroutine of the developed algorithm when these assumptions are not used.)

Let us consider the case of the unit arriving first at a fire at a random point in zone i. From the given unit locations (where $y_j = 1$), we identify the "closest" unit as the one with the smallest expected travel time to the zone centroid (i.e., smallest t_{ij}). The unit's travel time to the fire, say T_i, consists of two components: (1) a travel time, say v_i, from the unit's station to the point of entry into the zone, and (2) a travel time, say, w_i, from the point of entry to the fire:

$$T_i = v_i + w_i \tag{8.A1}$$

The expected value of T_i is

$$E(T_i) = \bar{v}_i + \bar{w}_i \tag{8.A2}$$

Thus the systemwide average travel time for the first arriving unit, $\bar{t}_1$, is

$$\bar{t}_1 = \sum_{i=1}^{m} p_i(\bar{v}_i + \bar{w}_i) \tag{8.A2}$$

where p_i is the proportion of fires in zone i.

We assume that v_i and w_i are statistically independent. Although they are travel times for travel on nonoverlapping road segments, this assumption is not strictly valid because some factors (e.g., weather) that affect travel time v_i also affect w_i, and vice versa. Also, if the zone has more than one closest unit (i.e., parts of the zone have different closest units than other parts), then there may be further dependencies between v_i and w_i. Nevertheless, for practical purposes, this assumption is reasonable.

The first unit's systemwide travel time variance var(t_1) is computed from

$$\text{var}(t_1) = E[(t_1 - \bar{t}_1)^2] \tag{8.A4}$$

$$= E(t_1^2) - \bar{t}_1^2 \tag{8.A5}$$

$$= \sum_{i=1}^{m} p_i E(T_i^2) - \bar{t}_1^2 \tag{8.A6}$$

$$= \sum_{i=1}^{m} p_i E[(v_i + w_i)^2] - \bar{t}_1^2 \tag{8.A7}$$

$$= \sum_{i=1}^{m} p_i E[(v_i^2) + 2E(v_i)E(w_i) + E(w_i^2)] - \bar{t}_1^2 \tag{8.A8}$$

$$= \sum_{i=1}^{m} p_i[\text{var}(v_i) + \bar{v}_i^2 + 2\bar{v}_i\bar{w}_i + \text{var}(w_i) + \bar{w}_i^2] - \bar{t}_1^2, \tag{8.A9}$$

where the statistical independence of v_i and w_i is utilized in (8.A8).

The computation procedure for the mean and variance of the second unit's travel time is similar to the above case.

The covariance of the first and second unit's travel times is, by definition,

$$\text{cov}(t_1, t_2) = E(t_1, t_2) - \bar{t}_1\bar{t}_2 \tag{8.A10}$$

$E(t_1, t_2)$ can be computed from

$$E(t_1, t_2) = \sum_{i=1}^{m} p_i E[(v_{1i} + w_{1i})(v_{2i} + w_{2i})] \tag{8.A11}$$

$$= \sum_{i=1}^{m} p_i[E(v_{1i}v_{2i}) + E(v_{1i}w_{2i}) + E(w_{1i}v_{2i}) + E(w_{1i}w_{2i})] \tag{8.A12}$$

where v_{1i}, w_{1i} are the travel times v_i, w_i associated with the first unit, and v_{2i} and w_{2i} those associated with the second unit.

If we assume that v_{1i} and v_{2i} are mutually independent[2] and are independent of w_{1i} and w_{2i}, then

$$E(t_1, t_2) = \sum_{i=1}^{m} p_i[\bar{v}_{1i}\bar{v}_{2i} + \bar{w}_{1i}\bar{v}_{2i} + \bar{w}_{2i}\bar{v}_{1i} + E(w_{1i}w_{2i})] \qquad \textbf{(8.A13)}$$

The remainder of this appendix focuses on methods of computing the components of equations (8.A3), (8.A9) and (8.A13).

The expected travel time from the unit's station to the zone boundary (v_i) can be computed from either travel time data or using Kolesar's (1975) travel time–distance relationship. Equation (8.38) in the body of the chapter gives this relationship for Albany, New York.

The variance of this travel time depends on the variance of the effective speed. Data from Hendrick et al. (1974) showed that the coefficient of variation of travel times in Denver, was only about 0.15. Since Denver has transportion characteristics similar to Albany's, we assume $\text{var}(v_i)$ as $0.0225\bar{v}_i^2$ in our application. In any case, an empirical estimate of $\text{var}(v_i)$ is required in this model.

To compute the travel time means, variances, and covariances within zones $[\bar{w}_{1i}, \bar{w}_{2i}, \text{var}(w_{1i}), \text{var}(w_{2i}), \text{and } E(w_{1i}w_{2i})]$, some analyses based on geometric probability are necessary. First, note that because of our assumption of homogeneous zones,

$$\bar{w}_{ki} = \frac{\bar{r}_{ki}}{s_i} \qquad k = 1, 2 \qquad \textbf{(8.A14)}$$

$$\text{var}(\bar{w}_{ki}) = \frac{1}{s_i^2} \text{var}(r_{ki}) \qquad k = 1, 2 \qquad \textbf{(8.A15)}$$

$$E(w_{1i}w_{2i}) = \frac{1}{s_i^2} E(r_{1i}r_{2i}) \qquad \textbf{(8.A16)}$$

where r_{ki} is the intrazonal travel distance for k^{th} unit and s_i is the speed of travel in zone i. Now to compute $\bar{r}_{ki}$ and $\text{var}(r_{ki})$, we must consider three cases: (1) fire-unit entry from the corner of the zone, (2) fire-unit entry from the edge midpoint of the zone, and (3) fire unit being located at the center of the zone. Figure 8.A1 illustrates these cases and Table 8.A1 shows how the means and variances are computed.

Even with the restriction of zone entry being at a corner or an edge midpoint, we have to consider several different patterns of the first and second units' entry points (see Fig. 8.A2) to compare $E(r_{1i}r_{2i})$. Figure 8.A3 illustrates the computation procedure for one such combination.

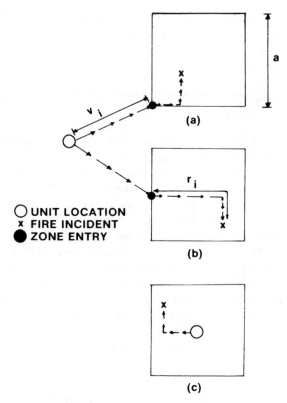

Figure 8.A1. Illustration of various cases of a unit's zone entry. (a) Entry at zone corner (also, v_i illustrated); (b) entry at zone edge midpoint (also, r_i illustrated); (c) unit's site at zone center.

This analysis has assumed that the first unit to respond to any fire in the zone is always the same and the second unit is always the same. That is, with regard to one-unit response, the entire zone is dedicated to one closest unit; and likewise, for a two-unit response the second unit is always the second-closest unit. This assumption is not always correct, especially with large zones, where parts of the zone may have a different closest unit than other parts. This nondedication of zones to units complicates the computation of the mean, variances, and covariances of the travel times. Reilly (1983) discusses these issues further and presents a procedural approach that partitions nondedicated zones into dedicated subzones to facilitate those computations.

Table 8.A1. Computation of Within-Zone Distance Means and Variances

Case	Probability Density Function[a]	Mean Distance	Distance Variance
1. Corner entry	$f_{xy}(x_0, y_0) = \begin{cases} \dfrac{1}{a^2} & 0 \leqslant x_0, y_0 \leqslant a \\ 0 & \text{otherwise} \end{cases}$	$\displaystyle\int_0^a\int_0^a (x+y)f_{xy}dxdy$ $= a$	$\displaystyle\int_0^a\int_0^a (x+y)^2 f_{xy}dxdy$ $- a^2 = 0.167a^2$
2. Edge mid-point entry	$f_{xy}(x_0, y_0) = \begin{cases} \dfrac{1}{a^2} & 0 \leqslant x_0, \leqslant a, \\ & -b \leqslant y_0 \leqslant b \\ 0 & \text{otherwise} \end{cases}$	$\displaystyle\int_{-b}^b\int_0^a (x+|y|)f_{xy}dxdy$ $= \dfrac{3a}{4}$	$\displaystyle\int_{-b}^b\int_0^a (x+|y|)^2 f_{xy}dxdy$ $-\left(\dfrac{3a}{4}\right)^2 = 0.104a^2$
3. Site at center	$f_{xy}(x_0, y_0) = \begin{cases} \dfrac{1}{a^2} & -b \leqslant x_0, y_0 \leqslant b \\ 0 & \text{otherwise} \end{cases}$	$\displaystyle\int_{-b}^b\int_{-b}^b (|x|+|y|)f_{xy}dxdy$ $= \dfrac{a}{2}$	$\displaystyle\int_{-b}^b\int_{-b}^b (|x|+|y|)^2 f_{xy}dxdy$ $-\left(\dfrac{a}{2}\right)^2 = 0.042a^2$

Note: For convenience, we use the notation $b \equiv a/2$.

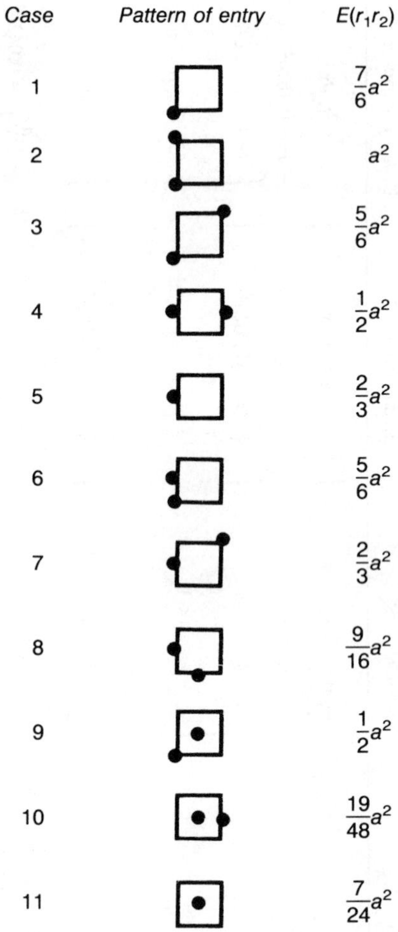

Case	Pattern of entry	$E(r_1r_2)$
1		$\frac{7}{6}a^2$
2		a^2
3		$\frac{5}{6}a^2$
4		$\frac{1}{2}a^2$
5		$\frac{2}{3}a^2$
6		$\frac{5}{6}a^2$
7		$\frac{2}{3}a^2$
8		$\frac{9}{16}a^2$
9		$\frac{1}{2}a^2$
10		$\frac{19}{48}a^2$
11		$\frac{7}{24}a^2$

Figure 8.A2. Computation of $E(r_1r_2)$ for alternate patterns of zone entry.

NOTES

1. Therefore equation (8.38) is not used for intrazonal travel. Recall that (8.38) is based on data for origin-destination trips. In this case, the unit is already moving when it enters the zone. If the speed in any zone had significant variation, including this variation would increase the computed variances and covariances of the intrazonal travel times based on the constant-speed model.

2. Note, again, that although this assumption is reasonable, it is not always valid. For one, the travel routes of the first and second units *to the zone boundary* (resulting in travel times v_{1i} and v_{2i}) may have common links that introduce some correlation

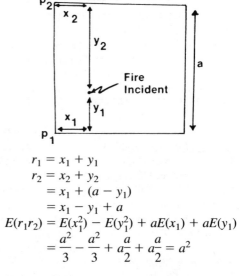

p_1 is the point of first unit entry

p_2 is the point of second unit entry

$$r_1 = x_1 + y_1$$
$$r_2 = x_2 + y_2$$
$$= x_1 + (a - y_1)$$
$$= x_1 - y_1 + a$$
$$E(r_1 r_2) = E(x_1^2) - E(y_1^2) + aE(x_1) + aE(y_1)$$
$$= \frac{a^2}{3} - \frac{a^2}{3} + a\frac{a}{2} + a\frac{a}{2} = a^2$$

Figure 8.A3. Computation of $E(r_1 r_2)$ for one case of zone entry.

between v_{1i} and v_{2i} (traffic delays on common links will affect both units' travel times in the same manner). Furthermore, exogenous events such as weather could contribute to additional correlation between v_{1i} and v_{2i}.

REFERENCES

Carter, G. M., E. J. Ignall, and W. E. Walker, 1975, *A Simulation Model of the New York City Fire Department: Its Use in Development Analysis,* P-5110-1, The New York City Rand Institute, New York.

Corman, H., E. J. Ignall, K. L. Rider, and K. A. Stevenson, 1976, Fire casualties and their relation to fire company response distance and demographic factors, *Fire Technology* **12**:193–203.

Daberkow, S. G. and G. A. King, 1977, Response time and the location of emergency facilities in rural areas: a case study, *American Journal of Agricultural Economics* **59**:467–476.

Dormont, P., J. Hausner, and W. Walker, 1975, *Firehouse Site Evaluation Model: Description and User's Manual,* R-1618/2, New York City Rand Institute, New York.

Fitzsimmons, J. A., 1973, Allocation of emergency ambulances to fire stations, *Fire Technology* **9**:112–118.

Gordon, G. and K. Zelin, 1970, A simulation study of emergency ambulance service in New York City, *New York Academy of Science Transactions* ser. II, **32**:414–427.

Halpern, J., G. Isherwood, and Y. Wand, 1979, Response time and fire property losses in single and double-family dwelling units, *INFOR* **17**:373–379.

Handler, G. Y. and P. B. Mirchandani, 1979, *Location on Networks: Theory and Algorithms,* MIT Press, Cambridge, Mass.

Hendrick, T. E., D. R. Plane, C. Tomasides, D. Monarchi, and F. W. Heiss, 1974, *Denver Fire Services Project Report.* Denver Urban Observatory, Denver.

Hirsch, W. Z., 1973, *Urban Economic Analysis,* McGraw-Hill, New York.

Hogg, J. M., 1971*a, A Model of Fire Spread,* Home Office Scientific Advisory Board, Great Britain.

Hogg, J. M., 1971*b,* The siting of fire stations, *Operational Research Quarterly* **19:**627–636.

Ignall, E., K. L. Rider, and R. Urbach, 1978, *Fire Severity and Response Distance: Initial Findings,* R-2013, New York City Rand Institute, New York.

Insurance Services Office, 1974, *Grading Schedule for Municipal Fire Protection,* Insurance Services Office, New York.

Keeney, R. L., 1973, A utility function for the response times of engines and ladders to fires, *Journal of Urban Analysis* **1:**209–222.

Keeney, R. L. and H. Raiffa, 1976, *Decisions with Multiple Objectives: Preference and Value Tradeoffs,* Wiley, New York.

Kolesar, P. 1975, A model for predicting average fire engine travel times, *Operations Research* **23:**603–613.

Larson, R. C., 1974, A hypercube queueing model for facility location and redistricting in urban emergency services, *Computers and Operations Research* **1:**67–95.

Larson, R. C., 1975, Approximating the performance of urban emergency service systems, *Operations Research* **23:**845–868.

Larson, R. C. and A. R. Odoni, 1981, *Urban Operations Research,* Prentice-Hall, New Jersey.

Nilsson, E., J. Swartz, and M. Westfall, 1974, *Application of Systems Analysis to the Operation of a Fire Department,* NBS Technical Note 782, Government Printing Office, Washington, D.C.

Public Technology, Inc., 1974, *Fire Station Location Package, Chief Executive's Report,* Washington, D.C.

Reilly, J., 1983, *Development of a Fire Station Placement Model with Consideration of Multiple Arriving Units,* Ph.D. diss., Rensselaer Polytechnic Institute, New York.

Rider, K. L., 1975, *A Parametric Model for the Allocation of Fire Companies,* R-1615, New York City Rand Institute, New York.

Savas, E. S., 1969, Simulation and cost-effectiveness analysis of New York's emergency ambulance service, *Management Science* **15:**B-608–627.

Schaenman, P., J. Hall, Jr., A. Schainblatt, J. Swartz, and M. Karter, 1977, *Procedures for Improving the Measurement of Local Fire Protection Effectiveness,* Urban Institute, Washington, D.C.

Schilling, D. A., C. ReVelle, J. Cohon, D. J. Elzinga, 1980, Some models for fire protection locational decisions, *European Journal of Operational Research* **5:**1–7.

Teitz, M. B. and P. Bart, 1968, Heuristic methods for estimating the generalized vertex median of a weighted graph, *Operations Research* **16:**955–961.

Toregas, C., R. Swain, C. ReVelle, and L. Bergman, 1971, The location of emergency service facilities, *Operations Research* **19:**1363–1373.

U. S. Department of Housing and Urban Development, 1977, *Deployment Methodology for Fire Departments,* Government Printing Office, Washington, D.C.

Volz, R. A., 1971, Optimum ambulance location in semi-rural areas, *Transportation Science* **5:**193–203.

Walker, W., J. Chaiken, and E. Ignall, 1979, *Fire Department Deployment Analysis,* North-Holland, New York.

9

LOCATION, DISPATCHING, AND ROUTING MODELS FOR EMERGENCY SERVICES WITH STOCHASTIC TRAVEL TIMES

Mark S. Daskin

Northwestern University

Providing emergency services has become one of the primary responsibilities of urban governments. Such services include emergency medical services and fire and police protection. In all three cases, performance is often rated along two dimensions: first, the speed with which the system can respond to emergencies (Keeney, 1973); and second, the ability of the responding personnel to handle the situation once they arrive on the scene. This chapter is concerned with response time and with means of reducing it, and not with techniques for improving the service quality once the emergency personnel are on the scene. Nevertheless, we recognize that the two performance dimensions are related, and often, improving the deployment of facilities primarily to reduce response time also results in ennhanced skill levels (Eaton et al., 1985).

The response time of emergency vehicles is composed of many elements (Fitzsimmons, 1973) including (1) the dispatching delay during which the caller describes the nature and location of the emergency to the dispatcher, (2) possible queueing delays during which the call must wait for an available unit to respond to the demand, and (3) the time required for the responding unit to travel to the emergency.

Dispatching delays depend on the nature of the service and the skill levels of the dispatchers. For example, the dispatching delay in multitiered emergency medical service systems (EMS), often exceeds that found in less advanced EMS services, since the dispatchers must ask the caller a series of questions to determine the most appropriate vehicle to dispatch. Basic life support vehicles are sent to non–life threatening incidents, while advanced life support vehicles and paramedics are reserved for life threatening events. Any deleterious effects associated with the larger dispatch delay are offset by the increased likelihood of having an advanced life support vehicle available when needed. Queueing delays depend on both the number of vehicles deployed throughout the system and on the dispatching policy employed, since the dispatching policy may delay low-priority calls in a queue when the system is congested and many vehicles are busy responding to demands.

This chapter focuses on reducing travel time delays, which depend on five factors:

1. The number of vehicles deployed.
2. The location of the vehicle bases.
3. The relocation policies employed during vehicle busy periods.
4. The dispatching policy that determines which vehicle will be sent to each emergency as a function of the number of busy vehicles, the locations of the available vehicles, and the nature of the emergency.
5. The routing policy that determines the paths used by vehicles traveling to the emergency scene.

As more vehicles are deployed, the mean travel time to an emergency decreases. With a fixed number of vehicles, the mean time can often be improved by selecting vehicle bases that are, on an average, closer to demands. When a call for emergency service is received, it is likely that some vehicles will be busy responding to existing demands. A vehicle relocation policy may be used during busy periods to reposition available vehicles to respond to future calls better. The dispatching policy is responsible for balancing the needs of the current call for service against the anticipated needs of calls that may come in the near future. Finally, the paths that emergency vehicles use in traveling to calls (Daskin and Haghani, 1982, 1984) as well as the routes used by vehicles being repositioned (Berman and Rahnama, 1983) have both been shown to affect system performance measures in general, and expected response times in particular.

Location, relocation, dispatching, and routing policies are all related, and all interact in determining the ability of the emergency system to respond quickly to demands. Most previous research in the analysis of emergency services has focused on individual components of the problem, without considering the interactions between various policies. Thus, for example, many location models ignore relocation options, take the dispatching policy as an input, and assume that vehicles travel along minimum paths.

This chapter formulates a multiobjective model that simultaneously determines the number of vehicles to deploy and their locations and identifies the appropriate dispatch policy (when all vehicles are available) and the routes vehicles should use in traveling to emergency locations. Two important components of the formulation are that it models the need to dispatch two vehicles to an emergency and it accounts for the effects of travel time variability on the system performance measures. The model does not explicitly consider three factors that are often important in designing emergency service systems: it does not model vehicle busy periods; it does not consider vehicle relocation policies; and it ignores the possibility of deploying a hierarchy of vehicle types. Nevertheless, the model formulation represents an attempt to integrate several of the important determinants of emergency system performance into a single optimization model. Incorporating additional factors is a task for future research. The formulation that results has a nonlinear objective function and linear constraints. The chapter also presents a simplified version of the model, outlines a heuristic solution procedure for the simplified model, and discusses preliminary results.

REVIEW OF RELATED LITERATURE

Much of the literature on emergency services relates to determining the number of vehicles to deploy and their locations. This chapter and most of the literature are concerned with locating vehicles on a network composed of nodes and links.

One of the earliest and simplest optimization models to be applied to this problem is the location set covering model initially formulated by Toregas et al. (1971). This model is formulated as

$$\text{Min} \sum_i X_i \tag{9.1}$$

subject to

$$\sum_{i \in N_k} X_i \geq 1 \quad \forall k \tag{9.2}$$

$$X_i = 0, 1 \quad \forall i \tag{9.3}$$

where

$X_i = 1$ if a vehicle is located at node i, 0 otherwise,
$N_k =$ set of vehicles sites i that can serve or cover demands at node k of the network.

The objective function (9.1) minimizes the number of vehicles deployed. Constraint (9.2) states that each demand node k must be served by at least one vehicle. A node is usually said to be served if it can be covered or reached by one of the vehicles in a user-specified travel time T. Constraint (9.3) states that the decision variables are integer.

This formulation stipulates that vehicles are to be located on the nodes of the network. However, Church and Meadows (1979) showed that restricting the set of available locations to the nodes of the network may result in suboptimal solutions both for this problem (i.e., more vehicles being located than are needed to cover all demands) and for the maximum covering location model discussed below. Church and Meadows showed that an optimal solution to these problems may be found by adding a finite number of "network intersect points" to the set of candidate vehicle locations.

The location set covering model has been used by a number of researchers. Berlin and Liebman (1971) used the model to identify EMS vehicle locations whose performance they then analyzed in greater detail using a simulation model. Jarvis, Stevenson, and Willemain (1975) used the model in determining EMS locations in rural areas. The manual solution technique they developed for the model is based on eliminating dominated locations and including nodes whose exclusion would result in uncovered demands. (Node A dominates node B if node A can cover all the demands covered by node B plus at least one additional demand location.) Walker (1974) used the set covering model to locate fire companies.

The set covering model suffer from a number of weaknesses. First, since many alternate optima are likely to exist, the model is not very discriminating. Several researchers have exploited this property in developing hierarchical objective set covering models in which the primary objective is to minimize the number of vehicles needed to cover all demands, and in which an additional term is added to the objective function (9.1) to account for the secondary objective. Plane and Hendrick (1977) formulated a hierarchical objective set covering model used in locating fire stations in Denver. Since the city already had an established fire department, the secondary objective in their model minimized the number of stations that needed to be moved to attain complete coverage. This secondary objective was a proxy for minimizing the cost of relocating existing fire stations. Daskin and Stern (1981) developed an alternate hierarchical objective set covering model to account approximately for a second weakness of the basic model, namely, its inability to account for vehicle busy periods. The secondary objective in this model maximizes a measure of back-up coverage, where the back-up coverage of node k is defined as the slack in constraint (9.2), that is, the number of times node j is covered in excess of the one required time.

Perhaps the greatest weakness of the location set covering model, however, are that it ignores differences in the demand levels associated with the nodes and it requires that each node be covered at least once. This results in a large number of vehicles being located on the network, with many of the vehicles being used to cover nodes with very small demands. Recognizing these weaknesses, Church and ReVelle (1974) formulated the maximum covering location model as follows:

$$\text{Max} \quad \sum_k D_k Y_k \qquad\qquad (9.4)$$

subject to

$$\sum_{i \in N_k} X_i - Y_k \geqslant 0 \qquad \forall k \qquad\qquad (9.5)$$

$$\sum_i X_i = L \qquad\qquad (9.6)$$

$$X_i = 0, 1 \qquad \forall i \qquad\qquad (9.7a)$$

$$Y_k = 0, 1 \qquad \forall k \qquad\qquad (9.7b)$$

where

$Y_k = 1$ if node k is covered, 0 otherwise,
$D_k =$ the expected demand at node k,
$L =$ the number of vehicles to be located

and all other variables are as defined for equations (9.1)–(9.3).

The objective function (9.4) maximizes the total number of covered demands. Constraint (9.5) states that node k is not covered unless a vehicle is located at one of the nodes that can cover node k. Constraint (9.6) stipulates that exactly L vehicles are to be located (on the nodes of the network). Finally, constraints (9.7a and b) are the integrality constraints. Eaton, Church and ReVelle (1977) discuss many of the properties of this model and review the greedy-adding and greedy-adding-with-substitution heuristics that are often used to solve this problem.

Eaton et al. (1979, 1980, 1985) used the maximum covering location model in determining EMS vehicle locations in Austin, Texas, and Bennett, Eaton and Church (1982) used it in locating rural health care workers in developing countries. Benedict (1983) and Hogan and ReVelle (1984) formulated hierarchical objective extensions of the maximum covering model that maximize a variety of measures of demand-weighted back-up coverage. These extensions are similar to the hierarchical objective set covering model proposed by Daskin and Stern (1981). Daskin (1982, 1983) formulated a maximum expected covering location model to account approximately for vehicle busy periods. The model incorporates a probability of finding a vehicle busy and maximizes the expected number of demands covered by nonbusy vehicles. The maximum covering model defines the relative benefit of coverage as being equal to 1 if a node is covered within a critical time T, and 0 if it is not covered. In many cases, the relative value of coverage depends on the response time. The relative value need not increase or decrease monotonically with the response time. Church and Roberts (1983) extended the maximum covering model to account for benefits that depend on the response time.

In general, the benefit of an emergency service system decreases with the response time. For example, the likelihood of surviving a life threatening medical emergency decreases with increases in the response time. Similarly, fire damage is likely to increase with response time. An alternate formulation has been employed—that of the median problem originally formulated by Hakimi (1964) as follows:

$$\text{Min} \quad \sum_k \sum_i D_k t_{ik} Z_{ik} \tag{9.8}$$

subject to

$$\sum_i Z_{ik} = 1 \qquad \forall k \tag{9.9}$$

$$Z_{ik} \leq Z_{ii} \qquad \forall i, k\ i \neq k \tag{9.10}$$

$$\sum_i Z_{ii} = L \tag{9.11}$$

$$Z_{ik} = 0, 1 \qquad \forall i, k \tag{9.12}$$

where

$Z_{ik} = 1$ if demands at node k are assigned to a vehicle located at node i, 0 otherwise,

$t_{ik} =$ response time for a vehicle located at node i in responding to demands at node k

and all other variables are as already defined.

The objective function (9.8) minimizes the total (or average) response time. Constraint (9.9) states that each demand node k is to be assigned to exactly one vehicle. Constraint (9.10) states that demands at node k cannot be assigned to a vehicle at node i unless demands at node i are also assigned to a vehicle at node i (meaning that a vehicle is located at node i). Constraint (9.11) requires exactly L vehicles to be located (on the nodes of the network). Constraint (9.12) states that the decision variables are integer variables. Note that the response time, t_{ik}, is taken along the minimum path from i to k and that the model will automatically assign demands to the vehicle with the smallest response time. Also, Hakimi (1964) showed that at least one optimal solution to this problem involves locating vehicles only on the nodes of the network. Thus, in this case, restricting attention to the nodes as candidate locations will not result in a suboptimal solution.

Berlin, ReVelle, and Elzinga (1976) adapted this model to EMS systems and incorporated into the formulation the travel time from the emergency scene to the hospital. Weaver and Church (1981), to account for vehicle busy periods in an approximate manner, extended the formulation to include the probability of demands being served by vehicles other than the closest one.

Whereas Weaver and Church (1981) and Daskin (1982, 1983) accounted for vehicle busy periods in an appropriate manner, Fitzsimmons (1973) used an $M/G/\infty$ queueing model. The mean response time in this model is computed conditional on the number of busy vehicles (the state of the system). Fitzsimmons embedded this model in a heuristic search procedure in an attempt to identify locations that minimize the systemwide mean response time. In fact, however, the response time depends not only on the number of busy vehicles, but on the identity of the busy (or available) vehicles. If the available vehicles are distributed randomly throughout the system, the expected response time to the next demand will be less than if all the available vehicles are clustered in one portion of the network. This observation motivated the development of the hypercube queueing model (Larson, 1974). In this model, the state of the system describes the identity of the available and busy vehicles and is denoted by an L-tuple of 0/1 elements. For a system with L vehicles, this results in 2^L linear equations. Larson (1975) also developed an approximate solution based on solving L nonlinear equations.

Jarvis (1975) employed the hypercube model in developing vehicle dispatching strategies and vehicle location plans. This work is significant because it is one of the few models that explicitly accounts for vehicle busy periods while simultaneously modeling location and dispatching strategies. The model allows for only single-vehicle responses to emergencies and assumes that vehicles travel along their minimum time paths.

Multivehicle dispatches have been examined by a number of authors. Chaiken (1971) developed a Markov model to estimate the number of busy vehicles at an emergency scene. Chelst and Barlach (1981) extended the hypercube model to account for demands that require two vehicles. They assumed that the service times of each of the two vehicles follow independent, identically distributed exponential distributions, and that link travel times are deterministic. They used the model to estimate a number of performance measures including the response times of the first and second arriving vehicles. Chelst and Jarvis (1979) proposed using the hypercube model to approximate the distribution of response times. The source of randomness in their model is the uncertainty regarding the identity of the available vehicle with the smallest response time. They too assume deterministic network travel times. In the following section, we outline a model that can be used to estimate the response time of the first arriving vehicle when two or more vehicles are dispatched to an emergency scene and link travel times are stochastic.

In all of the models outlined above, travel times on the network are assumed to be deterministic. Mirchandani and Odoni (1979) extended the median problem to account for stochastic travel times. Their model assumes travel times to be known when a demand for service arises; however, the state of the system (as described by the link travel times) changes over time according to a Markov process. They formulated this problem on an interger linear program and showed that at least one optimal solution to the problem involves locating only on the nodes of the network. This work is significant because it shows that the optimal locations obtained when the distribution of link travel times is considered differ from the locations that would be identified by a deterministic median problem using the mean link travel times. Weaver and Church (1983) discuss a lagrangian relaxation approach to solving this problem.

The discussion of models of emergency services would not be complete without mentioning the vehicle dispatching results obtained by Carter, Chaiken, and Ignall (1972). They considered the vehicle dispatching problem with fixed vehicle locations and showed that when service to anticipated future demands is considered, it may be suboptimal to dispatch the nearest available vehicle. In the following section we show that when link travel times are stochastic and multiple vehicle dispatches are required, dispatching the nearest available vehicles may be suboptimal even in terms of service to the current demand (without considering the effects of the current dispatch on the availability of vehicles to respond to future demands.)

A MULTIVEHICLE RESPONSE-TIME MODEL WITH STOCHASTIC TRAVEL TIMES

At the heart of the combined location, dispatching, and routing model discussed in the next section is a model of response times when link travel times are stochastic and multiple vehicles are dispatched to the emergency scene. With given vehicle paths to the emergency scene, the model determines the approximate distribution of the arrival time of the first arriving vehicle, a measure that is closely related to the perceived quality of the service (Keeney, 1973). The model is both descriptive and prescriptive, in that it is also used to suggest changes in vehicle routing and dispatching policies. This section reviews earlier work by the author in developing such a model (Daskin and Haghani, 1982, 1984) and suggests extensions that are important in formulating the combined model.

The model makes three key assumptions:

1. Travel times on nonoverlapping links are independent.
2. The variance of the link travel times is proportional to the mean link travel time. The proportionality constant, γ, will be called the unit variance. That is, the link variance will equal γ times the mean link travel time for all links.
3. All vehicles are dispatched simultaneously.

In addition, this chapter assumes that link travel times are normally distributed. Assuming normally distributed link times makes the analysis considerably more tractable at the risk of losing some realism because the normal distribution admits the possibility of negative link travel times. Tests of the normal assumption against a more realistic, although considerably less tractable. Erlang-k distribution indicate that the normal approximation is quite good in most cases (Daskin and Haghani, 1982).

Recognizing that the route or path that a vehicle takes in traveling from its location to an emergency scene will be composed of several links, the path travel times are also normally distributed under these assumptions. The mean path travel time will equal the sum of the mean link times for those links that comprise the path. Similarly the variance of the path travel time will equal the sum of the link variances, again for those links on the path. Alternatively, the path time variance will equal the mean path travel time multiplied by γ, the unit variance. Finally, the covariance of two path travel times will equal the sum of the variances on the links that the two paths share in common. The covariance will also equal the sum of the mean link travel times for the common links, multiplied by the unit variance. In short, path times follow a multivariate normal distribution with means, variances, and covariances that are readily computable from the means and variances of the link travel times.

Let the random variable T_i, denote the travel time for path i, and let μ_i, and σ_i^2 be the mean and variance of the travel time on path i. Let σ_{ij}^2 be the travel time covariance between paths i and j and let μ_{ij} be the mean time on the links shared by paths i and j. Finally, let Y be a random variable equal to the minimum of the path travel times. To find the distribution of y, we need to determine

$$\text{prob}(Y \leq y) = \text{prob}[\min_i(T_i) \leq y] \qquad (9.13)$$

Clark (1961) developed a recursive procedure for finding the approximate distribution of the maximum of a set of jointly distributed normal random variables. Adapting his formulae to find the distribution of the minimum of a set of random variables, we obtain

$$E[\min(T_1, T_2)] = \mu_2 - (\mu_2 - \mu_1)\Phi(\alpha) - a\,\phi(\alpha) \tag{9.14}$$

$$E\{[\min(T_1, T_2)]^2\} = \mu_2^2 + \gamma\mu_2 + (\mu_1^2 + \gamma\mu_1 - \mu_2^2$$
$$- \gamma\mu_2)\Phi(\alpha) - (\mu_1 + \mu_2)a\phi(\alpha) \tag{9.15}$$

$$\mathrm{var}[\min(T_1, T_2)] = E\{[\min(T_1, T_2)]^2\} - E^2[\min(T_1, T_2)] \tag{9.16}$$

$$\mathrm{cov}[\min(T_{i,}\, T_2), T_3] = \gamma\,[\mu_{23} + (\mu_{13} - \mu_{23})\,\Phi(\alpha)] \tag{9.17}$$

where

$$\alpha^2 = (\sigma_1^2 + \sigma_2^2 - 2\,\sigma_{12}^2) = \gamma(\mu_1 + \mu_2 - 2\mu_{12}),$$

$$\alpha = \frac{\mu_2 - \mu_1}{a},$$

$\phi(\alpha)$ = the standard normal distribution evaluated at α,

$\Phi(\alpha)$ = the standard cumulative normal distribution evaluated at α.

Equations (9.14)–(9.17) may be used recursively to find the distribution of Y. Note that if two paths are perfectly correlated, a^2 will equal 0 and α will be ill defined. Daskin and Haghani (1984) give special equations for this case.

As the mean travel time on the common links increases (holding the individual mean path times and the unit variance constant), the properties of the mean and variance of the minimum path time provide insight into desirable routing and dispatching policies. For example, consider the two alternate configurations shown in Fig. 9.1. In both cases, the mean path time for the vehicle located at node A is 21 units, and the travel time for the vehicle based at B is 20 units. However, in case (a) the mean time on the common links is 18 units, whereas in case (b) it is only 2 units. Thus, the covariance of the path times is considerably greater in case (a) than it is in case (b). Using equations (9.14)–(9.16) we find that for case (a) the expected value of the minimum of the two path times is 19.857, while the variance is 5.899; in case (b) we obtain 19.111 and 4.436 for the mean and variance of the minimum path time, resepctively. In general, we can show that if the mean path times and the unit variance are held constant, then the mean and variance of the minimum path time increase with the expected time in common between two paths (Daskin and Haghani, 1984).

Let us define the probability that a demand is covered in T_c time units to be the probability that the first vehicle will arrive within T_c time units. We can also show that if T_c is greater than or equal to the mean of the minimum

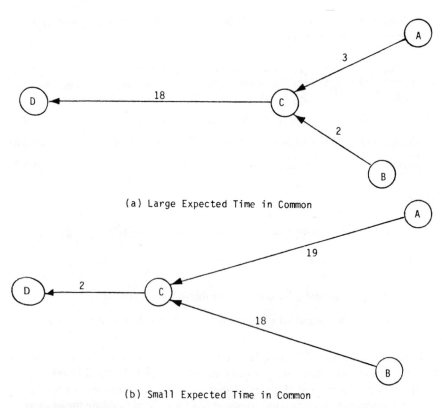

(a) Large Expected Time in Common

(b) Small Expected Time in Common

Figure 9.1. Two alternative routing schemes varying the expected time in common between the paths. A, B = vehicle locations; C = intermediate node; D = emergency location.

path time—the situation in which we are likely to be most interested—then the probability that a demand is covered decreases as the time in common between two paths increases. These properties are illustrated in Figs. 9.2 and 9.3, which plot the mean and variance of the minimum path time as a function of the expected time on the common links, holding $\mu_1 = 21$ and $\mu_2 = 20$ for three values of γ. Fig. 9.4 plots the probability that the demand is covered in 21 time units as a function of the expected time in common between two paths.

These figures suggest that it may be possible to increase the probability that a demand is covered by reducing the path time covariances, even if doing so requires increasing the mean path times slightly. This notion is formalized in the concept of an equivalent independent path. Consider Fig. 9.5. With

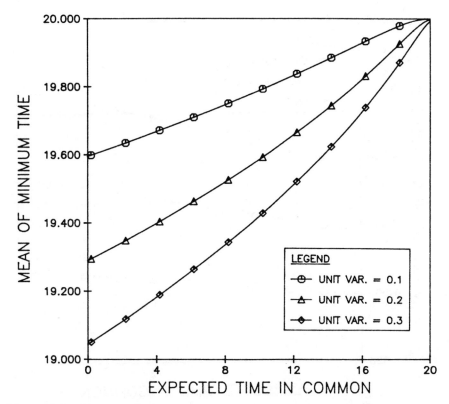

Figure 9.2. Mean of the minimum time versus the expected time in common between two paths.

vehicles again located at nodes A and B and the demand at node D, Fig. 9.5(a) illustrates an equivalent independent path that results from dispatching the more remote vehicle from an alternate node E; Fig. 9.5(b) illustrates an equivalent independent path for the more remote vehicle that does not share any links in common with the path traveled by the vehicle based at A. To find the allowable expected travel time on the equivalent independent path, we take the coverage probability that would result from routing the vehicles along the paths that share links in common and equate it with the coverage probability that results from routing the vehicles along independent paths.

Formally, if we have two paths, let μ_Y and σ_Y^2 be the mean and variance of the medium path time for the original (shared-link) routes. Define Z_e to be a random variable equal to the travel time on the equivalent independent path,

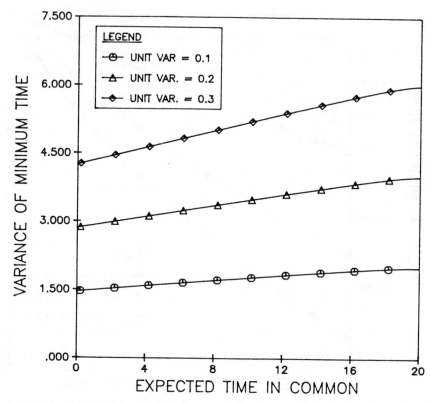

Figure 9.3.   Variance of the minimum time versus the expected time in common between the paths.

and let μ_e be the mean time on the equivalent independent path. To find μ_e, we solve the following equation

$$P_c = \text{prob(covered)} = \Phi\left(\frac{T_c - \mu_Y}{\sqrt{\gamma\mu_Y}}\right)$$

$$= 1 - [\text{prob}(T_2 \geq T_c)][\text{prob}(Z_e \geq T_c)]$$

$$= 1 - \left[1 - \Phi\left(\frac{T_c - \mu_2}{\sqrt{\gamma\mu_2}}\right)\right]\left[1 - \Phi\left(\frac{T_c - \mu_e}{\sqrt{\gamma\mu_e}}\right)\right] \tag{9.18}$$

The solution to (9.18) is given by

$$\mu_e = \frac{2T_c + \beta^2\gamma \mp \sqrt{(2T_c + \beta^2\gamma)^2 - 4T_c^2}}{2} \tag{9.19}$$

where

$$\beta = \Phi^{-1} \left[1 - \frac{1 - P_c}{1 - \Phi \left(\frac{T_c - \mu_2}{\sqrt{\gamma \mu_2}} \right)} \right]$$

and

$\Phi^{-1}(X)$ = inverse standard cumulative normal distribution evaluated at X.

The positive root of (9.19) is used if $\beta \leq 0$, and the negative root is used if $\beta \geq 0$.

Fig. 9.6 plots the mean equivalent independent path time as a function of

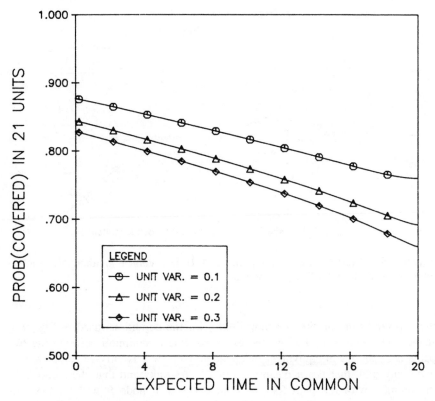

Figure 9.4.   Coverage probability versus expected time in common between the paths.

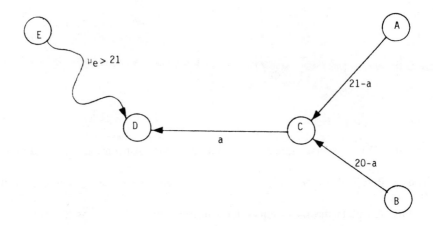

(a) Dispatching From a More Remote Location

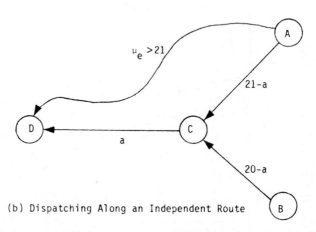

(b) Dispatching Along an Independent Route

Figure 9.5. Alternate independent paths. A, B, E = vehicle locations; C = intermediate node; D = emergency location.

the expected time on the common link(s) for the conditions used in Fig. 9.4. If the unit variance is 0.3 and the expected time in common between the two paths is 2, the mean equivalent independent path time is only 21.263. That is, using any path that has a mean time of 21.263 units and that shares no links in common with the path for the vehicle based at node *B* will increase the probability that demand is covered. If the expected time in common between the two original paths is 18, however, the mean equivalent independent path

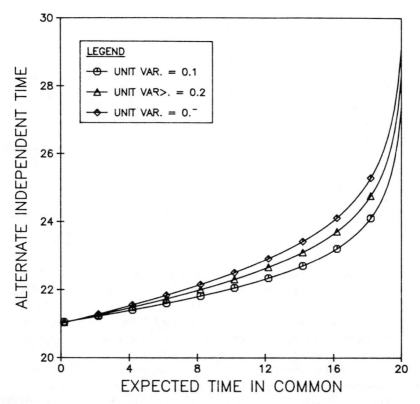

Figure 9.6. Mean time on the alternate independent path versus expected time on common links for the original routing.

time increases to 25.134. Now, any path that is independent of the path traveled by the vehicle based at node *B* and that has a mean time of less than 25.134 will increase the probability that demand is covered; and 25.134 represents a 19.7% increase over the original expected time of the vehicle based at *A*.

In deriving the mean equivalent independent path time, we implicitly assumed that the vehicle with the smaller mean travel time would be routed along its minimum path, and that either the other vehicle would be routed along an independent path, or an alternate vehicle that could travel along an independent path would be dispatched. However, it is not always optimal to dispatch either vehicle along its minimum expected path.

Consider the network shown in Fig. 9.7, where both vehicles are located at node *A* and the emergency is located at node *B*. Table 9.1 summarizes the three routing options of interest. If both vehicles are routed along the expected

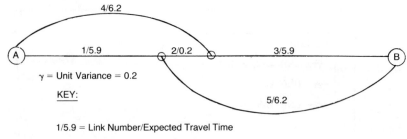

γ = Unit Variance = 0.2

KEY:

1/5.9 = Link Number/Expected Travel Time

All Links Directed From A Toward B.

A: Vehicle Location

B: Emergency Location

Figure 9.7. Example in which neither vehicle should use its minimum path.

minimum path, the expected value of the minimum path time is 12, and its variance is 2.4 when the unit variance equals 0.2. If the second vehicle is routed along one of the longer paths, with a mean travel time of 12.1 and an expected time in common of 5.9, the expected value of the minimum path time decreases to 11.423 and the variance is reduced to 2.019. Finally, if both vehicles are routed along the longer (but different) paths with mean travel times of 12.1 and an expected time along the common links of 0.0, the mean of the minimum path time is further reduced to 11.222 and the variance equals 1.65. For $T_c \leqslant 5.032$, dispatching both vehicles along the minimum expected paths (case 1) maximizes the coverage probability. For $5.032 \leqslant T_c \leqslant 9.329$, dispatching one vehicle along the minimum path and one along one of the longer paths (case 2) maximizes the coverage probability. Finally, if $T_c \geqslant 9.329$, dispatching along the two independent paths, both of which exceed the expected minimum path time of 12, will maximize the coverage probability. Thus, it may not be optimal to dispatch either vehicle along the expected minimum path if maximizing the coverage probability is desired. Note that if the objective is to minimize the expected arrival time of the first vehicle, it is never optimal to dispatch either vehicle along the minimum expected path for the example shown in Fig. 9.7.

Finally, the discussion above has aimed at determining the distribution and properties of the arrival time of the first vehicle to arrive at an emergency scene. Some cases may also call for the distribution of the arrival time of the last vehicle to arrive at an emergency location. The distribution of this time can be readily found by straightforwardly applying Clark's (1961) formulae for the distribution of the maximum of a set of jointly distributed normal random variables.

Table 9.1. Routing Options and Implications for Example Network

Case	Vehicle	Path (Links)	Mean Path Time	Mean Common Time	Arrival Time of First Vehicle		Maximizes Coverage Probability for
					Mean	Variance	
1	1	1-2-3	12.0	12.0	12.0	2.4	$T_c \leqslant 5.032$
	2	1-2-3	12.0				
2	1	1-2-3	12.0	5.9	11.423	2.019	$5.032 \leqslant T_c \leqslant 9.329$
	2	4-3 or 1-5	12.1				
3	1	4-3	12.1	0.0	11.222	1.650	$T_c \geqslant 9.329$
	2	1-5	12.1				

A COMBINED LOCATION, DISPATCHING, AND ROUTING MODEL FOR EMERGENCY SERVICES

As indicated in the earlier literature review, few models deal with the location of emergency services when multiple vehicles must be dispatched to the emergency scene. Also, almost all location models assume that the nearest vehicle is dispatched to the scene and that it is routed along the minimum path. In fact, the minimum path travel times are generally taken as model inputs in emergency vehicle location models. This approach is justified, in part, because these models assume that link travel times are deterministic.

However, link travel times are almost certainly stochastic. Many sources of variability exist, including local traffic conditions, the inability or unwillingness of other drivers to yield to the emergency vehicles, and such stochastic events as train crossings. Also, many emergency services, such as almost all fire-related dispatches, simultaneously send multiple vehicles to the emergency. Under these conditions, the multivehicle response-time model of the third section indicated that (1) it may not be optimal to dispatch the nearest available vehicles and (2) it may be best not to route the vehicles along their minimum paths. The impacts of the routing and dispatching findings on the optimal vehicle locations can only be determined through the formulation and solution of a simultaneous location, dispatching, and routing model. After a brief discussion of the appropriate location objectives, this section formulates such a model. The model allows either one or two vehicles to be dispatched to each emergency. Emergencies that require a single vehicle are referred to as single-response demands; situations requiring a two-vehicle response are referred to as double-response demands. Link travel times are treated as independent normally distributed random variables as in the preceding section.

Two principal objectives have been employed in locating emergency vehicles: the maximum covering location objective and the median or minimum average time objective. The formulation to follow incorporates four objectives:

1. Minimize the expected arrival time of the first vehicle for single-response demands.
2. Minimize the expected arrival time of the first vehicle for double-response demands.
3. Maximize the expected number of covered single-response demands.
4. Maximize the expected number of covered double-response demands.

In some emergency services, when multiple vehicles are required, there is a disbenefit associated with having the first vehicle arrive too soon before the other vehicles. For example, in early 1980, both a police patrol car and an EMS vehicle were dispatched to medical emergencies in certain parts of Chicago. The police were needed to protect the EMS attendants who felt threatened in these neighborhoods. In fact, the EMS attendants often refused to enter buildings in these areas of the city unless escorted by the police. Clearly, the attendants did not want to arrive on the scene much before the police escort, as they would be subject to considerable abuse from the residents for refusing to attend to the emergency until the police arrived. Consequently, this section formulates a fifth objective of minimizing the difference between arrival times of the first and second (last) vehicle. Such an objective might be appropriate in any situation in which effective service can not begin until all vehicles are at the scene.

The model formulation simultaneously determines vehicle locations, the assignment of demands to vehicles (the dispatching policy), the routes to be used by the vehicles for double-response demands, and the expected travel times for these routes. Note that under the assumptions regarding the distributions of link travel times, it is always optimal to route a vehicle responding to a single-response demand along the minimum path computed in terms of expected link travel times. Thus, for single-response demands, the expected travel time from any candidate vehicle location to the emergency scene will be an exogenous model input; for double-response demands, the expected travel times from the vehicle locations to the emergency scene are endogenously determined decision variables. The complete set of decision variables for the model includes the following:

Location Decision Variables

$X_i = 1$ if a vehicle is located at node i, 0 otherwise.

Dispatching Decision Variables

$Y_{ik} = 1$ if a vehicle at node i is dispatched to single-response demands at node k, 0 otherwise.

$Z_{ik} = 1$ if a vehicle at node i is dispatched to double-response demands at node k, 0 otherwise.

Routing Decision Variables

$W_{ijk} = 1$ if a vehicle located at i traveling to node k for a double-response demand uses link j, 0 otherwise.

$V_{jk} = 1$ if link j is used by both double-response routes in going to node k, 0 otherwise.

Travel Time Decision Variables

$R_{ik} =$ expected travel time from node i to node k for double-response demands,

U_k = expected travel time in common for the two paths for double-response demands at node k.

In addition to these decisions variables, a number of input constants and input functions need to be specified. They are defined in Tables 9.2 and 9.3, respectively.

We begin by defining the constraints on the optimization problem and then turn to the formulation of the various objectives. The entire location problem is subject to a budget constraint:

$$\sum_i C_i X_i \leq L \qquad (9.20)$$

Each single-response demand must be assigned to exactly one vehicle, which can be formulated as

$$\sum_i Y_{ik} = 1 \qquad \forall k \qquad (9.21)$$

Similarly, all double-response demands must be assigned to exactly two vehicles, which can be formulated as

$$\sum_i Z_{ik} = 2 \qquad \forall k \qquad (9.22)$$

Table 9.2. Definitions of Input Constants

D_k = total demand at node k,

α_k = fraction of demand at node k requiring a double response,

t_{ik} = expected minimum path travel time from i to k (used for single-response demands),

τ_j = expected travel time on link j,

$A(s)$ = set of links outbound from (after) node s,

$B(s)$ = set of links inbound to (before) node s,

T_1, T_2 = critical coverage times for single and double response demand respectively,

C_i = cost of locating at node i,

L = budget,

γ = unit variance,

n = number of nodes.

Table 9.3. Definitions of Input Functions

Double-Response Demands

$E^1_{hik} = E^1_{hik}(R_{hk}, R_{ik}, U_k, \gamma)$

= expected value of the *minimum* of the travel times for vehicles located at h and i traveling to node k with expected path travel times of R_{hk} and R_{ik}, respectively, and an expected time on the common links of U_k. This function is defined to equal 0 if either R_{hk} or R_{ik} is 0.

$E^2_{hik} = E^2_{hik}(R_{hk}, R_{ik}, U_k, \gamma)$

= expected value of the *maximum* of the travel times for vehicles located at h and i traveling to node k with expected path travel times of R_{hk} and R_{ik}, respectively, and an expected time on the common links of U_k. If either $R_{hk} = 0$ or $R_{ik} = 0$, this function will be defined to equal the nonzero expected time; if both $R_{hk} = R_{ik} = 0$, the function will equal 0.

$V^1_{hik} = V^1_{hik}(R_{hk}, R_{ik}, U_k, \gamma)$

= variance of the *minimum* of the travel times for vehicles located at h and i traveling to node k with expected path travel times of R_{hk} and R_{ik}, respectively, and an expected time on the common links of U_k. This function is also defined to equal 0 if either $R_{hk} = 0$ or $R_{ik} = 0$.

$P^1_{hik} = P^1_{hik}(E^1_{hik}, V^1_{hik}, T_2)$

= probability that vehicles located at h and i can cover demands at node k in time T_2 when the expected value and variance of the minimum of the path times are E^1_{hik} and V^1_{hik}, respectively. If $E^1_{hik} = 0$, this function is defined to equal 1.

Single-Response Demands

$P^1_{ik} = P^1_{ik}(t_{ik}, \gamma, T_1)$

= probability that a vehicle at node i can cover demands at node k in time T_1.

The assignment variables, Y_{ik} and Z_{ik}, must be further constrained since a demand at node k can only be assigned to a vehicle based at node i if the model locates a vehicle at node i. These constraints may be represented as

$$Y_{ik} \leq X_i \quad \forall i, k \tag{9.23}$$

$$Z_{ik} \leq X_i \quad \forall i, k \tag{9.24}$$

The routes used by vehicles responding to double-response demands must be determined endogenously. For demand nodes k that are assigned to a vehicle based at a node i other than k ($i \neq k$), we require the following at each node s of the network:

$$\sum_{j \in A(s)} W_{ijk} - \sum_{j \in B(s)} W_{ijk} = \begin{cases} Z_{ik} & \text{for } s = i \\ 0 & \text{for } s \neq i, s \neq k \\ -Z_{ik} & \text{for } s = k \end{cases}$$
$$\forall s; \forall i, k \ i \neq k \tag{9.25}$$

Constraint (9.25) states that for each node s of the network, the number of vehicles leaving node s that originate at node i and serve demands at node k minus the number of such vehicles entering node s must equal one of three values: 1 if node s is also node i, the vehicle location node; -1 if node s is node k, the emergency location; and 0 if node s is any other node in the network. If demands at a node are to be served by a vehicle at that node (i.e., if $Z_{ii} = 1$), then we assume that no links of the network are used and we require

$$W_{iji} = 0 \qquad \forall i, j \qquad (9.26)$$

The W_{ijk} variables indicate whether link j is used by vehicles going from i to k. In addition, we must identify those links j that are used by both vehicles assigned to double-response demands at node k. If link j is used by both vehicles dispatched to demands at k ($\sum_i W_{ijk} = 2$), then $V_{jk} = 1$; if not ($\sum_i W_{ijk} = 0$ or 1), then we require $V_{jk} = 0$. The following two constraints guarantee this condition:

$$V_{jk} \geq \sum_i W_{ijk} - 1 \qquad \forall j, k \qquad (9.27)$$

$$\sum_i W_{ijk} \geq 2V_{jk} \qquad \forall j, k \qquad (9.28)$$

Constraint (9.27) ensures that $V_{jk} = 1$ if link j is used by both vehicles, but it also allows V_{jk} to equal 1 if either one or none of the vehicles going to k use link j. Since V_{jk} is an integer variable, constraint (9.28) prevents V_{jk} from exceeding 0, except when link j is in fact used by both vehicles going to an emergency at node k (when $\sum_i W_{ijk} = 2$). This constraint is only needed when the objective of minimizing the expected difference between the response times of the first and second arriving vehicles is included in the model. Both of the other two objectives related to double-response demands—maximizing the total expected coverage and minimizing the expected response time of the first arriving vehicle averaged over all demands—are degraded as the expected travel time in common between the two paths increases, all else held constant, as shown in the preceding section. Thus, if only these objectives are employed, constraint (9.28) will not be needed.

Finally, the expected travel times for double-response demands are determined endogenously as well. The expected travel time for a vehicle located at node i to an emergency at node k is given by

$$R_{ik} = \sum_j \tau_j W_{ijk} \qquad \forall i, k \qquad (9.29)$$

The expected travel time on the common links for the two vehicles responding to double-response demands at node k is given by

$$U_k = \sum_j \tau_j V_{jk} \qquad \forall k \qquad\qquad (9.30)$$

Finally, we have the following integrality and nonnegativity constraints:

$$
\begin{aligned}
X_i &= 0, 1 & \forall i \\
Y_{ik} &= 0, 1 & \forall i, k \\
Z_{ik} &= 0, 1 & \forall i, k \\
V_{jk} &= 0, 1 & \forall j, k \\
W_{ijk} &= 0, 1 & \forall i, j, k \\
R_{ik} &\geq 0 & \forall i, k \\
U_k &\geq 0 & \forall k \qquad\qquad (9.31)
\end{aligned}
$$

The five objectives can also be formulated in terms of the decision variables, input constants, and input functions defined earlier. The two objectives related to single-response demands can be formulated as linear functions of the decision variable Y_{ik} as follows:

$$\text{Min} \sum_k \sum_i (1 - \alpha_k) D_k t_{ik} Y_{ik} \qquad\qquad (9.32)$$

$$\text{Max} \sum_k \sum_i (1 - \alpha_k) D_k P^1_{ik}(t_{ik}, \gamma, T_1) Y_{ik} \qquad\qquad (9.33)$$

Objective function (9.32) is a median objective that minimizes the total (or average) travel time to single-response demands. The term $(1 - \alpha_k) D_k$ represents the number of single response demands expected at node k. The travel time, t_{ik}, is the minimum path time from node i to node k. Since only one of the Y_{ik} terms will equal 1, by constraint (9.21), the inner summation represents the total travel time in responding to single-response demands at node k.

Objective function (9.33) maximizes the expected number of single-response demands covered in time T_1. While the probability term, $P^1_{ik}(t_{ik}, \gamma, T_1)$, is technically an input *function*, its arguments are all known inputs and its value may be computed a priori for each candidate location i and each demand node k. Thus, this objective is also linear in the decision variable Y_{ik}. The inner summation represents the expected number of covered single-response demands at node k.

Before formulating the three objectives related to double-response demands, it is useful to review the way in which the key double-response input functions are evaluated. The functions are E^1_{hik} and E^2_{hik}, the expected arrival times of the first and second vehicles, respectively, to arrive at node k when vehicles are dispatched from h and i; and P^1_{hik}, the probability that node k is covered by vehicles dispatched from h and i.

For all three objective functions, it will prove convenient to write the objective function in terms of a triple summation, with the inner two summations covering all candidate vehicle locations h and i. In doing so, however, we will inadvertently add extraneous terms that will have to be removed. This procedure is shown in Table 9.4, which summarizes the way in which each of the three functions is defined for a given emergency location k and all possible combinations of vehicle locations h and i. We designate the vehicle locations actually assigned to demands at node k by $\hat{h}$ and $\hat{\imath}$, respectively. That is,

$$Z_{\hat{h}k} = Z_{\hat{\imath}k} = 1$$

and

$$Z_{ik} = 0 \qquad \forall \, i \neq \hat{h} \text{ and } i \neq \hat{\imath}$$

Note that since the assignment variables, Z_{ik}, are 0 or 1 and $\Sigma_i Z_{ik} = 2$, the vehicle locations $\hat{h}$ and $\hat{\imath}$ cannot be the same for any demand node k. also, note that R_{ik} (or R_{hk}) will equal zero whenever either (1) a vehicle at node i (or h) is not assigned to demands at node k, or (2) node i (or h) is the same as node k.

The elements in the first row of Table 9.4 are the only ones we want to have remain in the objective function, and they may be selected in one of two ways. First, we could write each of the objectives in terms of the product of the appropriate input function and $Z_{ik}Z_{hk}$. Thus, for example, minimization of the demand-weighted expected response time of the first arriving vehicle could be written as

$$\text{Min} \sum_k \sum_{h \neq i} \sum_i \alpha_k D_k E^1_{hik}(R_{hk}, R_{ik}, U_k, \gamma)Z_{ik}Z_{hk} \qquad \textbf{(9.34)}$$

This formulation introduces significant nonlinearities, as it involves the product of two integer variables and a nonlinear function whose arguments are decision variables. The formulation has the advantage of being easy to understand because it implies that we should only evaluate the input functions for cases in which both vehicles are assigned to the demand. That is, for each demand node k, we need only evaluate E^1_{hik} for the single case in which $Z_{ik}Z_{hk} = 1$. Such a formulation may be useful in developing heuristics.

However, from an optimization perspective, the nonlinearities associated with (9.34) should be avoided. Therefore, we adopt a second approach that eliminates the need for the product $Z_{ik}Z_{hk}$. Specifically, we first sum the appropri-

Table 9.4. Summary of Evaluations of Objective-Function Elements for Double-Response Demands for a Particular Demand Node k

For		Number of Cases	E^1_{hik}	E^2_{hik}	P^1_{hik}
$h =$	$i =$		minimum to be used in obj. fn.	maximum to be used in obj. fn.	coverage probability
$\hat{h}$	$\hat{i}$	1	R_{ik}	R_{ik}	$P^1_{ik}(R_{\hat{i}k}, \gamma, T_2)$
$\hat{h}$	$\neq \hat{i}, \hat{h}$	$n-2$	0	R_{ik}	1
$\hat{h}$	$\hat{h}$	1	$R_{\hat{h}k}$	$R_{\hat{h}k}$	$P^1_{\hat{h}k}(R_{ik}, \gamma, T_2)$
$\hat{h}$	$\neq \hat{i}, \hat{h}$	$n-2$	$R_{\hat{h}k}$	$R_{\hat{h}k}$	1
$\neq \hat{i}, \hat{h}$	$\neq \hat{i}, \hat{h}$	$n^2 - 2n + 1$	0	0	1
Sum of all elements except first row		n^2	$R_{ik} + R_{\hat{h}k} = \sum_i R_{ik}$	$(n-1)(R_{ik} + R_{\hat{h}k}) = (n-1)\sum_i R_{ik}$	$\sum_i P^1_{ik}(R_{ik}, \gamma, T_2) + n^2 - n - 1$

ate function over all values of h and i. We then subtract those extraneous elements that need not appear in the objective function.

Using this approach, minimizing the demand-weighted expected response time of the first arriving vehicle may be written as

$$\text{Min} \sum_k \sum_h \sum_i \alpha_k D_k E^1_{hik}(R_{ik}, R_{hk}, U_k, \gamma) - \sum_k \sum_i \alpha_k D_k R_{ik} \quad (9.35)$$

Table 9.4 indicates that E^1_{hik} will equal 0 in all cases except when the following conditions hold: (1) $h = \hat{h}$ and $i = \hat{i}$, (2) $h = i = \hat{i}$, or (3) $h = i = \hat{h}$. Case (1) is the one we need to include in the objective function. Cases (2) and (3) are removed by the second term of (9.35), which will be nonzero only when $i = \hat{i}$ or $i = \hat{h}$.

Maximization of the expected number of covered double-response demands may be written as

$$\text{Max} \sum_k \sum_h \sum_i \alpha_k D_k P^1_{hik}(E^1_{hik}, V^1_{hik}, T_2)$$

$$- \sum_k \alpha_k D_k \left[\sum_i P^1_{ik}(R_{ik}, \gamma, T_2) + n^2 - n - 1 \right] \quad (9.36)$$

where E^1_{hik} and V^1_{hik} in the first term are functions of R_{hk}, R_{ik}, U_k, and γ, as indicated in Table 9.3. As shown in Table 9.4, P^1_{hik} is evaluated as 1 when either $R_{hk} = 0$ or $R_{ik} = 0$ (which occur when $h \neq \hat{h}$ or $i \neq \hat{i}$, respectively). When $h = i = \hat{h}$ (or $h = i = \hat{i}$), the expected response time of the first arriving vehicle is $R_{\hat{h}k}$ (or $R_{\hat{i}k}$) and the coverage probability equals $P^1_{\hat{h}k}(R_{\hat{h}k}, \gamma, T_2)$ [or $P^1_{\hat{i}k}(R_{\hat{i}k}, \gamma, T_2)$]. Thus,

$$\sum_i P^1_{ik}(R_{ik}, \gamma, T_2) = n - 2 + P^1_{\hat{i}k}(R_{\hat{i}k}, \gamma, T_2) + P^1_{\hat{h}k}(R_{\hat{h}k}, \gamma, T_2)$$

The summation over h and i in the first term of (9.36) includes an additional $n^2 - 2n + 1 + n - 2 = n^2 - n - 1$ terms equal to 1, which are subtracted in the second term of (9.36).

Finally, we can formulate the minimization of the demand-weighted difference between the expected response times of the first and second arriving vehicles as follows:

$$\text{Min} \sum_k \sum_h \sum_i \alpha_k D_k (E^2_{hik} - E^1_{hik}) - \sum_k \alpha_k D_k \left[\sum_i (n - 2) R_{ik} \right] \quad (9.37)$$

where E^2_{hik} and E^1_{hik} are functions of R_{hk}, R_{ik}, U_k, and γ. The justification for the second term of (9.37) is similar to that outlined for the second term of (9.35), and it follows from Table 9.4.

The complete formulation of the multiobjective model is summarized in Table 9.5. Each of the five objectives has been weighted by an objective-function weight, $\beta_m(m = 1, \ldots, 5)$.

SIMPLER FORMULATIONS

The multiobjective formulation presented in the preceding section jointly determines vehicle locations, the assignment of vehicles to demand nodes for both single- and double-response demands, and the optimal vehicle routes in the case of double-response demands. With n nodes and m one-way links, the model has $(n^2 + n)m + 2n^2 + n$ integer variables, $n^2 + n$ continuous variables, and $(n + 1)^3 + 2mn - n^2$ constraints excluding the definitional constraints (9.26) and (9.31). For a small network with only 21 nodes and 38 links, the formulation would have 18,459 integer variables, 462 continuous variables, and 11,803 constraints. A simpler, more compact formulation is clearly needed, as such a large nonlinear mixed-integer programming problem is not likely to be solvable with existing codes and algorithms.

Most of the constraints, (9.25)–(9.30), and decision variables, W_{ijk}, V_{jk}, R_{ik}, and U_k, are associated with finding the optimal routes for double-response demands. Elimination of these constraints and decision variables would remove $(n^2 + n)m$ integer variables, $n^2 + n$ continuous variables, and $n^3 + n + 2nm$ constraints. If we consider only the coverage objectives, (9.33) and (9.36), the optimal double-response demand routes for any pair of vehicle locations h and i and any demand location k will be those routes that maximize P_{hik}^1. In principle, these routes may be found exogenously, just as minimum path travel times are determined exogenously for the median problem. Similarly, if we only consider the median objectives, (9.32) and (9.35), the optimal routes for double-response demands may be determined exogenously for all combinations of vehicle and demand locations. This statement is also true if we only consider objective (9.37), minimization of the expected difference in response times for the first and second arriving vehicles. Only when we consider the objectives simultaneously must we include the routing-related decision variables and constraints; the routes that (1) maximize expected coverage, (2) minimize the expected response time of the first vehicle, and (3) minimize the expected difference between the first- and second-vehicle response times may differ, and endogenously determined trade-offs may need to be made.

Model Formulation

In this section we present a simpler formulation of the model previously outlined. The simplifications are based on the separation of the double-response objectives, thereby allowing the optimal double-response routes to be determined exoge-

Table 9.5. Formulation of the Combined Location, Dispatching, and Routing Model

$$\text{Min } \beta_1 \quad \sum_k \sum_i (1 - \alpha_k) D_k\, t_{ik} Y_{ik} \tag{9.32}$$

$$\text{Max } \beta_2 \quad \sum_k \sum_i (1 - \alpha_k) D_k\, P^1_{ik}(t_{ik},\, \gamma,\, T_1) Y_{ik} \tag{9.33}$$

$$\text{Min } \beta_3 \quad \sum_k \sum_h \sum_i \alpha_k D_k E^1_{hik} - \sum_k \sum_i \alpha_k D_k R_{ik} \tag{9.35}$$

$$\text{Max } \beta_4 \quad \sum_k \sum_h \sum_i \alpha_k D_k P^1_{hik}$$
$$- \sum_k \alpha_k D_k \left[\sum_i P^1_{ik}(R_{ik},\, \gamma,\, T_2) + n^2 - n - 1 \right] \tag{9.36}$$

$$\text{Min } \beta_5 \quad \sum_k \sum_h \sum_i \alpha_k D_k (E^2_{hik} - E^1_{hik})$$
$$- \sum_k \alpha_k D_k \left[\sum_i (n - 2) R_{ik} \right] \tag{9.37}$$

subject to

$$\sum_{i,} C_i X_i \qquad\qquad \leq L \tag{9.20}$$

$$\sum_i Y_{ik} \qquad\qquad = 1 \qquad \forall k \tag{9.21}$$

$$\sum_i Z_{ik} \qquad\qquad = 2 \qquad \forall k \tag{9.22}$$

$$-X_i + Y_{ik} \qquad\qquad \leq 0 \qquad \forall i,\, k \tag{9.23}$$

$$-X_i + Z_{ik} \qquad\qquad \leq 0 \qquad \forall i,\, k \tag{9.24}$$

$$\sum_{j \in A(s)} W_{ijk} - \sum_{j \in B(s)} W_{ijk} = \begin{cases} Z_{ik} & \text{for } s = i \\ 0 & \text{for } s \neq i,\ s \neq k \\ -Z_{ik} & \text{for } s = k \end{cases}$$
$$\forall s;\ \forall i,\, k;\ i \neq k \tag{9.25}$$

$$W_{iji} = 0 \qquad \forall i,\, j \tag{9.26}$$

$$\sum_i W_{ijk} - V_{jk} \leq 1 \qquad \forall j,\, k \tag{9.27}$$

$$-\left(\sum_i W_{ijk} \right) + 2V_{jk} \leq 0 \qquad \forall j,\, k \tag{9.28}$$

$$\sum_j \tau_j W_{ijk} - R_{ik} = 0 \qquad \forall i,\, k \tag{9.29}$$

$$\sum_j \tau_j V_{jk} - U_k = 0 \qquad \forall k \tag{9.30}$$

$$X_i = 0,\, 1 \qquad \forall i \tag{9.31}$$

$$Y_{ik} = 0,\, 1 \qquad \forall i,\, k \tag{9.31b}$$

$$Z_{ik} = 0,\, 1 \qquad \forall i,\, k \tag{9.31c}$$

$$V_{jk} = 0,\, 1 \qquad \forall j,\, k \tag{9.31d}$$

$$W_{ijk} = 0,\, 1 \qquad \forall i,\, j,\, k \tag{9.31e}$$

$$R_{ik} \geq 0 \qquad \forall i,\, k \tag{9.31f}$$

$$U_k \geq 0 \qquad \forall k \tag{9.31g}$$

nously. The model also assumes that a fixed number of vehicles are to be located on the nodes of the network. Thus, the budget constraint (9.20) is reformulated as

$$\sum_i Y_{ii} = L \qquad (9.38)$$

where we have assumed that $t_{ii} < t_{ji} \; \forall i, j, j \neq i$, and where L now equals the number of vehicles to be located rather than the budget. The single-response assignment constraint (9.21), requiring each demand node to be assigned to a single vehicle, remains unchanged:

$$\sum_i Y_{ik} = 1 \qquad \forall k \qquad (9.21)$$

We again stipulate equation (9.23)—that single-response demands at node k can only be assigned to a vehicle at i if a vehicle is located at node i:

$$Y_{ik} \leq Y_{ii} \qquad \forall i, k, i \neq k \qquad (9.23)$$

To formulate the double-response assignment constraints, we define a new decision variable, Z_{hik}, as follows:

$Z_{hik} = 1$ if double-response demands at node k are assigned to vehicles located at h and i, 0 otherwise.

Constraints (9.22) and (9.24) are replaced by

$$\sum_h \sum_i Z_{hik} = 1 \qquad \forall k \qquad (9.39)$$

$$Z_{hik} \leq Y_{hh} \qquad \forall h, i, k \qquad (9.40a)$$

$$Z_{hik} \leq Y_{ii} \qquad \forall h, i, k \qquad (9.40b)$$

Constraint (9.39) states that each demand node k will be assigned to only one *pair* of vehicles for double-response demands. Constraints (9.40a) and (9.40b) are analogous to constraint (9.23).

Subject to these constraints, we define two objective functions: (1) maximizing a weighted sum of the expected number of covered single- and double-response demands, and (2) minimizing a weighted sum of the expected total (first arriving) response times for a single- and double-response demands. These two objectives can be formulated as follows:

$$\text{Max } \beta_2 \ \sum_k \sum_i (1 - \alpha_k) D_k P^1_{ik} Y_{ik} \ +$$

$$\beta_4 \ \sum_k \sum_h \sum_i \alpha_k D_k P^1_{hik} Z_{hik} \tag{9.41}$$

$$\text{Min } \beta_1 \ \sum_k \sum_i (1 - \alpha_k) D_k t_{ik} Y_{ik} \ +$$

$$\beta_3 \ \sum_k \sum_h \sum_i \alpha_k D_k E^1_{hik} Z_{hik} \tag{9.42}$$

The single-response and double-response coverage probabilities (P^1_{ik} and P^1_{hik}, respectively) as well as the single- and double-response travel times for the first arriving vehicle (t_{ik} and E^1_{hik}, respectively) can be found exogenously. All other variables are as already defined. Thus, each objective is linear in the decision variables Y_{ik} and Z_{hik}. Table 9.6 summarizes the revised formulations.

The formulations involve $n^3 + n^2$ integer variables and $2n^3 + n^2 + n + 1$ constraints. For a problem with 21 nodes, we would have 9,702 variables and 18,985 constraints—still a very large problem. However, it is now linear in

Table 9.6. Revised Formulations with Exogenously Determined Routes

$$\text{Max } \beta_2 \ \left[\sum_k \sum_i (1 - \alpha_k) D_k P^1_{ik} Y_{ik} \right] \ +$$

$$\beta_4 \ \left[\sum_k \sum_h \sum_i \alpha_k D_k P^1_{hik} Z_{hik} \right] \tag{9.41}$$

$$\text{or Min } \beta_1 \ \left[\sum_k \sum_i (1 - \alpha_k) D_k t_{ik} Y_{ik} \right] \ +$$

$$\beta_3 \ \left[\sum_k \sum_h \sum_i \alpha_k D_k E^1_{hik} Z_{hik} \right] \tag{9.42}$$

subject to

$\sum_i Y_{ii}$	$= L$		(9.38)
$\sum_i Y_{ik}$	$= 1$	$\forall k$	(9.21)
$Y_{ik} - Y_{ii}$	≤ 0	$\forall i, k; i \neq k$	(9.23)
$\sum_h \sum_i Z_{hik}$	$= 1$	$\forall k$	(9.39)
$-Y_{hh} + Z_{hik}$	≤ 0	$\forall h, i, k$	(9.40a)
$- Y_{ii} + Z_{hik}$	≤ 0	$\forall h, i, k$	(9.40b)
Y_{ik}	$= 0, 1$	$\forall i, k$	(9,43a)
Z_{hik}	$= 0, 1$	$\forall h, i, k$	(9.43b)

both the constraints and the objective function. Furthermore, the number of constraints of the form of (9.40a) and (9.40b) can be reduced by reformulating:

$$\sum_k \sum_i Z_{hik} \leq nY_{hh} \qquad \forall h$$

$$\sum_k \sum_h Z_{hik} \leq nY_{ii} \qquad \forall i$$

where n is the number of nodes. Similarly, constraint (9.23) can be rewritten as

$$\sum_k Y_{ik} \leq nY_{ii} \qquad \forall i$$

Rewriting these constraints in this way will reduce the number of constraints to $5n + 1$. However, experience with similar problems suggests that the linear programming (LP) relaxation of the revised form is less likely to result in an all-integer solution; so the reduction in the number of constraints is likely to come at the expense of greater effort in a branch-and-bound procedure needed to force an integer solution.

Both formulations may be viewed as extensions of the L-median problem. [We use the term *L-median*, rather than *p-median*, since we are locating L vehicles as indicated by constraint (9.38).] Consider the model with objective function (9.42). The first term of (9.42) is simply the standard L-median objective. This term, together with constraints (9.38), (9.21), (9.23), and (9.43a) constitute an L-median formulation. The second term of (9.42) and constraints (9.39), (9.40a), (9.40b), and (9.43b) extend the L-median model to account for double-response demands. Similarly, if objective function (9.41) is rewritten in terms of minimizing the expected number of *uncovered* demands, it too may be viewed as an extension of the L-median formulation.

Model Solution

To solve the model we have formulated, we must compute the single- and double-response coverage probabilities [in the case of objective (9.41)] or the expected response times of the first arriving vehicle [in the case of objective (9.42)] must be computed. The single-response probabilities and times are simple to evaluate, since it will always be optimal to dispatch a vehicle to a single-response demand along the minimum expected time path. However, as noted, for double-response demands it may not be optimal to route the vehicles along the minimum paths.

In principle, we would need to evaluate all possible combinations of paths for the two vehicles for each triplet of two vehicle locations and one emergency location. Enumerating all such paths would be a prohibitive exercise. Therefore, we constrain the vehicle with the smaller minimum expected travel time to travel along its minimum expected time path, and we consider only alternate paths for the vehicle with the larger minimum path expected travel time. With this simplification, there will be a trade-off between (1) reducing the covariance (or expected time in common) between the two path times and (2) increasing the expected travel time of the second, more remote, vehicle. The points on this trade-off curve must be identified for each pair of vehicle locations and each emergency location.

These trade-off curves were identified by using the following algorithm for each combination of vehicle and emergency locations:

1. Find the minimum expected time paths for each vehicle.
2. Find the common links for the minimum time paths, and record the expected travel time of the more remote vehicle and its expected time in common with the nearer vehicle.
3. If the expected common time is 0, stop.
4. "Eliminate" the common links from the network and find the new minimum path for the more remote vehicle.
5. If no path exists (i.e., the more remote vehicle and the emergency are in separate subnetworks) go to step 7; otherwise, record the expected travel time of the more remote vehicle and the time in common (0) with the path of the nearer vehicle.
6. If the expected times for the more remote vehicle found in steps 1 and 5 are identical, stop; all paths on the trade-off curve have been identified. Otherwise, go to step 7.
7. "Restore" all links to the network.
8. For each possible subset of the common links:
   a. "Remove" the links in the subset from the network.
   b. Find the new minimum path for the more remote vehicle.
   c. If a path is found in step 8a and if the resulting expected travel time and time in common are not dominated by an existing solution, record the expected travel time and the time in common and remove any solutions that are dominated by this solution from the emerging trade-off curve.

Step 8 warrants additional explanation. Let $E(n)$ and $C(n)$ be the expected time for the more remote vehicle and the expected time in common for the two paths in the nth solution. If $E(n) \geq E(m)$ and $C(n) \geq C(m)$, then solution m dominates solution n, and solution n is discarded. (If strict equality holds for both inequalities, the solutions correspond to the same point on the trade-off curve and only one need be retained.) Step 8 may be expedited by using an implicit enumeration scheme rather than explicitly enumerating solutions for all subsets of the set of common links. Such a scheme was adopted in the solution procedure for the example problem discussed in the next section.

Once the nondominated points on the trade-off curve are determined for each combination of vehicle and emergency locations, the model of multivehicle response times with stochastic travel times can be used to find the point on each curve that maximizes the coverage probability or minimizes the expected response time of the first arriving vehicle. These values will be P_{hik}^1 and E_{hik}^1, respectively.

Once the values of P_{hik}^1 and E_{hik}^1 are known, the model can be solved in a variety of ways, including using branch-and-bound techniques on the LP relaxation. Recognizing that the model includes an embedded L-median formulation, it may be possible to adapt one of the optimal techniques commonly used to solve such problems. However, owing to the large size of the problem for even small networks, the model was solved heuristically using a vertex substitution algorithm similar to that used by Tietz and Bart (1968) in solving the L-median problem and by Eaton, Church, and ReVelle (1977) in solving the maximum covering problem.

Finally, note that the model formulation shown in Table 9.6 does not explicitly specify the routing rule used to determine the coverage probabilities, P_{hik}^1, and the expected response time of the first arriving vehicle, E_{hik}^1. In particular, it is not necessary that the double-response vehicles be routed along paths that optimize these quantities. For example, we can compute both the optimal coverage probabilities and those that result from routing both vehicles along their minimum expected time paths. By solving the model of Table 9.6 with these two values, we can determine the extent to which the optimal routes increase the expected coverage of double-response demands, and we can also find out whether or not the optimal vehicle locations are sensitive to the routes used by the vehicles in traveling to the emergency scene. This approach was employed for the test results to be discussed next.

Sample Model Results

The model and the solution procedure were tested using the 21-node network shown in Fig. 9.8. The network was derived from the 25-node problem used by Berman and Rahnama (1983) by (1) eliminating nodes 22 through 25, which form a tail, and (2) renormalizing the node weights so that the remaining 21 weights sum to 100.

We computed two sets of double-response coverage probabilities and expected response times for the first arriving vehicle. The first set assumed that both vehicles were routed along their minimum expected travel time paths. In fact, in cases in which alternate minimum paths were available, the model tended

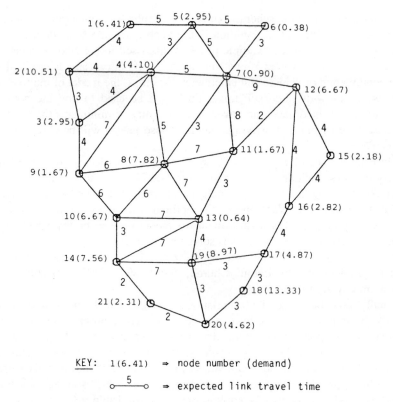

KEY: 1(6.41)   ⇒ node number (demand)

o——5——o   ⇒ expected link travel time

Figure 9.8.   21-node network example.

to select the path that *maximized* the expected travel time in common between the two paths.

The second set of coverage probabilities and expected response times were based on constraining the nearer of the two vehicles to travel along its minimum expected path. The more remote vehicle was routed along the path that corresponded to the optimal point on the trade-off curve of increased expected travel time for the more remote vehicle and reduced path time covariance. Such a trade-off curve is shown in Table 9.7 for the case of vehicles located at nodes 1 and 2 and an emergency at node 14. The unit variance, γ, is set equal to 0.3, and the critical coverage time, T_2, is set to 18. Vehicle 2 is the nearer vehicle, with an expected travel time of 16. Three alternate paths—with expected travel times of 20, 22, and 27—are identified for the vehicle based at node 1. The second path significantly reduces the expected travel time in common with the vehicle based at node 2. This path maximizes the coverage probability and minimizes the expected response time of the first arriving vehicle.

Table 9.7. Example Trade-Off Curve

Expected Second- Best Time	Expected Time in Common	Second- Best Path	Coverage Probability	Expected Response Time
20	16	1-2-3-9-10-14	0.819	16.000
22	3	1-2-4-8-10-14	0.827	15.969
27	0	1-2-4-8-13-14	0.820	15.999

Notes: 1. Vehicles located at nodes 1 and 2; emergency located at node 14.
2. Unit variance equal to 0.3; critical coverage time equal to 18.
3. Vehicle based at node 2 is closer, with an expected travel time of 16 using a path of 2-3-9-10-14.

In all cases, the weights, β_m, on the objective function were set equal to 1, and all values of α_k were equal. Under these assumptions, the objective functions (9.41 and 9.42) can be rewritten as

$$\text{Max } (1 - \alpha)\sum_k \sum_i D_k P^1_{ik} Y_{ik} + \alpha \sum_k \sum_h \sum_i D_k P^1_{hik} Z_{hik} \tag{9.41'}$$

$$\text{Min } (1 - \alpha)\sum_k \sum_i D_k t_{ik} Y_{ik} + \alpha \sum_k \sum_h \sum_i D_k E^1_{hik} Z_{hik} \tag{9.42'}$$

In each case, the first term represents the value of the objective function for single-response demands multiplied by $1 - \alpha$, and the second term is the value of the objective function for double-response demands multiplied by α.

For each of the two objective functions and each routing scheme, the vertex substitution algorithm was solved for 11 different values of α from 0.0 to 1.0, in increments of 0.1. Table 9.8 presents the results of minimizing the expected response time of the first arriving vehicle using the two different routing schemes for the more remote vehicle. As the value of α increases, the weight placed on serving double-response demands increases. The table indicates that the locations of the vehicles are insensitive to the value of α. Also, the expected value of the response time of the first arriving vehicle for double-response demands is only slightly smaller than the single-response-time value using either the minimum expected path for the more remote vehicle or the path that optimizes the objective function. This is so because the second vehicle tends to be considerably farther from the emergency than the first vehicle, and so the minimum expected time depends heavily on the expected response time of the nearer vehicle.

Finally, the table indicates that there is little difference in this case between using the expected minimum travel time or the path that optimizes the objective function. This condition results from three factors. First, in most cases the expected minimum response time for double-response demands is determined

**Table 9.8 Minimizing the Expected Response Time
of the First Arriving Vehicle**

Number of Vehicles	αValue		Locations	Single-Response Time	Double-Response Time
	Low	High			
(a) More remote vehicle using minimum expected path					
1	0.0	1.0	10	951.282	undefined
2	0.0	1.0	2, 20	503.846	503.124
3	0.0	1.0	2, 12, 20	370.513	360.765
4	0.0	1.0	2, 8, 12, 20	284.103	277.954
(b) More remote vehicle using "best" path					
1	0.0	1.0	10	951.282	undefined
2	0.0	1.0	2, 20	503.846	503.060
3	0.0	1.0	2, 12, 20	370.513	360.765
4	0.0	1.0	2, 8, 12, 20	284.103	277.465

Note: Times given are the demand-weighted expected value of the response time of the first arriving vehicle.

largely by the expected response time of the nearer vehicle. Second, minimum paths of the two vehicles generally have very few links in common. Had the model allowed multiple vehicles to be stationed at a single vehicle base (e.g., several fire engines housed at a single station), there would probably have been greater overlap in the paths selected and a greater difference between the results of using the minimum expected paths and those that optimize the objective function. Third, it is important to recognize that neither case constrains the model to selecting the pair of vehicles that are "closest" to the emergency scene. Rather, the model selects the pair of vehicles that, under the given routing scheme, can best serve each demand node. However, in almost all cases, this feature results in selecting the two vehicles with the smallest expected response times.

Table 9.9 presents the results of maximizing the expected number of covered demands. Both critical coverage times, T_1 and T_2, were set to 9. The qualitative conclusions are similar to those discussed above for minimizing the expected response tme of the first arriving vehicle. However, two points are worth noting. First, the locations that maximize the expected number of covered demands differ from the locations that minimize the expected response time of the first arriving vehicle. Second, when locating four vehicles, the expected number of covered demands using the locations found for the routes that maximize coverage [part (b) of the table] is *less* than the number of covered demands using the locations found for the case in which both vehicles use the expected minimum paths [part (a)]. This finding is purely a function of the heuristic nature of the

Table 9.9. Maximizing the Expected Number of Covered Demands

Number of Vehicles	α Value		Locations	Single-Response Coverage	Double-Response Coverage
	Low	High			
(a) More remote vehicle using minimum expected path					
1	0.0	1.0	13	53.333	0.0
2	0.0	1.0	4, 19	86.909	87.654
3	0.0	0.758	2, 13, 17	96.203	97.356
	0.758	1.0	2, 13, 18	96.098	97.390
4	0.0	0.825	2, 8, 16, 21	99.636	99.819
	0.825	1.0	2, 8, 16, 20	99.568	99.834
(b) More remote vehicle using "best" path					
1	0.0	1.0	13	53.333	0.0
2	0.0	1.0	4, 19	86.909	87.676
3	0.0	1.0	2, 13, 17	96.203	97.423
4	0.0	1.0	2, 10, 11, 17	98.980	99.290

Note: In some cases, the vertex substitution heuristic failed to find the solution indicated for the given value of α. This failure results from the inability of the heuristic to consider multiple exchanges simultaneously. However, the solution indicated was found for a different value of α and the indicated solution dominates the solution found by the heuristic. For example, for 4 vehicles with α = 0.9 and the more remote vehicle using the "best" path, the heuristic identified locations 2, 8, 13, 17, with a combined α-weighted objective function of 98.845. However, with α = 0.8, the heuristic found the indicated solution, which has a weighted objective function of 99.259 at α = 0.9.

vertex substitution algorithm and does not reflect on the routes chosen. If the vehicles based at the locations identified in part (a) of Table 9.9 used the routes that maximized coverage, the expected coverage would exceed that shown in the table.

CONCLUSIONS AND RECOMMENDATION FOR FUTURE STUDY

This chapter has reviewed a multivehicle routing model for emergency service with stochastic travel times (Daskin and Haghani, 1982, 1984). The model can be used to estimate the mean and variance of the travel time of the first vehicle to arrive at an emergency scene, accounting for possible covariances between vehicle paths. These values can then be used to estimate the probability that a demand can be covered by a pair of vehicles within a specified time.

The model suggests that the performance of the service—as measured either by the expected travel time of the first arriving vehicle or the coverage probability—may be enhanced by reducing the path time covariance between vehicle routes, even if doing so necessitates slightly increasing the expected travel time of one or more vehicles. The implications of this result on vehicle routing and dispatching strategies were discussed and the notion of alternate independent paths was reviewed.

A multiobjective location, dispatching, and routing model was formulated for the case in which two vehicles are dispatched to some demands and link travel times are stochastic. The model has linear constraints and a nonlinear objective function for the double-response demands. The formulation is not likely to be solvable with existing algorithms due to the nonlinear nature of the objectives and the large number of integer decision variables. By isolating specific objectives, we showed that we could remove the routing-related variables and constraints and thereby reduce the size of the problem. The model was shown to be an extension of the L-median problem.

The model was solved heuristically using a vertex substitution algorithm, and it was tested on a small 21-node network. The results tentatively suggest that

1. Vehicle locations are insensitive to the fraction of demand requiring a two-vehicle response.
2. The improvement in either objective function that results from using the routes that account for path time covariances is very small.

These results must be viewed with caution for several reasons. First, the model does not account for the queueing aspects of emergency services. Accounting for queueing phenomena would tend to enhance any apparent gains realized by using the routes that account for path time covariances. The vehicle routing model suggests it may be desirable to dispatch a more remote vehicle if doing so sufficiently reduces the path time covariance with nearer vehicles dispatched to the scene. In short, the model suggests dispatching a more dispersed set of vehicles than would be used if only the nearest vehicles were dispatched. This policy, in turn, would result in a more dispersed set of *available* vehicles for subsequent calls and smaller queueing-related delays. Incorporation of queueing-related phenomena is a major direction for future study (Berman, Larson, and Chiu, 1985; Chiu, Berman, and Larson, 1985).

Second, the models presented in this paper do not distinguish between vehicles and stations. In many cases, multiple vehicles are housed at a single station so that the fixed costs of station construction can be spread over more vehicles. Although the economics of station construction suggests the need for fewer stations, the routing and dispatching model indicates the desirability of more stations more widely dispersed. A trade-off between the economics and performance of the emergency service is likely to arise. Incorporating station location as well as vehicle location decisions into the formulation is a second important extension.

Third, the model is solved only heuristically. In light of the apparently small differences between the two routing schemes discussed in the preceding section, it would be desirable to solve the model optimally to assess the true impact of using the covariance-based routes rather than routing all vehicles on their minimum expected time paths.

Finally, the multivehicle routing model assumes independent travel times on nonoverlapping links of the network. Means should be developed to relax this assumption to account for stochastic events that might affect a region of the network.

REFERENCES

Benedict, J. M., 1983, *Three Hierarchical Objective Models Which Incorporate the Concept of Excess Coverage to Locate EMS Vehicles or Hospitals,* M.S. thesis, Department of Civil Engineering, Northwestern University, Evanston, Ill.

Bennett, V. L., D. J. Eaton, and R. L. Church, 1982, Selecting sites for rural health workers, *Social Science and Medicine* **16:**63–72.

Berlin, G. N. and J. C. Liebman, 1971, Mathematical analysis of emergency ambulance location, *Socio-Economic Planning Sciences* **8:**323–328.

Berlin, G. N., C. S. ReVelle, and D. J. Elzinga, 1976, Determining ambulance-hospital locations for on-scene and hospital services, *Environment and Planning, A* **8:**553–561.

Berman, O., R. C. Larson, and S. S. Chiu, 1985, Optimal server location on a network operating as an M/G/1 queue, *Operations Research* **33:**746–771.

Berman, O. and M. R. Rahnama, 1983, Optimal path of a single service unit on a network to a "nonemergency" destination, *Transportation Science* **17:**218–232.

Carter, G. M., J. M. Chaiken, and E. Ignall, 1972, Response area for two emergency units, *Operations Research* **20:**571–594.

Chaiken, J., 1971, *The Number of Emergency Units Busy at Alarms Which Require Multiple Servers,* New York City Rand Institute, Report Number R-531-NYC/HUD, New York.

Chelst, K. R. and Z. Barlach, 1981, Multiple unit dispatches in emergency services: models to estimate performance, *Management Science* **27:**1390–1409.

Chelst, K. R. and J. P. Jarvis, 1979, Estimating the probability distribution of travel times for urban emergency service systems, *Operations Research* **27:**199–204.

Chiu, S. S., O. Berman, and R. C. Larson, 1985, Locating a mobile server queueing facility on a tree network, *Management Science* **31:**764–772.

Church, R. and M. Meadows, 1979, Location modelling utilizing maximum service distance criteria, *Geographical Analysis* **11:**358–373.

Church, R. and K. L. Roberts, 1983, Generalized coverage models and public facility location, *Regional Science Association Papers* **53:**117–135.

Church, R. and C. ReVelle, 1974, The maximal covering location problem, *Regional Science Association Papers* **32:**101–118.

Clark, C. E., 1961, The greatest of a finite set of random variables, *Operations Research* **9:**145–162.

Daskin, M. S., 1982, Application of an expected covering model to EMS system design, *Decision Sciences* **13**:416–439.

Daskin, M. S., 1983, A maximum expected covering location model: formulation, properties and heuristic solution, *Transportation Science* **17**:48–70.

Daskin, M. S. and A. Haghani, 1982, *Routing and Dispatching Multiple Vehicles to an Emergency Scene,* Transportation Center Report TP-DAS-82.04, Northwestern University, Evanston, Ill.

Daskin, M. S., and A. Haghani, 1984, Multiple vehicle routing and dispatching to an emergency scene, *Environment and Planning, A* **16**:1349–1359.

Daskin, M. S. and E. H. Stern, 1981, A hierarchical objective set covering model for EMS vehicle deployment, *Transportation Science* **15**:137–152.

Eaton, D., R. Church, and C. ReVelle, 1977, *Location Analysis: A New Tool for Health Planners,* Methodological Working Document No. 53, Sector Analysis Division, Bureau of Latin America, Agency for International Development.

Eaton, D. J., et al., 1979, *Location Techniques for Emergency Medical Service Vehicles,* vol. 1–4, Policy Research Report 34, Lyndon B. Johnson School of Public Affairs, University of Texas, Austin.

Eaton, D. J., et al., 1980, *Analysis of Emergency Medical Services in Austin, Texas,* vol. 1–2, Policy Research Report 41, Lyndon B. Johnson School of Public Affairs, University of Texas, Austin.

Eaton, D. J., M. S. Daskin, B. Bulloch, and G. Jansma, 1985, Determining emergency medical service vehicle deployment in Austin, Texas, *Interfaces* **15**:96–108.

Fitzsimmons, J. A., 1973, A methodology for emergency ambulance deployment, *Management Science* **19**:627–636.

Hakimi, S. L., 1964, Optimum locations of switching centers and the absolute centers and medians of a graph, *Operations Research* **12**:450–459.

Hogan, K. and C. ReVelle, 1984, *Concepts and Applications of Backup Coverage,* Operations Research Group Report Series Number 84–05, Johns Hopkins Unviersity, Baltimore.

Jarvis, J. P., 1975, *Optimization in Stochastic Service Systems with Distinguishable Servers,* Technical Report 19–75, Operations Research Center, Massachusetts Institute of Technology, Cambridge, Mass.

Jarvis, J. P., K. A. Stevenson, and T. R. Willemain, 1975, *A Simple Procedure for the Allocation of Ambulances in Semi-Rural Areas,* Technical Report 13–75, Operations Research Center, Massachusetts Institute of Technology, Cambridge, Mass.

Keeney, R. L., 1973, A utility function for the response times of engines and ladders to fires, *Urban Analysis* **1**:209–222.

Larson, R. C., 1974, A hypercube queueing model for facility location and redistricting in urban emergency services, *Computers and Operations Research* **1**:67–95.

Larson, R. C., 1975, Approximating the performance of urban emergency service systems, *Operations Research* **23**:845–868.

Mirchandani, P. B. and A. R. Odoni, 1979, Locations of medians on stochastic networks, *Transportation Science* **13**:85–97.

Plane, D. R. and T. E. Hendrick, 1977, Mathematical programming and the location of fire companies for the Denver fire department, *Operations Research* **25**:563–578.

Teitz, M. B. and P. Bart, 1968, Heuristic methods for estimating the generalized vertex median of a weighted graph, *Operations Research* **16**:955–961.

Toregas, C., R. Swain, C. ReVelle, and L. Bergmann, 1971, The location of emergency service facilities, *Operations Research* **19**:1363–1373.

Walker, W. E., 1974, Using the set-covering problem to assign fire companies to fire stations, *Operations Research* **22**:275–277.

Weaver, J. R. and R. L. Church, 1981, Average response time and workload balance: two criteria for ambulance station location, in *Systems Science in Health Care*, C. Tilquin, ed., Pergamon, Montreal, pp. 975–983.

Weaver, J. R. and R. L. Church, 1983, Computational procedures for location problems on stochastic networks, *Transportation Science* **17**:168–180.

Part III

METHODOLOGICAL ISSUES IN LOCATION-ALLOCATION MODELING

10

INTERPERIOD NETWORK STORAGE LOCATION-ALLOCATION (INSLA) MODELS

**Samuel J. Ratick, Jeffrey P. Osleeb,
Michael Kuby, and Keumsook Lee**

Boston University

Goods-oriented location problems have traditionally been formulated as mathematical programs in a static setting, with the implicit assumption that none of the parameters varies significantly over time. However, for many commodity distribution problems, supply, demand, costs, capacities, and even network characteristics can change over time. This chapter focuses on dynamic situations in which parameter values vary temporally in a cyclical fashion, that is, on a daily, weekly, yearly, or on any other periodic basis. For many of these distribution problems, a critical feature in satisfying demand at the minimum possible cost is the use of storage at selected locations on the network.

There are numerous examples of network location problems with cyclical parameters. Residential energy demand exhibits yearly cycles: consumption of home heating oil peaks in the winter, while demand for steam coal to generate electricity may exhibit one peak in the winter and, where summers are warm, a second peak during the air-conditioning season. In the case of grain distribution, demand is fairly constant over time, but supply varies seasonally as well as geographically. Grains in different regions are harvested at different times and sell at different prices. Renewable energy sources such as solar and wind power, which require both fixed facilities and transmission networks, vary both seasonally and daily. Furthermore, for any type of good, the underlying transportation network characteristics can change in a cyclical fashion. Rail transportation in many areas is more expensive during the harvest season owing to congestion, and freezing can either close a river off to traffic or cause water transportation modes to take a more circuitous and expensive route.

Problems such as these, with so many interrelated decisions to be made, have been solved using mathematical programming techniques—particularly mixed-integer programming for handling the discrete choice of locations for facilities. Commodity distribution problems have traditionally been formulated as static, single-period models. However, interperiod models have two main advantages over single-period models. First, by addressing more than one time period they can offer different solutions regarding flows and assignments in response to the changing environment, thus influencing the locational choices. Second, multiple time periods make possible the storage of commodities between time periods. The availability of storage is the key element that distinguishes INSLA problems from other dynamic location problems, and that makes the modeling of such problems a more difficult task.

Fig. 10.1 shows several examples of cyclical supply and demand in which the total supply over the entire cycle is enough to satisfy the total demand. There are time periods in which the amount demanded exceeds the amount the system is able to supply. Storage makes it possible to use the surplus (shaded area) to cover the shortfall (dotted area) during deficit periods. Even when there are no deficit periods, storage enables least-cost sources and routes to be used to the greatest possible extent.

While the idea for an interperiod storage model was motivated by a need for economic and logistical realism in commodity distribution problems, the modeling technique draws on several areas in operations research. In the following section, INSLA models are placed in their proper context within the field of operations research, with a brief background on distribution problems and dynamic storage problems. The section after that discusses the commodity distribution INSLA problem and the models for solving it. The discussion shows that for the transportation/transshipment/storage logistics system to be accurately

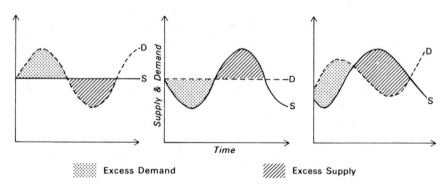

Figure 10.1.   Seasonal fluctuations of supply and demand.

represented by the model, the mathematical program for INSLA models must be combined with a special network design. The section following that one demonstrates these considerations with a sample INSLA problem. The final section briefly introduces potential enhancements of the basic model—such as multiple commodities, temporary storage, economies of scale, and stochastic environments—and briefly reviews solution techniques that have been used in solving INSLA models.

LITERATURE REVIEW

Interperiod network storage location-allocation models draw on several areas within the field of operations research. INSLA models are a certain type of dynamic location problem that basically combines network distribution models developed in the location and transportation science literature with mathematical programming storage models that have been developed primarily in the water resources and location literature. This section first outlines the development and important concepts of these two separate fields and then describe how they are combined to form an INSLA model.

Network Distribution Models

Network distribution models began with the classic transportation problem (Hitchcock, 1941), in which the cost of supplying a set of demand nodes from a set of supply nodes is minimized given the cost of transporting the commodity on routes linking the demand nodes with the supply nodes, the supply capacities, and the demand requirements. The assumption in these and most other transportation models is that transportation cost is a linear function of the amount shipped. The transportation problem can be formulated as follows:

$$\text{Min} \sum_i \sum_j C_{ij} X_{ij} \qquad (10.1)$$

subject to

$$\sum_j X_{ij} \le Y_i \qquad \forall i \epsilon Y \qquad (10.2)$$

$$\sum_i X_{ij} = D_j \qquad \forall j \epsilon D \qquad (10.3)$$

$$X_{ij} \ge 0 \qquad \forall i, j \qquad (10.4)$$

where

X_{ij} = amount transported from node i to node j,
C_{ij} = cost per unit for transport from i to j,
Y_i = amount supplied at i,
D_j = amount demanded at j,
Y = set of all supply nodes,
D = set of all demand nodes.

The simplest transportation problem was expanded by Beckman and Marshak (1955), and Orden (1956) to consider the transshipment of goods en route. Transshipment networks can be fully connected or partially connected. In fully connected networks, all supply nodes are connected to all transshipment nodes, which are connected to all demand nodes, but direct links from supply to demand do not exist. In partially connected networks, there can be any number of intermediate nodes or none at all between a supply node and a demand node. The latter type of network is the more general case, and the model can be formulated as follows:

$$\text{Min} \sum_{i \in N_j} \sum_{j} C_{ij} X_{ij} \tag{10.5}$$

subject to

$$\sum_{j \in M_i} X_{ij} \le Y_i \qquad \forall i \in Y \tag{10.6}$$

$$\sum_{i \in N_j} X_{ij} = D_j \qquad \forall j \in D \tag{10.7}$$

$$\sum_{i \in N_j} X_{ij} - \sum_{k \in M_j} X_{jk} = 0 \qquad \forall j \in J \tag{10.8}$$

$$X_{ij} \ge 0 \qquad \forall i \in N_j, j \tag{10.9}$$

where

M_i = set of nodes that can be supplied by node i,
N_j = set of nodes that can supply node j,
J = set of all transshipment nodes.

Equation (10.8) is the additional mass or materials-balance constraint: it requires that the amount shipped out of a transshipment node equal the amount shipped into it. With the flexibility provided by the notation using the sets M_i and N_j, this model can be applied to any type of network. Numerous methods exist for solving these network problems.

Transshipment of commodities usually requires the use of facilities and equipment that must be purchased and located at specific points on the network. Balinski (1961, 1965) first formulated the plant location problem in which a fixed charge (investment cost) must be incurred before the transshipment equipment can be used. An integer variable that can only take on values of zero or one is used to represent the decision on whether to build a facility at a node. A value of one indicates that the model solution indicates that the facility should be built, whereas zero indicates that it does not. The integer variables are multiplied by the investment cost in the objective function. Requiring the location variables to be integers prevents choosing partially opened facilities that pay only a portion of the investment costs. Gomory (1960) and Effroymson and Ray (1966) were instrumental in pioneering solution techniques such as cutting planes and branch-and-bound for mixed-integer mathematical programs.

$$\text{Min} \sum_{i \in N_j} \sum_j C_{ij} X_{ij} + \sum_j F_j W_j \qquad (10.10)$$

where

$W_j = 1$, if a facility is opened at node j, 0 otherwise,
F_j = investment cost of opening a facility at node j.

In the context of transshipment problems, the facilities needed are transshipment equipment such as port terminals, rail depots, handling equipment, and the like. If transshipment is restricted to nodes at which facilities are built, then the following constraint is added for each transshipment node:

$$\sum_{i \in N_j} X_{ij} \le M W_j \qquad \forall j \in D \qquad (10.11)$$

where M is a large number, generally set larger than the largest possible flow through j.

In the capacitated transshipment problem, the large number M is replaced by the capacity of the facility at node j:

$$\sum_{i \in N_j} X_{ij} \le B_j W_j \qquad \forall j \in J \qquad (10.12)$$

where B_j is capacity at node j.

The constraints and variables introduced thus far are the basic building blocks of any commodity distribution problem with facility location. Using these concepts, model formulations can be tailored to almost any type of distribution system. A model of this type was formulated for planning the U.S. coal export logistics system (Osleeb and Ratick, 1981, 1983a, b, c; Ratick and Osleeb, 1983). The Coal Logistics System (COLS) was designed for multiple transport modes, different facility types for the various modes, multiple sizes for each facility that incorporate economies of scale, multiple coal qualities (see the final section of this chapter), and multiple facility sites at any port with different channel depths. (For another application of the capacitated transshipment problem, see Schapiro, 1975).

Water Resource Models: Reservoir Sizing and Management

Although a number of disciplines address the problems of modeling storage facilities, we will concentrate here on the operational mathematical programming models applied to reservoir sizing and management because of their potential applicability to the INSLA problem. Some of the pioneering studies in this area include the work of ReVelle, Joeres, and Kirby (1969), ReVelle and Kirby (1970), ReVelle and Gundelach (1975), Loucks (1970), Eastman and ReVelle (1973), Gundelach and ReVelle (1975), and Houck, Cohon, and ReVelle (1980). Since the inflow of water to a reservoir fluctuates widely and somewhat regularly by season, these models were designed in a dynamic cyclical framework. The objective of these reservoir models is to build the most efficiently sized (least expensive) reservoir, and to operate it so that it meets various water resource constraints. These constraints include maintaining minimum and maximum releases of water; providing enough free capacity in the reservoir for flood control; meeting industrial, agricultural, and residential demands for water; and remaining within the chosen reservoir capacity. The continuity constraints given by the following equations constitute another type of water resource constraint that has been directly incorporated into the INSLA-type models:

$$S_t = S_{t-1} + r_t - X_t \qquad \forall t = 1, 2, \ldots, T$$
$$S_0 \le S_t \tag{10.13}$$

where

$r_t =$ inflow in period t,
$X_t =$ release in period t,
$S_t =$ amount in storage at end of period t.

These constraints link activities from one time period to the next.

The use of decision rules for the operation of storage relies on the assumption that the investigator can derive an efficient functional relationship between (1) the amount of the commodity that should be released from storage in that period—to help meet demand or for other purposes—and (2) the amount of the commodity (water) in storage (the reservoir) at the beginning of the period. In the reservoir models referenced, this relationship considered that the amount to be released from the reservoir in any time period, R_t, was linearly related to the amount in storage at the end of the previous period, S_{t-1}, the linear decision-rule relationship (LDR) is given by the following:

$$S_{t-1} = r_t - b_i \qquad \forall t = 1, 2, \ldots, T; \ i = 1, 2, \ldots, T/N \qquad \textbf{(10.14)}$$

where

> b_i = decision rule parameter for period i,
> $i = t$ Module N,
> N = number of periods per cycle,
> T = total number of time periods in the historical record.

The index i represents the repeating cyclical period in the data (i.e., days, months, years) and N represents the number of intervals in the cycle (e.g., $N = 12$ if i represents months). The (LDR) parameter b_i is the same for all repeating cyclical periods (e.g., the same each January if i represents months).

The use of decision rules serves two main functions. The computed values for the decision-rule parameters at the optimal solution of the models capture the information contained in the data used to develop the model, and they are therefore used to provide a method for operating the reservoir in each period i. By algebraic substitution, the LDR approach also allows the size of the models to be greatly reduced, concomitantly reducing solution times and costs without sacrificing any information contained in the original data. The decision-rule models have been formulated and used for both deterministic and stochastic environments.

Applying the decision-rule concepts to INSLA models is not straightforward. In the reservoir models, both the location of the reservoir and the route of the commodity (water) along the network (the river) are predetermined. The only decisions are how large a reservoir to build and how much to release in each time period. The absence of integer location variables (or associated fixed costs) allow for solution by linear programming techniques. Linking the concepts in the reservoir models with the dynamic goods-oriented location problems, in which both integer locational choices and related routing of commodities are decision variables in the model, complicates the use of storage/release decision rules.

Reinert (1983) gives an intermediate development incorporating some of the decision-rule concepts into the commodity distribution INSLA framework. His model is used for the location, sizing, and operation of wind-energy production and storage systems for both deterministic and stochastic situations. The choice of locations and configurations of different-sized wind turbines in wind farms are decisions made within the model. The size and operation of the single-battery storage system are also decision choices; however, the location of the electric storage battery is preselected. The model solution for each period includes a determination of how much of the electricity generated from each wind farm to deliver directly to storage, and how much either to deliver to storage (to store excess supply) or how much to release from storage (to overcome a shortfall).

The decision-rule parameters were not used to reduce the size of the model, since the amount of electricity delivered to the storage system depended on the optimal wind-turbine locational choices. However, the decision-rule parameters did provide a schema for operating the existing storage system during each period. Using representative 3-hour averages of wind speed, Reinert demonstrated the trade-offs between the reliability of meeting a fixed demand pattern and the fixed and operating costs for the wind turbines and storage system.

Dynamic Location Problems

In dynamic location problems, flow and allocation variables are indexed by time so that there can be a different solution for each time period in response to changes in the environment. One can identify two general classes of dynamic location problems that are distinguished by their treatment of time and the integer location variables. In certain dynamic situations (not discussed here), parameter changes over time are characterized by a trend extending out to a planning horizon. These situations have been treated by Wesolowsky (1973), Wesolowsky and Truscott (1975), Roodman and Schwarz (1975), and Van Roy and Erlenkotter (1982), among others. With such problems, the focus is on phasing facilities in or out in response to the changing environment, and therefore the integer location variables must also be indexed by time. However, when parameter changes are characterized by cycles that are usually not longer than a year, the most realistic assumption is that facilities are either built or not built for the entire cycle, a critical assumption underlying the INSLA models.

A few INSLA models for commodity distribution have appeared in the literature. One of the first, by Hilger, McCarl, and Uhrig (1977), was a model for locating grain subterminals in northwestern Indiana. Supplies and demands were input for each month, and storage variables linked the various months together. Integer variables represented the decision whether to build subterminals and ensured that the full investment costs were incurred before a terminal was used. Subsequently, Barnett et al. (1982) developed a world grain model with four time periods and three grain types for investigating the effects of U.S.-port-capacity constraints on national and world grain movements. Since Barnett et al. were concerned with current capacities, they did not use any integer variables for opening new facilities. Glover et al. (1979) formulated an INSLA model as a "dynamic transshipment network" similar to White's (1972, described further on) with the addition of integer variables and some important network design considerations. Their model reportedly saved Agrico Chemical Co. over $18 million during its first three years of implementation. These INSLA models will be explored in detail in the section following the next one.

INSLA MODELS FOR COMMODITY DISTRIBUTION

Temporal Continuity Constraints and Dynamic Transshipment

Incorporating storage into a goods-oriented network location problem requires that the model must "keep track" of goods over space as well as over time. Goods can be transported from one node to another, be stored there for any number of time periods, and then be released for shipment elsewhere. In any time period, shipments out of a node can be no greater than the sum of the amount supplied at the node plus the amount transported to the node during the current time period plus the amount in storage at the node from the previous time period. Likewise, the total amount transported to a node must be either transshipped, stored, or allocated to demand. If more than one commodity is being studied, this accounting system must be done separately. Keeping track of goods over space and time has been accomplished in the literature in two (seemingly) different ways. Both models begin with the classical transshipment formulation (10.5 and 10.9).

Hilger, McCarl, and Uhrig (1977) formulated a mixed-integer program for locating grain subterminals. Their model is a direct combination of the capacitated transshipment problem and the reservoir storage problem. All of their flow variables and costs were indexed by month t, with storage variables S_{jt} linking the months. Storage was permitted only at transshipment nodes.

The objective function (10.15) minimizes total costs that consist of transportation, handling costs, and storage costs—all of which can vary overtime—and investment costs. Supply constraints (10.16) ensure that supply is not exceeded and demand constraints (10.17) ensure that demand is satisfied. The materials-balance constraint is combined with the flow continuity constraint from the water resources literature to form a new continuity constraint (10.18a). This constraint defines storage at node j at the end of time period t to be the amount in storage at the beginning of the time period plus the sum of flows into j in time period t minus the sum of flows out of j in period t. Investment cost constraints (10.19) ensure that no goods can be shipped to or stored at node j unless a distribution center is opened there. Open storage capacity (10.20) must not be exceeded. Finally, there are integrality (10.21) and nonnegativity (10.22) constraints.

$$\text{Min} \sum_{i \in N_j} \sum_j \sum_t C_{ijt} X_{ijt} + \sum_j F_j W_j \qquad (10.15)$$

subject to

$$\sum_{j \in M_i} X_{ijt} \leq Y_{it} \qquad \forall\ j \in Y;\ t = 1, 2, \ldots, T \qquad (10.16)$$

$$\sum_{i \in N_j} X_{ijt} = D_{jt} \qquad \forall\ j \in D;\ t = 1, 2, \ldots, T \qquad (10.17)$$

$$S_{jt} = S_{jt-1} + \sum_{i \in N_j} X_{ijt} - \sum_{k \in M_j} X_{jkt} \qquad \forall\ j \in J;\ t = 2, 3, \ldots, T \qquad (10.18a)$$

$$S_{j1} = S_{jT} + \sum_{i \in N_j} X_{ij1} - \sum_{k \in M_j} X_{jk1} \qquad \forall\ j \in J \qquad (10.18b)$$

$$\sum_{i \in N_j} \sum_t X_{ijt} + \sum_t S_{jt} \leq MW_j \qquad \forall\ j \in J \qquad (10.19)$$

$$S_{jt} \leq B_j W_j \qquad \forall\ j \in J;\ t = 1, 2, \ldots, T \qquad (10.20)$$

$$W_j = (0, 1) \qquad \forall\ j \in J \qquad (10.21)$$

$$S_{jt}, X_{ijt} \geq 0 \qquad \forall\ i, j, t \qquad (10.22)$$

where

X_{ijt} = amount transported from i to j in period t,
C_{ijt} = cost per unit of transport from i to j in period t,
S_{jt} = amount in storage at j at the end of period t,
T = number of period,
Y_{it} = amount supplied at i in period t,
D_{jt} = amount demanded at j in period t.

Constraint (10.18b) defines storage at the end of the cycle to be the amount stored at the beginning of the cycle, ensuring that if the parameters were to keep cycling in the same way, the system could continue operating indefinitely.

Another method for keeping track of goods over space and time is to define a dynamic transshipment network, mentioned earlier, which is a network through space and time, where a node j is defined as an ordered pair (h, t): location h at time period t. White (1972) used this concept for planning the distribution of empty railroad cars: a railroad car at node $i = (g, t)$ could be shipped to a node $j = (h, t + u)$, where g and h are locations and u is the amount of time it takes to travel from g to h (see Fig. 10.2). Storage at node h from period t to $t + 1$ is represented as a link on the network from the node (h, t) to the node $(h, t + 1)$. In the type of routing problems White considered, careful scheduling was critical. Since rail cars can conceivably be needed at any time, White assumed that a node can be defined at any location for any time t. So, although all nodes are at discrete moments in time, the time horizon over which nodes can be defined is continuous.

The type of commodity distribution problems considered here are not concerned with scheduling and the routing of individual vehicles; the model is dynamic because conditions vary in different periods. Therefore, in an INSLA model the time index does not represent a moment in time but rather a period of time. Goods are shipped between locations via any number of intervening transshipment points during the time periods; the amount of time the journey requires is not considered important because commodities are continually being shipped. Storage is considered as a shipment between time periods at one location, as shown in Fig. 10.3.

Since the network incorporates storage and time, the problem can be solved as a transshipment problem. Storage capacity can be incorporated as an upper bound on the appropriate flow variable, and the materials-balance constraints (10.8) serve as both materials-balance and continuity constraints. Notice in Fig. 10.3 that there is a storage link defined between each node (j, T) and $(j, 1)$ to ensure that the optimal system would not degrade over a number of cycles. In a sense, it is a matter of nomenclature whether storage is viewed as a flow between periods at a single site or as an inventory at the site. The

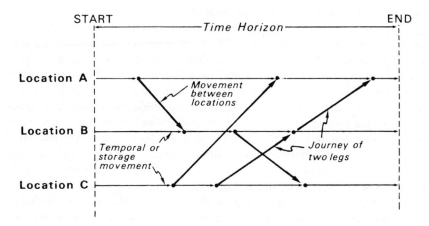

Figure 10.2. A dynamic transshipment network with a continuous time horizon.

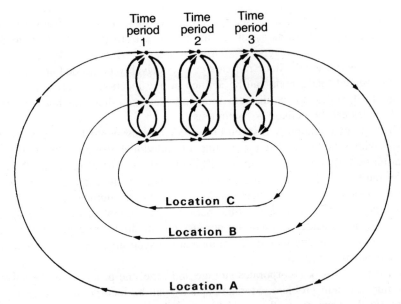

Figure 10.3.   A dynamic transshipment network with discrete time periods.

formulation for the dynamic transshipment network is less obvious because the time element is implicit, but one advantage of White's formulation is that the model can be solved using efficient network algorithms such as the out-of-kilter algorithm (White, 1972), the flow-augmentation algorithm (Jensen and Bhaumik, 1977), and others (see the final section).

White did not require any integer location variables in his empty-rail-car distribution model, but Glover et al., (1979) added integer variables to a dynamic transshipment network for optimizing the distribution and storage of agricultural chemicals. Since storage is represented as a flow variable, the entire problem can be solved as a capacitated transshipment problem, as in (10.5)–(10.10). Since there are now only integer and flow variables, the objective function consists of the sum of fixed costs and "transportation" costs. Facilities are no longer located at single nodes j on the network. Rather, they are located at sites indexed by h that correspond to T different nodes on the dynamic transshipment network. Constraint (10.11) must be modified so that nothing will be shipped to or stored at a site during any time period unless a distribution center has been opened. With a dynamic transshipment network, all capacities, including supply and demand, can also be represented by upper or lower bounds on links, as follows:

$$\text{Min} \sum_{i \in N_j} \sum_{j} C_{ij} X_{ij} + \sum_{h} F_h W_h \qquad (10.23)$$

subject to

$$\sum_{i \in N_j} X_{ij} - \sum_{k \in M_j} X_{jk} = 0 \qquad \forall \, j \in J \qquad (10.24)$$

$$LB_{ij} \leq X_{ij} \leq UB_{ij} \qquad \forall \, j, \, i \in N_j \qquad (10.25)$$

$$X_{ij} \leq M W_h \qquad \forall \, (i, j) \in A, \, j \in J_h, \, h \in H \qquad (10.26)$$

$$W_h \in (0, 1, 2, \ldots, G_h) \qquad (10.27)$$

where

> W_h = an integer, if one or more facilities are opened at loction h, 0 otherwise,
> H = set of all locations requiring a facility if there is any flow through them (all transshipment locations),
> J_h = all nodes j corresponding to location h in the various periods,
> F_h = investment cost for a facility at h,
> LB_{ij} = lower bound on flow from i to j,
> UB_{ij} = upper bound on flow from i to j,
> G_h = maximum number of facilities allowed at h,
> A = set of links that require a facility to be opened if there is any flow on them.

These constraints are adequate if the only possible facility type is a comprehensive distribution center that performs storage and transshipment functions. In the next section, this and other restrictive assumptions are made more realistic and flexible.

A Flexible Network Design for INSLA Problems

Desirable Features in an INSLA Model.   The model can be modified in several ways so it can more realistically embody the actual logistics system and also identify potential cost-saving strategies that previous model formulations do not address. However, a trade-off exists between model realism and model solution speed. This section develops a more detailed model; in general it will have a larger number of constraints and variables than the models previously described, and thus it will be more expensive to solve. Some decision makers may prefer the ability to run many cost-effective, less detailed scenarios over the ability to model the logistics system in detail. The purpose and person for whom the work is intended should dictate the scope and details of such models. However, advances in computer technology and new algorithms such as Karmarkar's (1984) may make solution speed a lesser concern than realistic application.

The model developed in this section addresses the following concerns:

1. Any combination of supply, demand, storage, and transshipment activities can occur at a site. For instance, coal can be stored at mines, ports, or power plants; and furthermore, power plants are sometimes located at mines and often at ports. Location at the same site implies that the different activities can share facilities that are jointly required. For instance, a port and a power plant can share a rail-car dumper and a storage pile, thus realizing significant investment cost savings.

2. Facilities performing different transshipment functions may be located independently. In the Hilger, McCarl, and Uhrig (1977) and Glover et al. (1979) models, the locational choice was whether to open a combination storage/transshipment facility. However, if a site is used only for transshipment, there is no need to build a storage facility. On the other hand, a storage facility will usually require transshipment equipment if only to unload and load the commodity for storage. (An exception to this rule is coal, which sometimes is stored in rail cars rather than in piles or silos.) Locating these facilities independently of each other can achieve significant savings in investment costs.

3. Similarly, modeling transshipment as functionally disaggregate activities and facilities—loading and unloading—improves the realism of the model in the following ways. First, different capacities and different operating and investment costs can be ascribed to the different activities. For instance, in coal handling, rail-car dumpers and ship loaders, while often paired together, have different hourly capacities. Second, when a site is used for transshipment and either supply or demand, certain shipments may be only unloaded or only loaded, but not both; in these cases, both operating costs should not be incurred. Third, superfluous facilities can be avoided. For instance, if a power plant site is a potential transshipment site but is not used for transshipment in the model solution, then only an unloader needs to be built at that site.

4. For accurate cost appraisal and capacity restrictions, the model formulation should allow the user to place a separate cost and capacity on each of the following activities:

   supply
   demand
   loading
   unloading

storage per time period
input to storage
release from storage
transportation

Supply will have a per-unit purchase or production cost and an upper limit on capacity; demand usually has no cost other than unloading, and the capacity limit is either a lower bound or an equality constraint. Loading and unloading will usually have different operating and investment costs, as well as different capacities. Often, a cost for storing a unit of commodity may be incurred each time period, such as refrigeration and other energy costs, maintenance costs, and the opportunity cost of holding the goods off the market. On the other hand, Hilger, McCarl, and Uhrig (1977) assumed that the opportunity cost was offset by an increase in price owing to hedging. Also, certain commodities degrade over time in storage; examples are energy loss in batteries or water evaporation in reservoirs. Such losses could be treated implicitly as a cost per unit stored per time period (or explicitly as a multiplicative fractional loss factor on the amount previously in storage, $St - 1$, in the continuity constraints). Other costs and capacities that are often ignored relate to input to and release from storage. With some goods, such as sand and gravel, the major costs are putting the goods on top of the pile and taking them off; once on the pile, such materials require very little maintenance. If these costs are ignored, storage of goods for a short period of time will appear much less costly than it actually is.

These four areas of improvement are important to help assure that there will be no unexpected costs or capacity bottlenecks when the system is constructed, and so that the model will be able to identify potential investment cost savings by omitting unnecessary facilities and planning for the sharing of unloading, loading, and storage facilities when possible.

Embedding Subnetworks at a Node.   The above considerations can be incorporated into an INSLA model by representing a node as a small transportation network nested within the overall transportation network (Osleeb and Ratick, 1985). We refer to the resulting network as a *logical network* because not all nodes and links correspond to real places or real transportation routes.

Each node, whether on a dynamic or regular transshipment network, is represented as a transportation network. Glover et al. (1979) divided each transshipment node into three nodes as shown in Fig. 10.4. The unloading cost and capacity are placed on the link from an unloading node to a B node that serves as a clearinghouse for the entire subnetwork. The loading cost and capacity are placed on the link from the B node to a loading node, and the storage cost and capacity on the link from the B node at time t to the B node at $t + 1$.

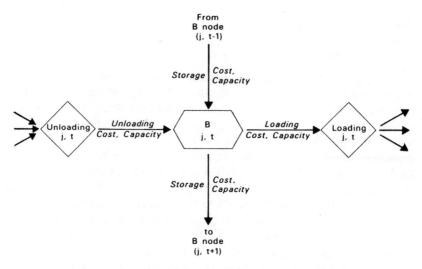

Figure 10.4.   A subnetwork consisting of three nodes.

Other locations on the network that can ship to this location are linked only to the unloading node, while all outward links from h are from the loading node. In this way, all arriving goods must first be unloaded before being loaded or stored. No shipments can enter the B node unless a distribution center is opened there, and thus nothing can be unloaded, loaded, or stored unless the fixed charge is incurred in the cost-minimizing objective function.

As formulated, this network design does not have supply and demand colocated with storage or transshipment, nor does it allow for separate facilities or input and release costs and capacities. However, this design can be expanded to incorporate all of these features. Fig. 10.5 shows the network representation of a site with all possible activities (Kuby et al., 1984). Supply, demand, and storage are placed in separate nodes, all connecting to the B node. Shipments are able to bypass storage, and the supply node can serve the demand node without requiring loading or unloading charges.

The approach suggested here provides a good deal of flexibility and allows for realistic model representations. The network for each potential or actual site can be tailored to conform to site characteristics. Nodes and links can be included or deleted as appropriate. The current formulation would require a B node to accompany every combination, a loading node always to accompany a supply node, and an unloading node always to accompany a demand node. The modeler would then be able to exercise judgement in designing the network: if a port and a demand are in the same city but are not able to share the same unloading facility, they should not be part of the same subnetwork. Each node

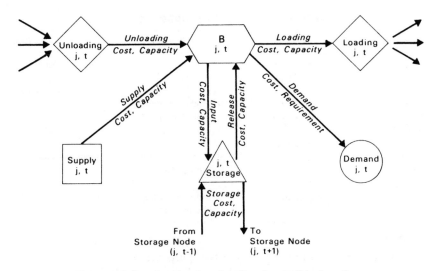

Figure 10.5.   A subnetwork with all possible functions.

on the original transshipment network will be represented as a subnetwork and will be linked via unloading and loading nodes to other sites on the network and via storage nodes to other periods on the network. This method can be used with either the temporal continuity-constrained model formulation, (10.15)–(10.22) or the dynamic transshipment formulation (10.24)–(10.27).

The constraints are essentially the same as those detailed earlier—whichever formulation is chosen—but they are applied to a different network. Capacities on the various links can be represented by two different methods. One way is to put upper and lower bounds on links as in the dynamic transshipment model (10.25). Table 10.1 describes the data for each type of link in the dynamic transshipment design. Each link (i, j) is for a specific period, and thus each of these parameters can vary over time. Glover et al. (1979) used both lower and upper bounds on transportation links: lower bounds to incorporate contractual agreements to ship a certain amount between locations; and upper bounds for physical capacities on bridges, pipelines, and the like.

An alternative formulation is to use actual facility capacities in constraints (10.5) and (10.26), instead of an arbitrary large number:

$$X_{ij} \leq B_{ij}W_h \qquad \forall\ (i, j)\epsilon A,\ h\epsilon H(i, j) \qquad\qquad \textbf{(10.28)}$$

$$\sum_{i\epsilon Nj} X_{ijt} \leq B_j W_j \qquad \forall\ j\epsilon L\cup U,\ t = 1, 2, \ldots, T \qquad \textbf{(10.29)}$$

where U is the set of unloading nodes and L is the set of loading nodes.

Table 10.1. Data-Input Description for Links

Link Type	Cost C_{ij}	Lower Bound LB_{ij}	Upper Bound UB_{ij}
Supply	Purchase or production cost	0 or minimum supply amount	Supply capacity or infinity
Demand	0	Demand requirement	Demand requirement
Loading	Loading cost	0 or other minimum	Loading capacity
Unloading	Unloading cost	0 or other minimum	Unloading capacity
Storage	Storage cost per time period	0 or other minimum	Storage capacity
Input to storage	Input cost	0 or other minimum	Input capacity
Release from storage	Release cost	0 or other minimum	Release capacity
Transport	Transport cost	0 or minimum transport amount	0 or transport capacity or infinity

The modified constraints thus serve two purposes: they serve as capacities or upper bounds on the flow variables, and they force facilities to be open when there is flow through them. This arrangement eliminates the need for upper bounds on all but the transport flow variables.

Certain storage facilities, such as silos or oil tanks, can be modular in that they are usually of a certain size, they do not show significant economies of scale, and there are often a large number of them at a single site. In these cases it may be more realistic to allow the model to choose the number of storage facilities rather than choose between various sizes of a single unit. Location variables would not be limited to values of zero and one but would be allowed to take any integer value up to a prespecified number G_j at each node j. This method is used mainly for the storage facilities, although the same logic could apply at times to loading and unloading facilities.

SAMPLE MODEL APPLICATION

This section presents an example of a commodity distribution problem with cyclically fluctuating parameter values. This example shows how a dynamic transshipment network is set up and how different sites are represented as embedded subnetworks. It also demonstrates some of the trade-offs made within the model in determining the least-cost solution. Fig. 10.6 shows the basic data for the problem. There are five sites, all of which are potential sites for storage facilities. *DV* and *DH* are inland sites, *SB* and *SE* are ports, and *QC* is an overseas demand. Winter, spring, summer, and fall are the four time periods. Fig. 10.7 shows that total supply peaks in the summer, while total demand peaks in the winter. Supply is 8000 units over the four-period cycle, and demand is 7500 units.

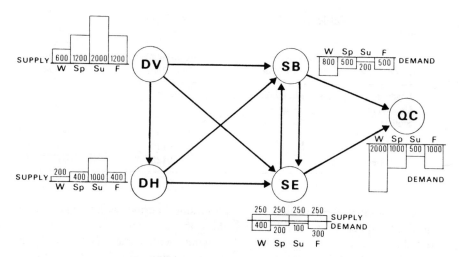

Figure 10.6.   Data for the sample problem.

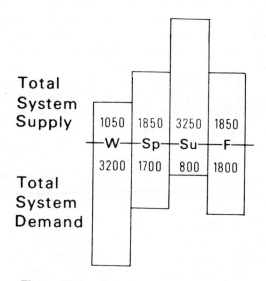

Figure 10.7.   System supply and demand over time.

Table 10.2 shows the costs and capacities of the three types of facilities. Since demand exceeds supply by 2150 units in the summer (see Fig. 10.7), it is obvious that at least three storage facilities will be necessary, since the release capacity of the storage facilities is given as 1000 units per period.

Table 10.2. Facility Costs and Capacities

Facility Type	Investment Cost per Year	Operating Cost per Ton	Capacity (tons × 10 per period)
Loading	$3,600,000	$0.50	1800
Unloading	3,600,000	0.50	1800
Storage	2,000,000		
storage		0.30	2000
input		0.125	1000
release		0.125	1000

Fig. 10.8 shows the expanded network for a single period, with the appropriate embedded subnetworks. Solid arrows show the transportation links between sites; dashed arrows show the logical networks within sites. *DV* is a supply site that can also have storage. It requires a clearinghouse *B* node, a storage node, and a loading node for shipping to other sites. The supply node has a directed link to the *B* node, from which goods can be put into storage (the link to the storage node) or loaded (the link to the loading node). A link in the reverse direction must be defined from the storage node to the *B* node so that goods can be released from storage. All links to other sites originate from the loading node.

DH is similar to *DV* except that it can receive shipments as well. Therefore, *DH* also requires an unloading node and an unloading link to the *B* node at *DH*. *SB* is similar to *DH* except that it has a demand node instead of a supply node. *QC* is only a demand node; there are no directed links from *QC* to any other location. *QC* thus requires storage and demand nodes and an unloading node. All links from other sites are directed to the unloading node. Finally, *SE* has all possible functions and all possible nodes. Everything must pass through the *B* node on its way to anywhere else—thus allowing any goods (1) from *DV* or *DH* or *SB*, (2) from its own supply at *SE*, or (3) from storage at *SE* to be loaded and shipped out to (1) satisfy demand at *SE* or (2) be stored at *SE*. Commodities in storage would not be released and put back into storage in the same period because there is a cost for input and release.

Assuring continuity between the four time periods can be accomplished either by indexing all of the flow variables by t and connecting the storage nodes by storage variables Sjt, or by setting up a dynamic transshipment network, as shown in Fig. 10.9. The network on each plane duplicates Fig. 10.8 for each of the seasons; however, if some of these routes were not open in one of the seasons, they could simply be omitted from the appropriate planes.

Fig. 10.10 shows the model solution for all four periods without the subnetworks within each site. The optimal solution has *SE*, *DV*, and *QC* as the chosen

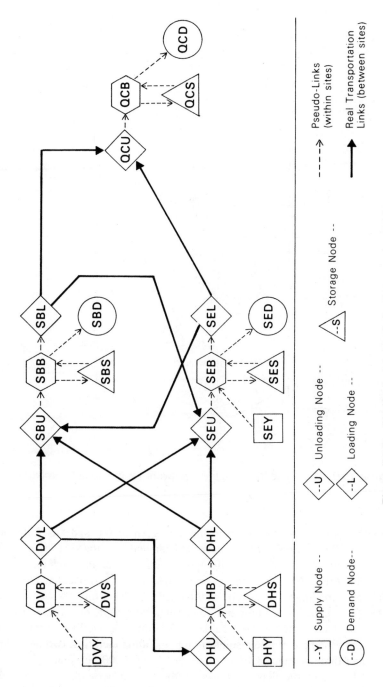

Figure 10.8. The logical network for the sample problem.

Supply Node -- --Y

Demand Node-- --D

Unloading Node -- --U

Loading Node -- --L

Storage Node -- --S

Pseudo-Links (within sites)

Real Transportation Links (between sites)

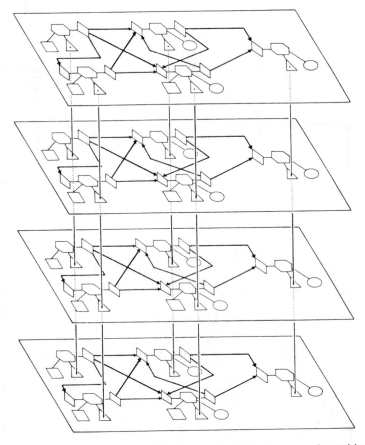

Figure 10.9.   Dynamic transshipment network for the sample problem.

storage sites. *SE* begins storing up in the spring, with *DV* and *QC* following
in the summer. Total supply at *SE* equals total demand over the year: by building
a storage facility, *SE* is able to self-supply without ever opening loading or
unloading facilities. No commodities are stored or transshipped at *DH*, and
therefore only a loading facility need be built at *DH*. *SB* transships all commodities
bound for *QC*, requiring loading and unloading equipment, but storage for the
demand at *SB* takes place at *DV*. Table 10.3 shows the facilities opened and
their costs in the sample problem solution: three loading facilities, two unloading
facilities, and three storage facilities, costing $24,000,000 per year.

Fig. 10.11 shows the detailed flows on the logical network, including the
subnetworks at each node for the summer and fall. We can follow a path

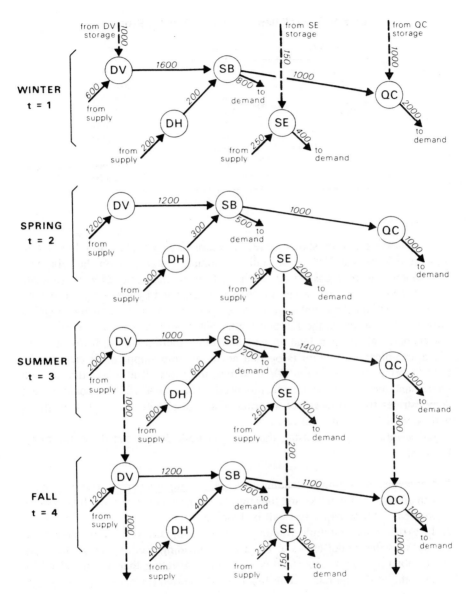

Figure 10.10. Solution to the sample problem for all four time periods.

from supply in one period to demand in another period through the various operations of supply, loading, transportation, unloading, input to storage, storage, release from storage, and demand. The materials-balance constraints ensure that no goods are lost or created except at supply and demand nodes.

Table 10.3. Facilities Opened in Sample Problem

Type of Facility	Location	Cost
Loading	DV	$ 3,600,000
	DH	3,600,000
	SB	3,600,000
Unloading	SB	3,600,000
	QC	3,600,000
Storage	DV	2,000,000
	SE	2,000,000
	QC	2,000,000
		$24,000,000

In another run (not shown here), transportation costs to *SB*—but not to *SE*—were increased 25% in the fall to simulate congestion on the railroads, and winter transport costs from *SB* to *QC* were increased 25% to simulate more-circuitous northern ocean transport routes caused by ice. To avoid paying the high costs on these links during these periods, all transshipment to *QC* is through *SE*, not just in the fall and winter but in all four periods, because it is not economical to open loading and unloading facilities at both *SB* and *SE*. Since an unloading facility is opened at *SE* for transshipment, it can also be used to bring in goods from *DV* to satisfy the shortfall at *SE* in the fall and winter, and therefore no storage facility needs to be built at *SE*. This configuration demonstrates the advantage of a formulation that allows independent facilities with separate transshipment functions.

Other interesting runs using the same network demonstrated the following features:

1. Capacities of loading, unloading, storage, and input and release from storage can all cause bottlenecks and thus influence routes and locations. It is difficult to predict which capacities will be binding.
2. In a run where multiple (modular) facilities of any type were allowed at any site, the model chose to build two unloading facilities at *SB* but only one loading and storage facility, thus demonstrating that it can indeed be important to disaggregate multifunction distribution centers into their component parts.
3. In a run with multiquality commodity types (see the next section), the higher-value commodity not only was shipped longer distances than the lower-value commodity but also was the only commodity put into storage: while the cost per ton for storing the two commodities is the same, the storage cost per unit value is less for the higher-value commodity.

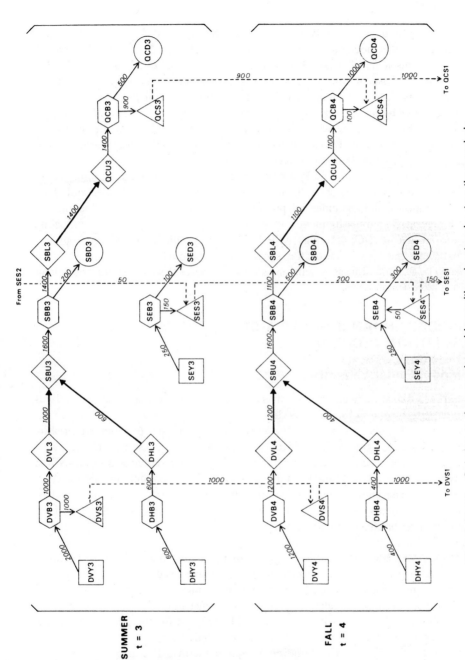

Figure 10.11. Solution to the sample problem with subnetworks, two time periods.

It is difficult to quantify the advantage that an INSLA model with logical subnetworks at nodes and with disaggregate facilities has over INSLA models without this level of detail. The objective-function value may actually increase because the costs of input and release were not included in previous formulations. Also, it is difficult to place a value on the model's ability to avoid logistical bottlenecks by having separate capacities for loading, unloading, input, release, and storage. One area where the savings are measurable is in the investment costs. The investment costs in the sample problem with disaggregate facility types was $24,000,000. However, investment costs increase by more than 60% to $38,800,000 in the sample problem if a single aggregated distribution center is used that must be open if any activity takes place at a node, and if the distribution center includes all possible facilities at a node (i.e., loading and storage at DV; transshipment and storage at DH, SB, and SE; and unloading and storage at QC). Savings will differ from problem to problem, but it is expected that transportation flows will shift to take advantage of the ability of supply, demand, transshipment, and storage purposes to share loading and unloading facilities.

MODEL ENHANCEMENTS AND SOLUTION TECHNIQUES

Model Enhancements

Earlier, we introduced the two basic programming formulations for solving interperiod network storage models, and described a subnetwork design for capturing logistical and economic characteristics unique to storage/transshipment systems. This section gives a brief overview of some enhancements of this model. The first, temporary storage, is not an essential part of an INSLA model, but is an additional feature that could be important for certain commodities. The others—multiple commodities, economies of scale, and stochastic environments—have been dealt with in other location-allocation problems but will be briefly summarized here.

Temporary Storage. It has been assumed thus far that goods must be stored in a storage facility of some kind. For some goods, such as coal, rock salt, or gravel, the normal method of storage may be to pile the material on the ground. However, even goods such as grain, which must normally be stored in silos, can be stored temporarily on the ground. The reason for temporary ground storage is that incremental capacity may be needed for a short period of time, but not enough to justify building a second facility. It can be assumed that temporary storage is more costly per bushel per month than permanent storage; otherwise short-term temporary storage might always be chosen first. While piling grain on the ground may incur less operating expenses than permanent storage, the losses from water damage, insects, and the like, surely make ground storage more expensive over time.

Hilger, McCarl, and Uhrig (1977) model temporary storage by allowing in-facility storage to exceed the storage capacity by the amount in temporary storage. An additional per-unit cost on temporary storage ensures that temporary storage is more expensive than in-facility storage and thus will be the first to be released. The authors' constraints do not, though, assure that grain will remain in temporary storage for one month only (see Osleeb and Ratick, 1985). However, adding a constraint requiring the amount in temporary storage to be released in the following period for loading or consumption, along with a tempo-rary-storage-capacity constraint, can make the representation of on-ground stor-age more realistic.

Multiple Commodities. Multicommodity network problems have been treated extensively in management science literature (see Ali et al., 1984, for an overview). We have isolated three classes of multicommodity problems in reference to modeling storage logistics. The categories are distinguished by whether the different commodities can be handled and stored by the same facili-ties, and by whether there are independent demands for each commodity type. The first type of multicommodity problem involves completely separate commodi-ties, such as oil and grain. Demands for these two goods are independent, and they cannot be stored or handled by the same facilities. The significant economic interactions between the two commodities are that they can compete for space at a site, they can combine to amortize large network improvements such as port dredging or offshore loading, or they can interact by causing congestion on the transportation network.

The second type of multicommodity problem involves commodities with independent demands but with similar handling capabilities. An example of this is different grain types, such as rice, wheat, and barley (assuming there is little or no substitution between these products). While these products cannot be stored together in the same silo, they can all be transshipped by the same unloading and loading equipment. Investment cost savings are possible by sharing such equipment. Barnett et al. (1982) refer to this type of problem as *mutually capacitated.*

The third class of multicommodity problems involves commodities such as coals, iron ores, or even strains of wheat that exist in multiple qualities or varieties. Demand is not generally in terms of an amount of each quality type, but rather for the generic commodity, for example, Btus of coal, pounds of pure iron, or perhaps calories of wheat. The different quality types are distin-guished by their content of various ingredients. For instance, Appalachian coal is high in Btus and low in sulfur, whereas Western coal is low in sulfur and heat content. The different coal types can be mixed together to satisfy demand for Btus as cheaply as possible without exceeding the sulfur restrictions (Osleeb and Ratick, 1983a, b, c). However, while demand is for Btus, pounds, or calories, tons of coal, ore, and wheat are what is shipped on the network. Multiquality commodities can usually be loaded and unloaded by the same facilities and can often be stored in the same facility simultaneously. For example, at a port, different coal types may have to be stored separately, but at the power plant coal types are often mixed in storage.

Depending on the type of multicommodity problem being considered, the network and model must be designed accordingly. In any of the three cases, all continuous variables must be indexed by commodity type, here indexed by q, and materials-balance constraints must be written for each commodity type q at all nodes except demand and supply nodes. At demand and supply nodes, demand and supply constraints must be written for each commodity type q, except for the demand constraint in the case of multiquality commodities, where all types are combined in a single demand equation with value factors for each type. Depending on whether each commodity requires a separate facility for storage and handling, the integer loading, unloading, and storage facility variables must be indexed by q, and one facility-capacity constraint must be written for each period t and each commodity type q. The case where a facility can store different commodities but not simultaneously can also be treated (see Osleeb and Ratick, 1984) for each of these enhancements.

Economies of Scale.   Economies of scale resulting from the maximal use of larger, more efficient bulk-handling facilities have long been recognized as an important feature in planning a distribution system (Balinski, 1965). Scale economies have traditionally been modeled with the choice of several facility sizes with increasing investment costs but decreasing operating costs so that they together define a piecewise linear concave cost function. In an INSLA problem with the subnetwork design, economies of scale can be incorporated by defining several unloading, loading, or storage links, where each link is associated with a different facility size. It is necessary to add constraints that ensure that the links are not used unless the associated investment costs are incurred, as well as constraints that govern the number of facilities that can be opened.

Stochastic Environment.   Supplies and demands for most commodities do not simply fluctuate in a deterministic cyclical fashion; they are also subject to strong stochastic influences. The demand for energy, the supply of grains, and the freezing of water transportation links all depend greatly on weather conditions, and often on the economic and political climate as well. This area of research demands attention at this point of INSLA model development. A great deal of work on stochastic environments has been done by Handler and Mirchandani (1979) for stochastic networks, by Larson (1974) in the context of service demands and queueing theory, and by ReVelle, Joeres, and Kirby (1969), ReVelle, and Kirby (1970), and ReVelle and Gundelach (1975) in the context of water inflows to reservoirs.

Chance-constrained programming (Charnes and Cooper, 1961; ReVelle, Joeres, and Kirby, 1969) appears to be an appropriate approach to handling this aspect of the problem. Much thought needs to be given to the question of what, if any, constraint types should be included. In their reservoir storage model, ReVelle, Joeres, and Kirby (1969) place minimum storage (for recreational purposes) and minimum freeboard (excess storage capacity for safety purposes), which may have equivalents in the commodity distribution problem. This area of research is fertile and could yield very provocative models.

Solution Techniques

All of the mathematical programs described in this chapter can be solved using conventional mixed-integer programming packages. However, a realistic model formulation for a problem of significant size may require other, faster solution techniques, several of which have been used to date. These other techniques include network solution algorithms, decomposition, and dual ascent procedures.

White (1972) presents an algorithm for optimizing a dynamic transshipment network, Barnett et al. (1982) and Ali et al. (1984) also use network-based algorithms. An obvious advantage of the dynamic transshipment formulation is that it allows the continuous portion of the problem to be solved in this manner, since the continuity constraints are represented as network materials-balance constraints. These procedures tend to be very fast because they do not require matrix inversions, but they also do not incorporate integer variables. Glover et al. (1979) solved their INSLA problem as a mixed-integer linear programming problem in which the linear program involved a large embedded network structure. These techniques have been advanced by Maier (1971), Glover et al. (1978), and Barr, Glover, and Klingman (1981).

A second, related approach involves the decomposition of large programs for faster solution. Bender's (1962) decomposition separates the integer variables from the continuous variables and solves the two problems separately, passing information from the integer problem to the continuous subproblem in the form of fixed values for the location variables, and in the other direction in the form of shadow prices (see also Geoffrion and Graves, 1974). Benders' decomposition, was used successfully by Hilger, McCarl, and Uhrig (1977) in solving an INSLA problem. Another decomposition technique, Dantzig and Wolfe's decomposition (1960), exploits the block structure of the continuous portion of a linear program. This method may be helpful in decomposing the problem into its various time periods and its various commodities. Conceivably, network algorithms could be used to solve each of these blocks within the Dantzig-Wolfe decomposition, all of which could be the subproblem of the Benders decomposition problem.

Dual ascent methods, as exemplified by Bilde and Krarup (1977), Erlenkotter (1978), and Van Roy and Erlenkotter (1982), may also be applicable to INSLA-type problems. Van Roy and Erlenkotter's procedures, known as DUALLOC (Erlenkotter, 1978) and DYNALOC, were developed for the dynamic facility location problem that selects the time-staged establishment of facilities, but they may be adaptable to INSLA problems that have cyclical parameters. Dual ascent methods generate an optimal solution or bounds for a branch-and-bound procedure by solving the dual of a linear program instead of the primal linear program. Dual ascent methods often yield an optimal solution directly. If not, they provide bounds for a branch-and-bound procedure that completes the search for an optimum. Van Roy and Erlenkotter have extended their procedures to problems with price-sensitive demands, linearized concave costs, interdependent projects, multiple commodities, and multiple stages.

The examples of the use of INSLA-type models referenced in the chapter include analyses of grain reserves, water storage and supply, the applicability of renewable but intermittent energy sources, the efficient distribution of agricultural chemicals, and the use of storage to ensure adequate supplies of fossil fuels. The use of INSLA models in many dynamic problems of location and transportation logistics may become more and more important as concern grows about the efficient and equitable use and allocation of scarce resources such as food, water, and energy, among others, and as computer-aided decision models become more acceptable to the people responsible for making resource-related decisions.

The ability of INSLA models to aid in the decision processes for which they are developed will depend on how well the models can be formulated to provide accurate, relevant and timely information. As mentioned previously, the formulation of INSLA models to date has been somewhat dictated by the trade-off between model realism and solution costs. Incorporating the enhancements mentioned above into the INSLA framework, in relevant situations, would improve the flexibility and utility of the models. However, such improvements could require the development of new solution techniques or modifications of existing techniques to allow for reasonable solution costs. Balanced research into the formulation and use of INSLA models will help ensure their future acceptance and use.

REFERENCES

Ali, I., D. Barnett, K. Farhangian, J. Kennington, B. McCarl, B. Patty, B. Shetty, and P. Wong, 1984, Multicommodity network problem: applications and computations, *American Institute of Industrial Engineers Transactions* **16:**127–134.

Balinski, M. L., 1961, Fixed-Cost Transportation Problem, *Naval Research Logistics Quarterly* **8**:41–54.

Balinski, M. L., 1965, Integer programming: methods, uses, computation, *Management Science* **12**:253–313.

Barnett, D., J. Binkley, B. McCarl, R. Thompson, and J Kennington, 1982, *The Effects of U.S. Part Capacity Constraints on National and World Grain Shipments*, Purdue University Agricultural Experiment Station Bulletin, No. 399, Lafayette, Ind.

Barr, R. S., F. Glover, and D. Klingman, 1981, A new optimization method for large scale fixed charge transportation problem, *Operations Research* **29**:448–463.

Beckmann, M. and T. Marschak, 1955, An activity analysis approach to location theory, *Kyklos* **8**:128–141.

Benders, J. F., 1982, Partitioning procedures for solving mixed-variables programming problems, *Numerische Mathematik* **4**:238–252.

Bilde, O. and J. Krarup, 1977, Sharp lower bounds and efficient algorithms for the simple plant location problem, *Annals of Discrete Mathematics* **1**:79–97.

Charnes, A. and W. Cooper, 1961, *Management Models and Industrial Applications of Linear Programming*, vol. 1, Wiley, New York.

Dantzig, G. B. and P. Wolfe, 1960, Decomposition principle for linear programs, *Operations Research* **8**:101–111.

Eastman, J, and C. ReVelle, 1973, Linear decision rule in reservoir management and design, 3: direct capacity determination and intra-seasonal constraints, *Water Resources Research* **9**:29–42.

Effroymson, M. and T. Ray, 1966, A branch and bound algorithm for plant location, *Operations Research* **14**:361–368.

Erlenkotter, D., 1978, A dual-based procedure for uncapacitated facility location, *Operations Research* **26**:992–1009.

Geoffrion, A. M. and G. W. Graves, 1974, Multicommodity distribution system design by Benders decomposition, *Management Science* **20**:822–844.

Glover, F., J. Hultz, D. Klingman, and J. Stutz, 1978, Generalized networks: a fundamental computer-based planning tool, *Management Science* **24**:1209–1220.

Glover, F., G. Jones, D. Karney, D. Klingman, and J. Mote, 1979, An integrated production, distribution, and inventory planning system, *Interfaces* **9**:21–35.

Gomory, R. E., 1960, *An Algorithm for the Mixed Integer Problem*, Rand Corporation, Santa Monica.

Gundelach, J. and C. ReVelle, 1975, Linear decision rule in reservoir management and design, 5: a general algorithm, *Water Resources Research* **11**:204–215.

Handler, G. and P. Mirchandani, 1979, *Location on Networks: Theory and Algorithms*, MIT Press, Cambridge, Mass.

Hilger, D. A., B. A. McCarl, and J. W. Uhrig, 1977, Facilities location: the case of grain subterminals, *American Journal of Agricultural Economics* **6**:74–684.

Hitchcock, F. L., 1941, The distribution of a product from several sources to numerous localities, *Journal of Mathematics and Physics* **20**:224–230.

Houch, M. H., J. Cohon, and C. ReVille, 1980, Linear decision rule in reservoir management and design, 6: incorporation of economic efficiency benefits and hydroelectric energy generation, *Water Resources Research* **16**:196–200.

Jensen, P. and G. Bhaumik, 1977, A flow augmentation approach to the network with gains minimum cost flow problem, *Management Science* **23**:631–643.

Karmarkar, N., 1984, *A New Algorithm for Linear Programming*, Bell Laboratories, New Jersey.

Kuby, M., K. Lee, J. Osleeb, and S. Ratick, A mathematical programming solution to the interperiod-network-storage-location-allocation problem, paper presented at Association of American Geographers Annual Meeting, Washington, D. C. April 1984.

Larson, R. C., 1974, A hypercube queuing model for facility location and redistricting in urban emergency services, *Computers and Operations Research* 1:67–95.

Loucks, D. P., 1970, Some comments on linear decision rules and chance constraints,'' *Water Resources Research* 6:668–671.

Maier, S., 1971, *A Compact Inverse Scheme Applied to a Multicommodity Network with Resource Constraints*, Technical Report no. 71–78, Operations Research House, Stanford University.

Orden, A., 1956, The transshipment problem, *Management Science* 2:276–285.

Osleeb, J. and S. Ratick, 1981, *Implications of Power Plant Coal Conversions on the Ports of New England, vols. 1 and 2*, U.S. Maritime Administration, NTIS:PB-82-136417, Washington, D.C.

Osleeb, J. and S. Ratick, 1983a, A mixed-integer and multiple objective programming model to analyze coal handling in New England, *European Journal of Operational Research* 12:302–313.

Osleeb, J. and S. Ratick, 1983b, The impacts of coal conversions on the ports of New England, *Economic Geography* 59:35–51.

Osleeb, J. and S. Ratick, 1983c, *Implications of World Coal Demands on U.S. Port Strategic Planning*, U.S. Maritime Administration, no. MA-DTMA-91-82-6-20032, Washington, D.C.

Osleeb, J. and S. Ratick, 1985, *Synergistic Gains from Storage on a Network: Development of an Interperiod Network Storage Location Allocation Model*, Final Report to the Geography and Regional Science Program, Award Number SES-8309173, National Science Foundation, Washington, D.C.

Ratick, S. and J. Osleeb, 1983, Optimizing freight transshipment: an evaluation of East Coast coal export options, *Transportation Research* 17A:493–504.

Reinert, K., 1983, Mathematical programming models for economic design and assessment of wind energy conversion systems, *Wind Engineering* 7:43–59.

ReVelle, C., E. Joeres, and W. Kirby, 1969, The linear decision rule in reservoir management and design, 1: development of the stochastic model, *Water Resources Research* 5:767–777.

ReVelle, C. and W. Kirby, 1970, Linear decision rule in reservoir management and design, 2: performance optimization, *Water Resources Research* 6:1033–1044.

ReVelle, C, and J. Gundelach, 1975, Linear decision rule in reservoir management and design, 4: a rule that minimizes output variance, *Water Resources Research* 11:197–203.

Roodman, G. M. and L. B. Schwarz, 1975, Optimal and heuristic facility phase-out strategies, *AIIE Transactions* 7:177–184.

Schapiro, T. F., 1975, Operations research models for energy planning, *Computers and Operations Research* 2:145–152.

Van Roy, T. J. and D. Erlenkotter, 1982, A dual-based procedure for dynamic facility location, *Management Science* 28:1091–1105.

Wesolowsky, G. O., 1973, Dynamic facility location, *Management Science* **19**:1241–1248.

Wesolowsky, G. O. and W. G. Truscott, 1975, The multiperiod location-allocation problem with relocation of facilities, *Management Science* **22**:57–65.

White, W. W., 1972, Dynamic transshipment networks: an algorithm and its application to the distribution of empty containers, *Networks* **2**:211–236.

11

SPATIAL INTERACTION BASED LOCATION-ALLOCATION MODELS

Morton E. O'Kelly
Ohio State University

NEAREST-CENTER OR PROBABILISTIC ALLOCATION?

Consider certain locationally fixed urban public facilities such as libraries, parks, and post offices. Recent research has addressed several problems related to the optimal provision of such facilities. Examples include the optimal location problem (Hansen, Peeters, and Thisse, 1983), the congestion of public facilities (Oakland, 1972, Leonardi, 1981), and the ability of jurisdictions to exclude some users from their facilities, together with related issues of spatial externalities (Ellickson, 1971; Papageorgiou and O'Kelly, 1984). A parallel set of problems also arises in the sphere of commercial locational analysis: facility siting (Erlenkotter, 1977, 1978), capacity constraints (Geoffrion and McBride, 1978), and trade-area delimitation (Batty, 1978). To a large extent, all of these problems can be readily solved if it is assumed that demand is allocated deterministically to the facilities.

In the absence of capacity constraints, an omnipotent planner could minimize the transportation costs of the system by forcing consumers to follow a least-cost allocation rule. The service areas could then be found by simply constructing proximal regions, and demand and congestion levels could be easily evaluated. Similarly, in the private facility location domain, locational strategies in a deter-

Acknowledgements: This material is based on work supported by the College of Social and Behavioral Sciences, Ohio State University, and the National Science Foundation under grant number SES 83-09590.

ministic world can be formulated. These solutions range in sophistication from the uniform spacing of centers in an ideal central place system to elaborate set covering rules in an emergency facility location system (Hogan and ReVelle, 1983).

REAL-WORLD TRAVEL PATTERNS

In reality, several services exist for which consumers choose a facility from which to obtain the service. The travel pattern of the consumers can produce a variety of allocations that differ from the simple nearest-center rule. Thus, although conventional location-allocation literature assumes that consumers travel to the nearest center offering a good to make their purchases, in reality individuals often travel further than necessary to purchase goods because, for example, other goods can be purchased at the same time (see Ambrose, 1968; Toyne, 1971; Warnes and Daniels, 1979; Fingleton, 1975; and Bacon, 1984).

The literature reports several empirical studies that explicitly test the nearest-center hypothesis (for a review see Hubbard, 1978). Davies (1976), for example found that shopping behavior in Coventry was either highly localized for convenience goods or involved less-frequent trips to the nearest large center. Other studies of consumer behavior in Britain found that the nearest-center hypothesis could be strongly refuted for both comparison shopping (Ambrose, 1968) and convenience items (Fingleton, 1975). Hodgson (1981) recently tested the hypothesis. He studied over 200,000 car license renewals in the city of Edmonton. Only 34% of the trips were to the facility nearest the licensee's home. Finally, a number of classic studies in the field have shown great variety in the types of factors that distort the basic travel pattern. Berry (1967) showed that interurban travel by a sample of rural Iowans exhibited greater dispersion of destination choices for fashion goods than for the more routinely purchased grocery and convenience goods. Similarly, Ray (1967) found that while the nearest-center allocation rule is important, cultural factors in the eastern Ontario area under study diminished the validity of the hypothesis. Compared to their French counterparts, English-speaking consumers had a propensity to make multipurpose trips, which diverted them from their nearest center.

The list of justification for these variations in behavior is a long one. Hanson (1980) discusses the role of diversification in consumer destination choice, and Hodgson (1978) and Devletoglou (1965) draw attention to the role of uncertainty about travel times and facility locations, (see also Golledge and Spector, 1978). Furthermore, consumers may be diverted from their nearest facility as a result of stochastic variations in attractiveness and congestion, (see, for example, Hodgson, 1981; Hubbard, 1978). For these reasons, several authors here suggested that the conventional models' assumption of travel to the nearest center should be replaced by a probabilistic allocation rule, and this suggestion has resulted in a series of studies of the role of consumer behavior in travel and interaction decisions (Clark and Rushton 1970; Clark 1968; and Rushton, Golledge, and Clark 1967).

IMPLICATIONS FOR FACILITY LOCATION

The implications of stochastic allocation of demand to facilities have had a long-standing appeal for theorists, (see, for example, Beckmann, 1983; Tapiero, 1974; Mirchandani and Oudjit, 1982; Webber, 1978). Some of these efforts are formulated in continuous space: Beaumont (1980), extending the work of Williams and Senior (1977), showed that the continuous probabilistic location-allocation problem reduces to the famous ice-cream-vendor location model. Webber (1979, pp. 327–340) used an information-minimizing model to obtain an agglomeration result (like Hotelling) under less restrictive assumptions. Probabilistic allocation of demand implies that expected inflows and congestion levels can be found, but these levels may be influenced more by the attraction of the facility than by its relative location. The extent to which users from outside a proximal area utilize a facility depends on the rate at which the probability of interaction declines with distance. If the probability of interaction is greatly diminished by a unit increase in distance, then travel across jurisdiction or trade-area boundaries will be slight. If, on the other hand, interaction is relatively unimpeded by distance there will be substantial overlap between jurisdictions, and in fact, some accessible facilities may attract users from many areas. Such a situation would show a large excess of visits to the facility over and above the number of trips originating near the facility, with a resultant problem of allocating costs for the service. This issue has been addressed by Linthorst and van Praag (1981), who model the interdependence between interaction patterns and service areas in the Netherlands.

These comments on the important role of the allocation mechanism in locating both private and public facilities form the background for the rest of the chapter. The next section presents an approach to probabilistic interaction modeling. That section is followed by some algorithmic developments for the probabilistic location-allocation model. Finally, a numerical example is developed for siting 20 facilities in the Hamilton, Ontario, region. The analysis uses a system of 181 neighborhoods.

SPATIAL INTERACTION BASED LOCATION-ALLOCATION MODELS

This part of the chapter presents Wilson's origin-constrained spatial-interaction model as an approach to stochastic allocation of demand to facilities. Next, this formulation is modified to incorporate integer variables denoting facility locations. Then the discussion states the well-known relationship between the probabilistic location-allocation model and the deterministic p-median problem and reviews a few numerical studies that use probabilistic allocation mechanisms in location models.

Spatial-Interaction Models of Allocation

A specific probabilistic allocation rule for the amount of flow between pairs of locations i and j can be written as a function of service demand that originates at i, the attraction at j, and the intervening distance. This classic *gravity* formulation allocates flows between pairs of locations depending on the locations' characteristics and their spatial arrangement. The idea has been used in such widely known models as those of Reilly (1931), Huff (1963; 1964) and Lakshmanan and Hansen (1965).

To begin formulation of a version of the model, an origin-constrained spatial interaction is defined as

$$S_{ij} = A_i O_i W_j \exp(-\beta C_{ij}) \tag{11.1}$$

where the terms A_i are

$$A_i = 1/\sum_j W_j \exp(-\beta C_{ij}) \tag{11.2}$$

O_i is the proportion of demand originating at i, W_j reflects the attractiveness of location j, and C_{ij} is a measure of the spatial separation between i and j. The parameter β is either calibrated to match some known interaction data or is defined exogenously. Note that A_i ensures that the sum of all the outflows from origin i add up to the amount of demand at that location.

It is easy to show that the expression in equation (11.1) is the solution to the following information-minimizing problem:

$$\text{Min} \sum_i \sum_j S_{ij}(\ln(S_{ij}/L_i W_j) - 1) \tag{11.3}$$

subject to

$$\sum_j S_{ij} = O_i \quad i = 1, \ldots, m \tag{11.4}$$

$$\sum_i \sum_j S_{ij} C_{ij} = \bar{C} \tag{11.5}$$

where $L_i = 1/m$; W_j is some known share of attraction at j; and in this example, $\bar{C}$ is the observed average trip length to be replicated by the model. (For a thorough development of this type of model see Wilson et al., 1981.) Following Wilson et al. (1981, p. 99), an allocation model such as (11.1) can also be generated as the solution to the following modification of the optimization problem above. Note that in the modified problem (11.6)–(11.7) constraint (11.5) has been relaxed and incorporated into the objective function and that the terms L_i and W_j have been assumed to be uniformly distributed across the locations.

$$\text{Min } (1/\beta) \sum_i \sum_j S_{ij} (\ln S_{ij} - 1) + \sum_i \sum_j S_{ij}C_{ij} \qquad (11.6)$$

subject to

$$\sum_j S_{ij} = O_i \qquad (11.7)$$

To minimize (11.6) subject to (11.7), form the Lagrangian

$$L = (1/\beta) \sum_i \sum_j S_{ij}(\ln S_{ij} - 1) + \sum_i \sum_j S_{ij}C_{ij} + \sum_i a_i \left(\sum_j S_{ij} - O_i \right) \qquad (11.8)$$

Then, setting

$$dL/dS_{ij} = (1/\beta) (1 + \ln S_{ij} - 1) + C_{ij} + a_i = 0 \qquad (11.9)$$

implying

$$S_{ij} = \exp[-\beta(a_i + C_{ij})] \qquad (11.10)$$

which can easily be translated to give the model

$$S_{ij} = \frac{O_i\exp(-\beta C_{ij})}{\sum_l \exp(-\beta C_{il})} \qquad (11.11)$$

which is in the form of a simple origin-constrained spatial-interaction model. This simple model generalizes the all-or-nothing assignment of demand typically embodied in location models.

So far, it has been assumed that facilities exist at all possible sites. In the rest of this section the probabilistic allocation model is used as a component of some simple facility location models. The discussion focuses on the formulation, properties, solution, and application of models of location *given* a spatial-interaction model such as (11.11).

The Location of Facilities: Problem Formulation

An origin-constrained spatial-interaction model with endogenous facility location can be derived as the solution to the following optimization problem:

$$\text{Min} \ (1/\beta) \sum_j Y_j \sum_i S_{ij} \ln(S_{ij} - 1) + \sum_i \sum_j Y_j S_{ij} C_{ij} \qquad \textbf{(11.12)}$$
$$(\text{S, Y})$$

subject to

$$\sum_j Y_j S_{ij} = O_i \qquad i = 1, \ldots, m \qquad \textbf{(11.13)}$$

$$\sum_j Y_j = p \qquad \textbf{(11.14)}$$

$$S_{ij} \leq Y_j \qquad i = 1, \ldots, M; j = 1, \ldots, n \qquad \textbf{(11.15)}$$

$$Y_j \text{ is 0 or 1} \qquad j = 1, \ldots, n \qquad \textbf{(11.16)}$$

where the decision variables include Y_j, which takes the value of one if a facility is located at j and is zero otherwise. The resulting model is

$$S_{ij} = A_i O_i Y_j \exp(-\beta C_{ij}) \qquad \textbf{(11.17)}$$

$$A_i = 1/\sum_l Y_l \exp(-\beta C_{il}) \qquad i = 1, \ldots, m \qquad \textbf{(11.18)}$$

Define J to be the set of locations with $Y_j = 1$. For a given J, the model has the property that the interactions S_{ij} satisfy the origin constraints.

Relationship Between the Models. The relationship between the optimization problem (11.12)–(11.16) and the conventional p-median problem must now be clarified. The value of the objective function at a solution to a p-median problem is given by

$$C(\text{min}) = \sum_i \sum_j O_i X_{ij} C_{ij} \qquad \textbf{(11.19)}$$

where X_{ij} allocates demand to the nearest of p available centers. $C(\text{min})$ is the minimum possible travel time in a system that satisfies all the demand and locates p facilities among the possible sites.

Turning to the probabilistic location-allocation model, consider the effects on an allocation array S as the impedance parameter β gets larger. As β increases, for a given set of facility locations, the term

$$\frac{Y_j \exp(-\beta C_{ij})}{\sum_l Y_l \exp(-\beta C_{il})} \qquad\qquad \textbf{(11.20)}$$

in equation (11.17) tends to X_{ij}, where $X_{ij} = 1$ if the travel time from i to j is smaller than the travel time from i to any other facility; and $X_{ij} = 0$ otherwise. Therefore, as the distance-impedance parameter β gets larger, the expression S_{ij} tends to $O_i X_{ij}$, and this model allocates the demand to the nearest facility in the same fashion as the p-median problem does. Thus the objective (11.12) reduces to the minimization of (11.19) when the impedance parameter β increases.

The nearest-center allocation rule, as a special case of the spatial-interaction model, has been used by Evans (1973), Wilson and Senior (1974), and Webber (1979, pp. 250–280) to show the connections between probabilistic allocation and linear programming.

Applications.   A number of issues have to be dealt with in applying location models. The number of centers p, for example, is very important in any application. The value of p might be dictated by a target level of customer service or by a marketing strategy. In theory the number of centers might be determined by the level of the service in a hierarchy of central functions (see, for example, Warnes and Daniels, 1979, p. 400). If the value of p is itself a decision variable, then the problem becomes more complicated. Usually, p is a decision variable in models with explicit measures of costs. Although the deterministic problem with facility costs has highly efficient solution methods (Erlenkotter, 1978), as yet only very simple versions of the corresponding probablistic problem have been solved. Two analyses along these lines (Webber, 1978; Leonardi, 1983) point to services being provided at all feasible locations.

Second, some limited numerical evidence as to the performance of the model is available. Hodgson (1978) took the results of a doubly constrained interaction model with endogenous facility location and compared them to a conventional p-median model. Ten possible locations were chosen at random from a square grid, and the distances between the points were calculated using the Euclidean metric. Intrazonal travel times were defined to be one-half the distance from a zone to its nearest neighbor. Three centers from among the ten possible sites were chosen, and values of $\beta = 0.5$, 1.0, and 1.5 were used. In six out of the ten runs with $\beta = 1.5$, the spatial-interaction model found the same solution values for the location as the deterministic model. The results did not coincide in all cases because the minimum feasible travel times (and hence the importance of β) vary between the runs. Further research is needed to show how these solutions depend on distance-deterrence terms *relative* to the minimum feasible travel time for the system.

Goodchild and Booth (1976) devised a probabilistic interaction model to describe the flow of population to 11 swimming pools in London, Ontario. Using this allocation model, two new locations were chosen to minimize the total distance traveled by the users of the system. The results of the analysis indicated some discrepancy between the municipality's plans for a new site and the estimate of the optimal location. Two important features of the analysis were a realistic estimate of demand based on age profiles throughout the city, and a demonstration that despite the inclusion of an elastic demand model, the usage of the system seemed to be largely independent of the supply of facilities. The analysts showed that in only a few cases did the number of nonusers exceed 10% of the users; and they concluded that "very few non-participants can be regarded as potential users of an increased supply of alternatives" for this particular recreational activity (Goodchild and Booth, 1976, p. 18).

Webber, O'Kelly, and Hall (1979) provided a more complex example of the interaction between facility location and the demand generated by residential locations. In the model, both residential and facility locations are endogenous. Further, the model reflects the behavior of consumers through a large number of allowable trip bundles; these bundles could contain 0, 1, or 2 visits to facilities. In the model's application to data from Hamilton, Ontario, the level of consumer demand proved to be relatively insensitive to facility location patterns.

Both of the studies mentioned found that the highly constrained nature of the interaction patterns made it impossible to induce variations in behavior by varying locations. Of course, because they use parameters calibrated from existing systems, there is really no way to judge from these results alone that demand can be assumed to be insensitive to facility locations. Nevertheless, two arguments support the assumption of inelastic demand as a first approximation. First, many services have a fixed demand that must be filled regardless of the location of facilities; and second, the conversion of the model to an elastic version seems to be relatively straightforward (Sheppard, 1980). Bearing these comments in mind, the next section proposes a solution strategy for a simple inelastic demand model. Some notes on extension to the elastic demand case are presented in the final section of the chapter.

A PROBABILISTIC LOCATION-ALLOCATION MODEL

This part of the chapter draws attention to the problem of finding solutions to location-allocation models that have imbedded spatial-interaction equations. The problem of finding global, as opposed to local, solutions to such mathematical programs is raised briefly in the next section. The section after that outlines a dual-based algorithm. The methods proposed here are analogous to the lagrangian relaxation technique developed for the p-median problem by Narula, Ogbu, and Samuelsson (1977).

Heuristic Versus Exact Methods

Problems such as (11.12)–(11.16) have been solved either by explicit enumeration or through some variant of the Teitz and Bart (1968) interchange heuristic (Hodgart, 1978). Leonardi (1983) has reported some numerical experience with a set of powerful heuristics that exploit the submodularity of the objective function. Hodgson (1978) and Beaumont (1980) have also solved probabilistic spatial-interaction models with heuristic methods. These researchers offer some suggestions as to the properties of facility locations in a system with realistic travel behavior, and the trial-and-error solution procedures of heuristics are acceptable for empirical applications. However, since the output from a heuristic is not guaranteed to correspond to a globally optimal solution, no sensitivity analysis can be performed. It is important therefore to develop an exact algorithm for the probabilistic location-allocation model. The following section suggests how to construct such a method.

Algorithmic Development

Suppose that several facilities are to be located. Indicate their location patterns by Y_j, where $Y_j = 1$ if a facility is located in j and is equal to 0 otherwise. Suppose that the frequency of demand for service at location i is known and is indicated by O_i. Finally, suppose that a known impedance factor β characterizes the effect of distance on interaction. Let the travel time from origin zone i to the facility at j be C_{ij}. Flows from origin zones to facilities are modeled as

$$S_{ij} = \frac{O_i Y_j \exp(-\beta C_{ij})}{\sum_l Y_l \exp(-\beta C_{il})} \tag{11.21}$$

where S_{ij} is the proportion of demand at i serviced by the facility at location j, and O_i is the proportion of all demand that occurs at i. Consider again the problem (11.12)–(11.16):

$$\text{Min } (1/\beta) \sum_i \sum_j Y_j S_{ij} (\ln S_{ij} - 1) + \sum_i \sum_j Y_j S_{ij} C_{ij} \tag{11.22}$$

subject to (11.13)–(11.16). Now relax constraint (11.13) to give the following lagrangian:

$$L = (1/\beta) \sum_i \sum_j Y_j S_{ij} (\ln S_{ij} - 1) + \sum_i \sum_j Y_j S_{ij} C_{ij}$$
$$+ \sum_i a_i \left(\sum_j Y_j S_{ij} - O_i \right) \tag{11.23}$$

which can be rearranged as

$$L' = (1/\beta) \sum_i \sum_j Y_j S_{ij} (\ln S_{ij} - 1 + \beta C_{ij} + \beta a_i) - \sum_i a_i O_i \quad \textbf{(11.24)}$$

Now, for all $Y_j = 1$,

$$dL'/dS_{ij} = (1/\beta) (1 + \ln S_{ij} - 1 + \beta C_{ij} + \beta a_i) \quad \textbf{(11.25)}$$

Equating (11.25) to zero gives

$$S_{ij} = \exp[-\beta (C_{ij} + a_i)] \quad \textbf{(11.26)}$$

substituting (11.26) back into L', we get

$$L' = (1/\beta) \sum_i \sum_j Y_j S_{ij} (-1) - \sum_i a_i O_i \quad \textbf{(11.27)}$$

which is the dual of the original problem. The minimization with respect to Y can be accomplished by selecting the p smallest values of

$$P_j(\mathbf{a}) = \sum_i (-1)\exp[- \beta(a_i + C_{ij})] \quad \textbf{(11.28)}$$

and then by setting the corresponding values of Y equal to one. This implicit treatment of the constraint on the number of facilities is suggested by Narula, Ogbu, and Samuelsson (1977) and is also used in Handler and Mirchandani (1979). The dual problem can thus be written

$$\underset{\mathbf{a}}{\text{Max}} \; FD = (1/\beta) \sum_j Y_j P_j (\mathbf{a}) - \sum_i a_i O_i \quad \textbf{(11.29)}$$

where the decision variable is the vector of multipliers $\mathbf{a}$.

The problem is to find the $\mathbf{a}$ vector that solves (11.29). As a preliminary result, note that for a given set of facility locations the optimal multipliers can be obtained analytically. Note that

$$\sum_j Y_j S_{ij} = O_i \quad \textbf{(11.30)}$$

and so

$$\sum_j Y_j \exp[-\beta (a_i + C_{ij})] = O_i \quad \textbf{(11.31)}$$

Therefore,

$$a_i = (-1/\beta) \ln[O_i \sum_j Y_j \exp(-\beta C_{ij})] \qquad (11.32)$$

It can easily be shown that for a given value of β, a lower (upper) bound on a_i can be found by setting facilities open in the p largest (smallest) entries in row i of C_{ij}. Next observe that for arbitrary values of **a**, and their associated set of facility locations, the primal constraints on the sum of the outflows from i may not be satisfied. Therefore—following Narula, Ogbu, and Samuelsson (1977), Handler and Mirchandani (1979), Held, Wolfe and Crowder (1974), and Kennington and Helgason (1980)—a subgradient search is employed for **a**. The primal constraint (11.30) requires that

$$f_i = O_i - \sum_j Y_j S_{ij} = 0 \qquad (11.33)$$

and an appropriate subgradient search sets the $(m + 1)$ value of a_i as

$$a_i(m + 1) = a_i(m) - q(f_i) \qquad (11.34)$$

where q is a step-length parameter determined by experiment. Notice that if $f_i > 0$, then $a_i(m + 1) < a_i(m)$, and so S_{ij} increases. Conversely, if $f_i < 0$, then $a_i(m + 1) > a_i(m)$ and S_{ij} decreases. Of course, if $f_i = 0$ for any i, the primal constraint is satisfied and no adjustment is made to the corresponding multiplier. In the next section, two alternative step-length rules are tested and a preferred method is described.

RESULTS

This part of the chapter reports some numerical results from an application of the algorithm. A 181-x-181 travel time array from the Hamilton, Ontario, metropolitan area is used as the basis for a test of the algorithm's performance. A set of 20 facilities is located in the 181 zones under a variety of characterizations of travel behavior. The next section discusses the choice of step-length procedure; in turn, the section after that discusses the performance of the algorithm over different parameter values. Finally, we compare the resulting locations of facilities under two different spatial-interaction assumptions.

Step-Length Procedure

Two different step-length procedures were tested. The results are quite sensitive to the choice between the two, so that detailed comments are provided to guide future research. The first step-length method sets the choice of q in equation (11.34) as

$$q = |a_i|a_i/0.1 \qquad (11.35)$$

The idea behind this method is that large corrections ought to be made whenever **a** is large (f takes care of the direction of the change). The resulting primal and dual objective values are shown in Table 11.1 for a variety of impedance parameters. The first column of the table shows the impedance parameter, and the second and third columns show the values of the primal (equation 11.22) and dual (equation 11.29) objectives, respectively. Notice that the dual objective provides a lower bound on the primal solution. There may be values of the primal less than those reported in column two, but they cannot be less than the corresponding entry in column three. The final column of the table shows the average travel time associated with the location-allocation "solution." For values of β less than 0.2, the dual bound is quite tight and a complete solution could be found with the addition of a branching routine. Rather than pursue a full branch-and-bound method, we decided to experiment with the step-length procedure.

Experiments with alternative step-length methods have shown the following adjustment to be preferable to the previous one:

$$q = \Delta/\beta \qquad (11.36)$$

where $\Delta = 4$ for iterations $1, \ldots, 5$; then Δ is multiplied by 0.75 every three iterations until $\Delta < 0.001$, when Δ is reset to 4. This method follows suggestions in Held, Wolfe, and Crowder (1974). The resulting objective-function values are shown in Table 11.2.

It is apparent that the primal-dual gap is smaller in general in Table 11.2 than in Table 11.1; furthermore the primal objective attains lower values in general using the second step-length method.

Table 11.1. Objective-Function Results, Step-Length Method 1

β	Primal Objective	Dual Objective	Average Travel Time
0.0125	−689.88	−690.07	10.2015
0.025	−339.84	−340.08	10.1282
0.050	−164.88	−165.27	9.9372
0.100	−76.86	−78.57	9.1529
0.200	−32.59	−37.41	10.0566
0.400	−11.17	−65.34	9.4756

Table 11.2. Objective-Function Results, Step-Length Method 2

β	Primal Objective	Dual Objective	Average Travel Time
0.0125	−689.88	−690.07	10.2015
0.025	−339.86	−339.99	10.1159
0.050	−164.90	−165.16	9.8930
0.100	−77.60	−77.96	9.1529
0.200	−34.56	−35.64	7.5015
0.400	−13.98	−28.06	5.5069

Table 11.3. Iteration Results for Primal and Dual, β = 0.025

K	F BKS	F I = K	ZINF	PCBAR	FD I = K	SUM1	SUM2	SUMSQ
1	−339.845	−339.845	−349.973	10.12	−340.081	−41.672	298.409	0.000101
2	−339.845	−339.845	−349.973	10.12	−340.065	−41.568	298.497	0.000090
3	−339.845	−339.845	−349.973	10.12	−340.051	−41.472	298.579	0.000080
4	−339.845	−339.845	−349.973	10.12	−340.039	−41.384	298.655	0.000071
5	−339.845	−339.845	−349.973	10.12	−340.029	−41.302	298.727	0.000063
6	−339.845	−339.845	−349.973	10.12	−340.018	−41.227	298.791	0.000057
7	−339.845	−339.845	−349.973	10.12	−340.012	−41.174	298.838	0.000052
8	−339.845	−339.845	−349.973	10.12	−340.005	−41.125	298.881	0.000048
9	−339.862	−339.862	−349.978	10.11	−339.999	−41.078	298.922	0.000036
10	−339.862	−339.862	−349.978	10.11	−339.997	−41.045	298.952	0.000034
11	−339.862	−339.862	−349.978	10.11	−339.993	−41.014	298.979	0.000032
12	−339.862	−339.862	−349.978	10.11	−339.991	−40.984	299.007	0.000031
13	−339.862	−339.862	−349.978	10.11	−339.989	−40.962	298.027	0.000029
14	−339.862	−339.862	−349.978	10.11	−339.986	−40.941	298.044	0.000028
15	−339.862	−339.862	−349.978	10.11	−339.984	−40.921	299.063	0.000027
16	−339.862	−339.862	−349.978	10.11	−339.983	−40.906	299.077	0.000026
17	−339.862	−339.862	−349.978	10.11	−339.982	−40.892	299.091	0.000026
18	−339.862	−339.862	−349.978	10.11	−339.981	−40.877	299.104	0.000025
19	−339.862	−339.862	−349.978	10.11	−339.980	−40.867	299.113	0.000024
20	−339.862	−339.862	−349.978	10.11	−339.979	−40.857	299.123	0.000024

F BKS : primal objective at best known solution (equation 11.22).

F I = K : primal objective at the kth iteration.

ZINF : $(1/\beta) \sum_i \sum_j Y_j S_{ij} (\ln S_{ij} - 1)$, from equation (11.22).

PCBAR : $\sum_i \sum_j Y_j S_{ij} C_{ij}$ average trip length, from equation (11.22).

FD I = K : dual objective at the current iteration (equation 11.27).

SUM1 : $(1/\beta) \sum_j Y_j A_j$ from equation (11.27).

SUM2 : $\sum_i a_i O_i$, from equation (11.27).

SUMSQ : sum of the squared errors, from equation (11.33).

Dual Gaps And the Value of Beta

Evidently the efficacy of the dual procedure depends on the value of the impedance parameter—notice that the gap between the primal and dual objectives remains large when β is large. There seems to be an explanation for this behavior of the algorithm. Tables 11.3 and 11.4 show the progress of the primal and dual objectives through 20 iterations for β values of 0.025 and 0.8, respectively.

Now we can see that the explanation for the gap between the primal and dual objectives is based on the large values of the multipliers produced by the dual method whenever β is large. This results in a large value of SUM2 (see Table 11.4) which, when subtracted from SUM1 leaves a large negative value

Table 11.4. Iteration Results for Primal and Dual, β = 0.80

K	F BKS	F I = K	ZINF	PCBAR	FD I = K	SUM1	SUM2	SUMSQ
1	−0.891	2.487	−9.724	12.21	—	—	—	—
2	−0.891	3.452	−9.944	13.39	−17034.270	−2497.	14536.582	—
3	−2.270	−2.270	−9.033	6.76	−14658.492	−94.32	14564.172	935.
4	−2.270	−2.197	−9.313	7.11	−14576.531	−9.774	14566.758	3.15
5	−3.190	−3.190	−9.426	6.23	−14570.516	−3.464	14567.055	0.25
6	−3.190	−3.176	−8.868	5.69	−14569.699	−2.572	14567.129	0.11
7	−3.190	−3.142	−9.400	6.25	−14569.359	−2.220	14567.141	0.06
8	−3.290	−3.290	−9.082	5.79	−14569.141	−1.977	14567.164	0.05
9	−3.290	−3.152	−9.363	6.21	−14568.988	−1.812	14567.180	0.03
10	−3.290	−3.111	−9.357	6.24	−14568.879	−1.698	14567.184	0.03
11	−3.356	−3.356	−9.172	5.81	−14568.797	−1.605	14567.195	0.02
12	−3.356	−3.161	−9.331	6.17	−14568.727	−1.529	14567.199	0.02
13	−3.356	−2.954	−9.246	6.29	−14568.672	−1.474	14567.199	0.02
14	−3.356	−2.983	−9.496	6.51	−14568.629	−1.430	14567.199	0.02
15	−3.564	−3.564	−9.056	5.49	−14568.582	−1.385	14567.199	0.02
16	−3.564	−2.959	−9.462	6.50	−14568.555	−1.356	14567.199	0.02
17	−3.676	−3.676	−9.078	5.40	−14568.520	−1.327	14567.195	0.01
18	−3.676	−2.834	−9.378	6.54	−14568.496	−1.301	14567.195	0.01
19	−3.676	−3.142	−9.371	6.22	−14568.473	−1.282	14567.191	0.01
20	−3.676	−2.919	−9.432	6.51	−14568.453	−1.263	14567.191	0.01

F BKS : primal objective at best known solution (equation 11.22).
F I = K : primal objective at the kth iteration.

ZINF : $(1/\beta) \sum_i \sum_j Y_j S_{ij} (\ln S_{ij} - 1)$, from equation (11.22).

PCBAR : $\sum_i \sum_j Y_j S_{ij} C_{ij}$ average trip length, from equation (11.22).

FD I = K : dual objective at the current iteration (equation 11.27).

SUM1 : $(1/\beta) \sum_j Y_j P_j$ from equation (11.27).

SUM2 : $\sum_i a_i O_i$ from equation (11.27).

SUMSQ : sum of the squared errors, from equation (11.33).

of the dual objective. At the same time, the value of the primal objective should be approaching the average trip length PCBAR. In the limit, as β gets extremely large the value of the primal objective should tend to PCBAR (see equations 11.19 and 11.20). (In the present case, however, when $\beta = 0.8$ the first component of the primal objective remains negative, simply because β is not set to a large number.) At the same time, the dual objective becomes highly negative as a result of the values of the multipliers. Thus, for large values of the impedance parameter, it is unlikely that the dual method presented here will ever produce a small gap.

Location-Allocation Results

The resulting patterns of locations and allocations for $\beta = 0.025$ are shown in Table 11.5, while similar data are presented in Table 11.6 for $\beta = 0.2$. These location-allocation tables are found using the second step-length method. The first column shows the location of the 20 facilities, and these locations are shown in Figs. 11.1 and 11.2. (These location codes are in the format *zznn*, where *zz* is the zone identifier and *nn* represents the neighborhood within that zone.) The inflow to each facility is the sum of the demands allocated to the

Table 11.5. Location-Allocation, $\beta = 0.025$

Location	Inflow	% Near	Average Travel Time
906	0.05008	10.69	9.99130
1009	0.05000	2.49	10.05423
1010	0.04965	13.70	10.36131
101	0.04995	12.90	10.20260
102	0.04996	0.43	10.21568
103	0.04985	1.02	10.31112
201	0.04995	0.44	10.21159
202	0.04953	3.67	10.52506
704	0.04990	1.73	10.23997
604	0.04986	0.94	10.27149
603	0.04956	1.28	10.50371
508	0.04982	2.10	10.27036
703	0.05010	2.36	10.05465
601	0.05085	2.55	9.47751
602	0.05084	1.53	9.47516
501	0.05084	1.03	9.43295
502	0.05032	0.84	9.81265
503	0.04993	2.60	10.07693
504	0.04957	27.11	10.30252
1803	0.04943	24.80	10.61086

Table 11.6. Location-Allocation, β = 0.20

Location	Inflow	% Near	Average Travel Time
101	0.04998	5.76	8.33937
102	0.04842	11.71	8.43474
306	0.04167	30.19	7.10565
601	0.05910	22.63	7.33365
602	0.05791	5.94	7.24831
501	0.05981	26.50	6.91516
502	0.05480	12.32	7.17670
504	0.05134	6.95	7.18778
403	0.05125	7.35	7.00691
404	0.05091	25.68	6.85981
405	0.04933	9.62	6.81975
406	0.04830	14.18	6.74652
304	0.04437	9.17	6.95107
1803	0.04782	0.00	8.51377
1805	0.04821	2.59	8.15379
1808	0.04399	6.45	8.17632
1814	0.04988	16.34	7.56769
1905	0.04810	48.94	7.32244
804	0.04856	7.74	8.28779
803	0.04624	38.44	8.25054

center. In the tables, the % Near column shows the percentage of the demand allocated to the facility, for which this is the *nearest* alternative. (The maps use heavier shading for the facilities that are nearest centers for a large portion of their users.) The column labeled Average Travel Time reports the average travel time (in minutes) for the users of each facility. Spatially, the two systems are quite different, with the small impedance parameter associated with a clustered set of facility locations (Fig. 11.1), and the large parameter showing a more dispersed set of locations (Fig. 11.2). The major contrast between the two tables is that the system with $\beta = 0.025$ exhibits a very uniform set of facilities, with little or no preference for the nearest facility. On the other hand, the system with a larger impedance parameter ($\beta = 0.2$) displays greater dispersion in facility locations as well as more marked use of the nearest facility and therefore shorter average travel times.

SUGGESTIONS FOR FURTHER RESEARCH

To conclude the chapter, this section raises several remaining research issues. Although brief, the discussion should convey a sense of the large number of potential extensions to the spatial-interaction-based location model.

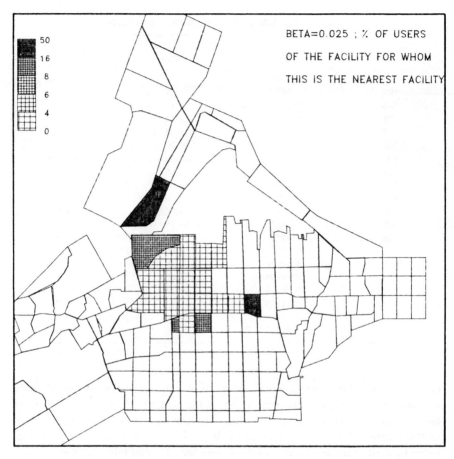

Figure 11.1. The location of facilities in the Hamilton, Ontario, region using an interaction model with β = 0.025.

Simultaneous Size and Location Models

The problem of simultaneously choosing facility locations and their sizes is a particularly difficult one. A fully interdependent model would determine locations endogenously together with the size of facilities at these locations. This general approach poses severe computational problems, given the current state of the art in mixed-integer programming. The general model involves the solution of a set of simultaneous equations describing facility size *and* the combinatorial problem of determining the set of facility sites.

Some have simplified this problem by separating the location and size problems. Goodchild (1978) chooses locations separately by solving a conventional

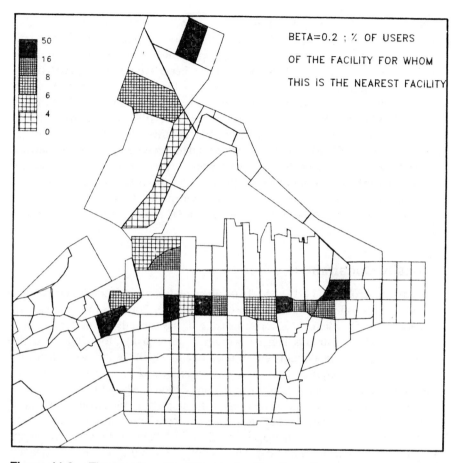

Figure 11.2. The location of facilities in the Hamilton, Ontario, region using an interaction model with β = 0.2.

location-allocation problem and then finds the facility sizes at these locations. The fundamental assumption underlying this method is that facility locations are chosen to minimize transportation costs, and that demand for the center can then be determined on the basis of a combination of distance, exogenous attraction parameters, and endogenous usage patterns. Obviously, there is some inconsistency between the behavioral rules governing the location and allocation phases, but it is an interesting approach to simplification of a difficult model.

Other theoretical advances have occurred, however, and it is now possible to describe a large collection of models that represent the size of retail facilities based on an imbedded spatial-interaction model. Batty (1978) describes an extension of Reilly's (1931) model of market areas and uses it to represent a retailing hierarchy. Wilson (1978) presents an explicit central place hierarchy that generalizes the traditional literature by including a probabilistic spatial-interaction model with endogenous facility location. Wilson and others have determined analytically some of the properties of the distribution of retail facilities when these facilities exhibit a dynamic size adjustment in response to utilization rates (Wilson et al., 1981; Wilson 1981). Allen and Sanglier (1979, 1981) extend the model to a dynamic form by devising equations that describe the changes in facility size as a function of market area population, and they have applied this framework to a regional simulation in Belgium. Ducca and Wilson (1976) describe a model of shopping-center distribution and apply the model to data from San Francisco. They consider only one level of service in their model. O'Kelly (1981, 1983a, 1983b) models different levels of a retailing system but keeps the facility locations fixed.

Capacity Constraints

Consider the problem (11.12)–(11.16) with an additional constraint that requires the sum of all inflows to a center to be less than or equal to the product K_jY_j (equation 11.15), where K_j is some measure of the capacity of the facility at j. If the capacity constraints can be written as an equality for all j, then the problem is effectively a doubly constrained one in which the inflows to facilities must match some required size—if a facility is located there. Hodgson (1978) has given some preliminary results from a related model. Many further difficulties remain, and it is important to determine the impact of such constraints on a system with several types of facilities. One particular issue is whether a facility would ever be optimally allocated an inflow less than its capacity. A highly attractive facility with limited capacity would probably be allocated a capacity inflow, but a less attractive facility with a large capacity could well be of excess size.

Constraints on Location

In practical applications of location models, some potential facility sites are already occupied by centers, while others are preempted by competing land uses. Such constraints can be incorporated in the model by restricting the range of the location indices j and k. More complex interaction between competing activities may require that the locational constraints be endogenous to the model, so that if a facility of a certain type is located in one area, then others may be restricted in choosing nearby sites for facilities. Furthermore, the addition of locational constraints enables the analyst to construct plans for incremental facility locations (of a myopic kind) by successively constraining the set of fixed locations. Goodchild and Booth (1976) give an example of this type of procedure. Other dynamic location-allocation schemes might follow the guidelines of Scott (1971) or Sheppard (1974). Finally, the set of feasible locations could be made endogenous to reflect municipal reorganization, expansion, or annexation. In this case, the size of the location-allocation problem could be increased over time.

Variations in Mobility

The effects of different levels of consumer mobility on the provision of facilities can be examined by varying the distance-impedance parameter β in the interaction model. (Mackett, 1980, and Sly and Tayman, 1980 report existing research on urban travel behavior and urban morphology.) The importance of consumer mobility has been recognized in several studies that draw attention to *immobile service-dependent households* (Wolch, 1981). For these groups, the travel decision may be so severely constrained that the concepts of efficient allocation and destination choice may be irrelevant. Such travel restrictions may perhaps be elucidated through a consideration of more disaggregated interaction patterns (say by income).

Access and Demand

The role of distance in intraurban spatial interaction is quite complicated. On the one hand, households with greater levels of accessibility are more likely to use a service; on the other hand, a distance-decay effect reveals a declining probability of interaction between i and j with increasing distance between them. The *rate* at which distance impedes interaction varies with the type of interaction, as does the rate at which demand declines with inaccessibility. Benson (1980) gives some locational implications of the sensitivity of demand to access, but further work is needed to extend these analyses to a model with realistic interaction effects.

A related problem arises with reference to applying the model to a rural health planning situation: planning for health facility location impinges on the spatial structure of a region—different plans may propose radically different spatial distributions of facilities. Although we may wish to perform an analysis of these locational patterns using interaction equations, the parameters of such models depend on the spatial arrangement of opportunities (Fotheringham, 1981). Therefore, it is not possible to use a calibrated spatial-interaction model from one distribution of facilities to assess a planned location pattern, for if the new scheme has different interfacility accessibility measures, then the calibrated model will be misleading. Fotheringham (1983a, 1983b) has developed a "competing destinations" model that avoids some of the pitfalls of conventional spatial-interaction formulations, and he has shown that impedance parameters calibrated using his model are relatively free of spatial-structure effects. An interesting research problem would be to use the competing destinations model in an interaction-based location-allocation problem. A major difficulty emerges in that the algorithm developed in this chapter relies on the calculation of $P_j(\mathbf{a})$ (equation 11.28) to determine "open" facilities; but in a competing destinations framework the term $P_j(\mathbf{a})$ would depend on which centers are open, and so the competing destinations location-allocation model cannot be solved using the methods developed here.

Hierarchical Levels of Service

Central place theorists have studied the relationships between hierarchical order and travel behavior (see also Narula and Ogbu 1979; O'Kelly and Storbeck 1984). The most relevent of these ideas is that high-order, infrequently demanded goods attract consumers over long distances. This notion suggests that the impact of accessibility on demand levels varies with the type of good. Such an idea cannot be fully explored without an explicit model of consumer interaction behavior. The probabilistic location-allocation model proposed here allows for different impedance parameters across different hierarchical levels. Some a priori expectations indicate that the interaction effects are likely to depend on the hierarchical level of the facilities. It is of some interest whether flows toward high-order facilities (e.g., major retail centers) are influenced by the size of the facility itself or by the size of the other facilities with which it interacts; also of interest is whether or not this relationship is symmetrical. At present, the role of distance in such flows is ambiguous; do neighboring large centers complement each other by generating a set of multistop, multipurpose trips between each other, or are such proximal locations likely to compete and draw sales away from each other? Casual observation of the current locational patterns of large suburban malls suggests that the former process is more likely, as many shopping centers are located close together. As yet though, no satisfactory explanation for the agglomeration of such large centers is available that is at the same time consistent with known patterns of spatial interaction. The methods described in this chapter provide some of the tools needed to pursue these research topics.

REFERENCES

Allen, P. M. and M. Sanglier, 1979, A dynamic model of growth in a central place system, *Geographical Analysis* **11:**256–272.

Allen, P. M. and M. Sanglier, 1981, A dynamic model of a central place system II, *Geographical Analysis* **13**:149–164.

Ambrose, P. J., 1968, An analysis of intra-urban shopping patterns, *Town Planning Review* **38**:327–334.

Bacon, R. W., 1984, *Consumer Spatial Behaviour*, Oxford University Press, New York.

Batty, M., 1978, Reilly's challenge: new laws of retail gravitation which define systems of central places, *Environment and Planning, A* **10**:185–219.

Beaumont, J. R., 1980, Spatial interaction models and the location-allocation problem, *Journal of Regional Science* **20**:37–50.

Beckman, M. J., 1983, On optimal spacing under exponential distance effect, in *Locational Analysis of Public Facilities*, J.-F. Thisse and H. G. Zoller, eds., North-Holland, Amsterdam, pp. 117–125.

Benson, B. L., 1980, Loschian competition under alternative demand conditions, *American Economic Review* **70**:1098–1105.

Berry, B. J. L., 1967, *Geography of Market Centers and Retail Distribution.* Prentice-Hall, New Jersey.

Clark, W. A. V., 1968, Consumer travel patterns and the concept of range. *Association of American Geographers Annals* **52**:386–396.

Clark, W. A. V. and G. Rushton, 1970, Models of intraurban consumer behavior and their implications for central place theory, *Economic Geography* **46**:486–497.

Davies, R. L., 1976, *Marketing Geography*, Methuen. London.

Devletoglou, N. E. 1965. A dissenting view of duopoly and spatial competition, *Economica* **35**:140–160.

Ducca, F. W. and R. H. Wilson, 1976, A model of shopping center location, *Environment and Planning, A* **8**:613–623.

Ellickson, B., 1971, Jurisdictional fragmentation and residential choice, *American Economic Review* **61**:334–339.

Erlenkotter, D., 1977, Facility location with price-sensitive demands: private, public, and quasi-public, *Management Science* **24**:378–386.

Erlenkotter, D., 1978, A dual-based procedure for uncapacitated facility location, *Operations Research* **16**:992–1009.

Evans, S. P., 1973, A relationship between the gravity model for trip distribution and the transportation problem in linear programming, *Transportation Research* **7**:39–46.

Fingleton, B., 1975, A factorial approach to the nearest centre hypothesis, *Institute of British Geographers Transactions* **65**:131–140.

Fotheringham, A. S., 1981, Spatial structure and distance-decay parameters, *Association of American Geographers Annals* **71**:425–436.

Fotheringham, A. S., 1983a, A new set of spatial interaction models: the theory of competing destinations, *Environment and Planning, A* **15**:15–36.

Fotheringham, A. S., 1983b, Some theoretical aspects of destination choice and their relevance to production-constrained gravity models, *Environment and Planning, A* **15**:1121–1132.

Geoffrion, A. M. and R. McBride, 1978, Lagrangean relaxation applied to capacitated facility location problems, *AIIE Transactions* **10**:40–47.

Golledge, R. G. and A. N. Spector, 1978, Comprehending the urban environment: theory and practice, *Geographical Analysis* **10**:403–426.

Goodchild, M. F., 1978, Spatial choice in location-allocation problems: the role of endogenous attraction, *Geographical Analysis* **10**:65–72.

Goodchild, M. F. and P. J. Booth, 1976, Modelling human spatial behaviour in urban recreation facility site location, **DP-007,** Department of Economics, University of Western Ontario, London.

Handler, G. Y. and P. B. Mirchandani, 1979, *Location on Networks: Theory and Algorithms.* MIT Press, Cambridge, Mass.

Hansen, P., D. Peeters, and J.-F. Thisse, 1983, Public facility location models: a selective survey, in *Locational Analysis of Public Facilities,* J.-F. Thisse and H. G. Zoller, eds., North-Holland, Amsterdam, pp. 223–262.

Hanson, S., 1980, Spatial diversification and multipurpose travel: implications for choice, *Geographical Analysis* **12**:245–257.

Held, M., P. Wolfe, and H. P. Crowder, 1974, Validation of sub-gradient optimization, *Mathematical Programming* **6**:62–88.

Hodgart, R. L., 1978, Optimizing access to public services: a review of problems, models, and methods of locating central facilities, *Progress in Human Geography* **2**:17–48.

Hodgson, M. J., 1978, Towards more realistic allocation in location-allocation models: an interaction approach, *Environmental and Planning, A* **10**:1273–1285.

Hodgson, M. J., 1981, A location-allocation model maximizing consumers' welfare, *Regional Studies* **15**:493–506.

Hogan, K. and C. ReVelle, 1983, Backup coverage concepts in the location of emergency services, *Modeling and Simulation* **14(4)**:1423–1428.

Hubbard, R., 1978, A review of selected factors conditioning consumer travel behavior, *Journal of Consumer Research* **5**:1–21.

Huff, D. L., 1963, A probabilistic analysis of shopping center trade areas, *Land Economics* **39**:81–90.

Huff, D. L., 1964, Defining and estimating a trading area, *Journal of Marketing* **28**:34–38.

Kennington, J. L. and R. V. Helgason, 1980, *Algorithms for Network Programming,* Wiley, New York.

Leonardi, G., 1981, A unifying framework for public facility location problems, *Environment and Planning, A* **13**:1001–1028, 1085–1108.

Leonardi, G., 1983, The use of random-utility theory in building location-allocation models, in *Locational Analysis of Public Facilities,* J.-F. Thisse and H. G. Zoller, eds., North-Holland, Amsterdam, pp. 357–383.

Lakshmanan, T. R. and W. G. Hansen, 1965, A retail market potential model, *American Institute of Planners Journal* **31**:134–143.

Linthorst, J. M. and B. van Praag, 1981, Interaction patterns and service areas of local public services in the Netherlands, *Regional Science and Urban Economics* **11**:39–56.

Mackett, R. L., 1980, The relationship between transport and the viability of central and inner urban areas, *Journal of Transportation and Economic Policy* **14**:267–294.

Mirchandani, P. B. and A. Oudjit, 1982, Probabilistic demands and costs in facility location problems, *Environment and Planning, A* **14:**917–932.

Narula, S. C., U. I. Ogbu, and H. M. Samuelsson, 1977, An algorithm for the *p*-median problem, *Operations Research* **25:**709–713.

Narula, S. C. and U. I. Ogbu, 1979, A hierarchical location-allocation problem, *Omega* **7:**137–143.

Oakland, W. H., 1972, Congestion, public goods and welfare, *Journal of Public Economics* **1:**339–357.

O'Kelly, M. E., 1981, A model of the demand for retail facilities incorporating multistop multipurpose trips, *Geographical Analysis* **13:**134–148.

O'Kelly, M. E., 1983*a*, Multipurpose shopping trips and the size of retail facilities, *Association of American Geographers Annals* **73:**231–239.

O'Kelly, M. E., 1983*b*, Impacts of multistop multipurpose trips on retail distributions, *Urban Geography* **4:**173–190.

O'Kelly, M. E. and J. E. Storbeck, 1984, Hierarchical location models with probabilistic allocation, *Regional Studies* **18:**121–129.

Papageorgiou, Y. Y. and M. E. O'Kelly, 1984, Guidelines for designing an abstract new town, *Geographical Analysis* **16:**97–120.

Ray, D. M., 1967, Cultural differences in consumer travel behavior in Eastern Ontario, *Canadian Geographer* **11:**143–156.

Reilly, W. J., 1931, *The Law of Retail Gravitation,* Putman, New York.

Rushton, G., R. G. Golledge, and W. A. V. Clark, 1967, Formulation and test of a normative model for the spatial alloction of grocery expenditures by a dispersed population, *Association of American Geographers Annals* **47:**389–400.

Scott, A. J., 1971, Dynamic location-allocation systems: some basic planning strategies, *Environment and Planning, A* **3:**73–82.

Sheppard, E. S., 1974, A conceptual framework for dynamic location-allocation analysis, *Environment and Planning, A* **6:**547–564.

Sheppard, E. S., 1980, Location and the demand for travel, *Geographical Analysis* **12:**111–128.

Sly, D. F. and J. Tayman, 1980, Changing metropolitan morphology and metropolitan service expenditure in cities and rings, *Social Science Quarterly* **61:**595–611.

Tapiero, C. S., 1974, The demand and utilization of recreational facilities: a probability model, *Regional Science and Urban Economics* **4:**173–183.

Teitz, M. P. and P. Bart, 1968, Heuristic methods for estimating the generalized vertex median of a weighted graph, *Operations Research* **16:**955–961.

Toyne, P., 1971, Customer trips to retail business, in W. L. D. Ravenhill and K. J. Gregory, eds., *Exeter Essays in Geography,* Exeter University Press, Exeter.

Warnes, A. M. and P. W. Daniels, 1979, Spatial aspects of an intrametropolitan central place hierarchy, *Progress in Human Geography* **3:**384–406.

Webber, M. J., 1978, Spatial interaction and the form of the city, in A. Karlquist, L. Lundquist, F. Snickars, and J. W. Weibull, eds., *Spatial Interaction Theory and Planning Models,* North-Holland, Amsterdam, pp. 203–226.

Webber, M. J., 1979, *Information Theory and Urban Spatial Structure,* Croom Helm, London.

Webber, M. J., M. E. O'Kelly, and P. D. Hall, 1979, Empirical tests on an information minimizing model of consumer characteristics and facility location, *Ontario Geography* **13:**61–80.

White, A. N., 1979, Accessibility and public facility location, *Economic Geography* **55:**18–35.

White, R. W., 1978, The simulation of central place dynamics: two-sector systems and the rank-size distribution, *Geographical Analysis* **10:**201–208.

Williams, H. C. W. L., and M. L. Senior, 1977, A retail location model with overlapping market areas: Hotelling's problem revisited, *Urban Studies* **14:**203–205.

Wilson, A. G., 1974. *Urban and Regional Models in Geography and Planning,* Wiley, New York.

Wilson, A. G., 1978, Spatial interaction and settlement structure; towards an explicit central place theory, in A. Karlquist, L. Lundquist, F. Snickars, and J. W. Weibull, eds., *Spatial Interaction Theory and Planning Models,* North-Holland, Amsterdam, pp. 137–156.

Wilson, A. G., and M. L. Senior, 1974, Some relationships between entropy maximizing models, mathematical programming models and their duals, *Journal of Regional Science,* **14:**207–215.

Wilson, A. G., 1981, *Catastrophe Theory and Bifurcation: Applications to Urban and Regional Systems.* Croom Helm, London.

Wilson, A. G., J. D. Coehlo, S. M. Macgill, and H. C. W. L. Williams, 1981, *Optimization in Locational and Transport Analysis.* Wiley, New York.

Wolch, J., 1981, The location of service-dependent households in urban areas, *Economic Geography,* **57:**52–67.

12

DATA AGGREGATION AND THE *p*-MEDIAN PROBLEM IN CONTINUOUS SPACE

Paul A. Casillas

Oregon State University

The solutions to many problems in locational analysis depend on, among other things, distances from demand points to service centers or to potential service centers. Because of this dependency, one expects that errors in distance measurement will introduce errors into these solutions.

One type of distance measurement error may occur owing to the aggregation of demand points in an area. This chapter presents a brief overview of this aggregation problem in locational analysis and then discusses the author's attempt to analyze errors in the solution of the *p*-median problem that are caused by data aggregation.

DATA AGGREGATION IN LOCATIONAL ANALYSIS

In this chapter, data aggregation refers to the assumption that the dispersed demand in an area is located at a single point. This point (whose weight is equal to the sum of the weights of the demand points in the area) is then used as the basis for measuring distance to all demand in the area. Because true distances to demand are no longer measured under this assumption, the solution of any problem involving these distances will seemingly be affected. Hillsman and Rhoda (1978) and Goodchild (1979) show that this type of data aggregation does, in fact, cause errors in the solutions of the *p*-median, minimax, and maximal covering problems.

Acknowledgement: The author thanks the University of Iowa, Department of Geography, for its aid in the preparation of this work.

In the first part of their article, Hillsman and Rhoda identify three sources of errors that result from using distance from an aggregation point to a service center to represent the true average distance between the service center and the demand points in an area. The source A error is the simple measurement error that results from measuring distance to a service center from an aggregation point instead of from the dispersed demand in an area. The source B error is a special case of source A error and arises when a service center is located at an aggregation point. In such a case, the distance to the service center from the demand assumed to be centered at the aggregation point is zero. However, this measurement underestimates the true distance from the demand to the service center, since the center is actually serving dispersed demand. Finally, when distances from aggregation points to service centers are used to assign demand to the nearest center, some demand may be assigned to the wrong center. This error is classified as the source C error.

In the second part of their article, Hillsman and Rhoda analyze how large these three types of errors are for different levels of aggregation of various uniformly distributed demand patterns. They conclude that aggregation causes measurement errors of up to 8% for the demand patterns they examined, and they assert that larger errors may be expected for real-world patterns.

Goodchild (1979) also uses a uniform demand-distribution pattern to examine the effects of different levels of aggregation, but he is concerned mainly with the effects of aggregation on the locations of the service centers given in the solution of the p-median problem. Along with presenting results on locations, he also discusses how aggregation can affect the solution of coverage and minimax problems. Goodchild's final conclusion is ". . . that the effects of aggregation error on median problems are substantial . . ." but that ". . . aggregation tends to produce much more dramatic effects on location than on the values of the objective function" (1979, p. 253).

Even though their investigative objectives differ, Hillsman and Rhoda and Goodchild agree that the problems caused by data aggregation warrant careful analysis. Hillsman and Rhoda write "We believe that [aggregation] error is large enough to potentially affect the methodological and substantive interpretations of the results of research on such spatial systems" (1978, pp. 85–86). In a similar tone, Goodchild asserts "The results of this paper show that solutions using aggregated data are open to extensive manipulation, and in fact cast some degree of doubt on the usefulness of many location-allocation models" (1979, p. 253). With these comments in mind, let us consider a range of responses to the data aggregation problem.

One response is to act as if aggregation does not cause errors and to use analyses based on aggregated data as if they were analyses based on unaggregated data. Although this is the easiest and sometimes the only possible response, the works cited above and the results of this paper show that it ignores a source of errors in the solutions of many location problems.

A second response is to never use aggregated data. This sounds good in theory, but two practical constraints often make this response untenable. First, aggregated data is often the only data available for solving a given location problem. Second, in terms of efficiency and accuracy, current methods of solution for many types of location problems cannot deal with large data sets (see Rushton, Goodchild, and Ostresh, 1973). Thus, even if an unaggregated data set is available, the practical limits of the solution method may prevent using it to solve a particular location problem.

A third response, which seeks a middle ground, is to analyze the errors caused by data aggregation. Initially, this analysis should, at the very least, point out the types, effects, and possible magnitudes of errors caused by aggregation. Ultimately, it should be complete and general enough to be used in developing a locational method that adjusts for aggregation errors. The work presented in this chapter—an extension of the work of Hillsman and Rhoda (1978) and Goodchild (1979)—addresses the first goal.

THE p-MEDIAN PROBLEM IN CONTINUOUS SPACE

The following sections deal with a study of the errors caused by data aggregation in the solution, for one particular demand pattern, of the p-median problem in continuous space. Before describing the methods and results of this study, however, it is worthwhile to discuss the reasons for studying this particular problem.

The p-median problem was chosen as the object of study for two reasons. First, this is the problem that Goodchild (1979) examined, so it will be possible to compare his findings with the findings here. Second, in many ways the p-median is a very basic location problem. Many other problems, such as maximal covering and maximum attendance, can be solved using edited data and the methods for the p-median problem (see Hillsman 1979, 1981). Consequently, an analysis of aggregation errors in the p-median problem may contribute useful information to the same analysis of other types of problems.

The decision to use only one fixed unaggregated demand pattern for the present study requires some explanation, particularly because a valid criticism of the study is that its results may all depend directly on this pattern.

When studying the effects of data aggregation on the solution of the p-median problem, the analyst can vary a number of factors: the number of unaggregated demand points, the pattern of the unaggregated demand points, the method of aggregation (including the type and location of the aggregated demand points), the level of aggregation, the system for measuring distances, and the number of service centers to be located. A study of the type here that systematically dealt with all of these factors would be voluminous to say the least. To keep the size reasonable, therefore, this chapter holds constant all factors except level of aggregation, location of aggregated demand points, and number of service centers to be located. By varying these three factors, the study's main goal is to see how different levels of aggregation affect the value of the objective function and the location of service centers for the p-median problem in continuous space. Because all other factors are constant, the chapter claims no universality of results but aims only to provide an initial attempt to analyze errors caused by the aggregation of nonuniformly distributed demand.

Method

The following paragraphs outline the main points of the method of the study.

1. A demand point data base consisting of 500 unaggregated demand points of unit weight was generated by randomly choosing positive x, y coordinates for each point according to the negative exponential distribution $F(t) = b*\exp(-bt)$ with $b = 0.6$. This method generated a nonuniform distribution pattern in which demand was clustered near the origin and thinned out along the positive x- and y-axes. Fig. 12.1 is a scattergram of a sample of 50 unaggregated demand points, and it provides a rough idea of how the unaggregated demand was distributed. (A nonuniform pattern was chosen in the hope of avoiding pathological results that may arise when analyzing uniform patterns.)

2. For a given level of aggregation, say to K points, K points were randomly chosen from the unaggregated data base. These K points served as reference centers. Using the standard Euclidian metric, each remaining demand point was then assigned to the reference center closest to it. The result of this assignment was the definition of K zones with the ith zone containing all of the demand points closest to the ith reference center.

3. The centroid of all the demand points in the ith zone (including the ith reference center itself) was computed, and this point was used as the aggregation point for the demand in the zone. The weight of the ith aggregation point was set equal to the number of unaggregated demand points in the ith zone. (Because of the way the reference centers were chosen, one expects that there will be more zones in areas of high demand density and fewer in areas of low demand density. Such a distribution commonly occurs in real life, and this aspect of the study may suggest an extension to the analysis of zoning errors done by Goodchild, 1979, and Gould, Nordbeck, and Rystedt, 1971.)

4. Using the Euclidian metric, multiple runs of the Tornqvist heuristic algorithm (Rushton, Goodchild, and Ostresh, 1973) and the demand patterns generated by steps 1, 2, and 3, the 1-, 2-, 4-, and 6-median problems were solved for the unaggregated demand and for each level of aggregation. (The study examined four levels of aggregation: to 200, 150, 100, and 50 points.)

5. For each level of aggregation steps 2, 3, and 4 were carried out again. As a result, the study examined two different patterns for each of 200, 150, 100, and 50 aggregated demand points.

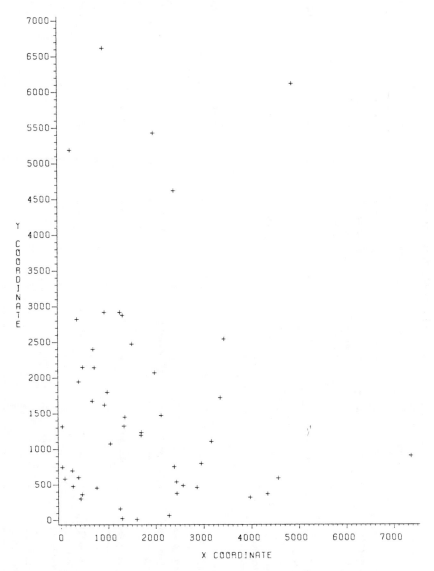

Figure 12.1. Locations of 50 unaggregated demand points.

Results

The basic numerical results of the study are presented in Tables 12.1–12.4. The first half of each table presents the results obtained from solving the indicated p-median problem for the first set of aggregation patterns, and the second half

Table 12.1. Values for the 1-Median Problem

	No Aggregation (500 Points)	Aggregation to 200 Points	Aggregation to 150 Points	Aggregation to 100 Points	Aggregation to 50 Points
First set of aggregation patterns					
(1) "Cost" and locations of optimal solution for given level of aggregation	$Z = 944,431$ (1305, 1264)	$Z = 942,845$ (1309, 1262)	$Z = 941,036$ (1308, 1263)	$Z = 939,889$ (1302, 1253)	$Z = 943,534$ (1299, 1221)
(2) True cost of serving unaggregated demand from locations in (1)	944,431	944,432	944,431	944,450	944,676
(3) Percent error between cost in (2) and optimal cost of serving unaggregated demand (optimality error)	—	0.00%	0.00%	0.00%	0.02%
(4) Percent of error between values in (1) and (2) (cost-estimate error)	—	−0.17%	−0.36%	−0.48%	−0.12%
Second set of aggregation patterns					
(1) "Cost" and locations of optimal solution for given level of aggregation	$Z = 944,431$ (1305, 1264)	$Z = 940,561$ (1311, 1268)	$Z = 941,755$ (1313, 1266)	$Z = 938,151$ (1331, 1256)	$Z = 929,997$ (1319, 1268)
(2) True cost of serving unaggregated demand from locations in (1)	944,431	944,434	944,437	944,510	944,450
(3) Percent error between cost in (2) and optimal cost of serving unaggregated demand (optimality error)	—	0.00%	0.00%	0.01%	0.00%
(4) Percent of error between values in (1) and (2) (cost-estimate error)	—	−0.41%	−0.28%	−0.67%	−1.53%

Table 12.2. Values for the 2-Median Problem

	No Aggregation (500 Points)	Aggregation to 200 Points	Aggregation to 150 Points	Aggregation to 100 Points	Aggregation to 50 Points
First set of aggregation patterns					
(1) "Cost" and locations of optimal solution for given level of aggregation	Z = 754,484 (1483, 3582) (1264, 818) 754,484	Z = 752,143 (1475, 3586) (1274, 814) 754,519	Z = 749,883 (1859, 3205) (1152, 783) 757,238	Z = 747,334 (1583, 3589) (1216, 827) 754,837	Z = 753,061 (3607, 920) (862, 1325) 765,409
(2) True cost of serving unaggregated demand from locations in (1)					
(3) Percent error between cost in (2) and optimal cost of serving unaggregated demand (optimality error)	—	0.00%	0.36%	0.05%	1.45%
(4) Percent of error between values in (1) and (2) (cost-estimate error)	—	−0.31%	−0.97%	−0.99%	−1.61%
Second set of aggregation patterns					
(1) "Cost" and locations of optimal solution for given level of aggregation	Z = 754,484 (1483, 3582) (1264, 818) 754,484	Z = 757,156 (3189, 908) (786, 1372) 760,964	Z = 751,492 (1561, 3464) (1223, 808) 754,810	Z = 747,240 (1515, 3485) (1248, 793) 754,681	Z = 739,228 (1167, 3229) (1332, 776) 758,033
(2) True cost of serving unaggregated demand from locations in (1)					
(3) Percent error between cost in (2) and optimal cost of serving unaggregated demand (optimality error)	—	0.86%	0.04%	0.03%	0.47%
(4) Percent of error between values in (1) and (2) (cost-estimate error)	—	−0.50%	−0.44%	−0.98%	−2.48%

Table 12.3. Values for the 4-Median Problem

	No Aggregation (500 Points)	Aggregation to 200 Points	Aggregation to 150 Points	Aggregation to 100 Points	Aggregation to 50 Points
First set of aggregation patterns					
(1) "Cost" and locations of optimal solution for given level of aggregation	$Z = 516{,}690$ (777, 577) (820, 2292) (3265, 854) (1707, 5602)	$Z = 518{,}812$ (524, 899) (1247, 3675) (2033, 617) (4405, 1161)	$Z = 510{,}309$ (776, 590) (808, 2369) (3232, 813) (1671, 5640)	$Z = 509{,}310$ (735, 557) (798, 2126) (3126, 833) (1796, 5220)	$Z = 481{,}942$ (626, 828) (1240, 3620) (2485, 651) (6727, 1547)
(2) True cost of serving unaggregated demand from locations in (1)	516,690	522,079	516,919	518,237	533,890
(3) Percent error between cost in (2) and optimal cost of serving unaggregated demand (optimality error)	—	1.04%	0.04%	0.30%	3.33%
(4) Percent of error between values in (1) and (2) (cost-estimate error)	—	−0.62%	−1.28%	−1.72%	−9.73%
Second set of aggregation patterns					
(1) "Cost" and locations of optimal solution for given level of aggregation	$Z = 516{,}690$ (3265, 854) (777, 577) (820, 2292) (1707, 5602)	$Z = 516{,}416$ (1267, 3692) (516, 944) (2053, 639) (4478, 1219)	$Z = 510{,}973$ (3170, 841) (754, 583) (857, 2336) (1863, 6078)	$Z = 513{,}537$ (1228, 3668) (469, 832) (1999, 675) (4400, 1351)	$Z = 484{,}329$ (1294, 3508) (529, 850) (2271, 604) (5494, 1454)
(2) True cost of serving unaggregated demand from locations in (1)	516,690	522,333	518,333	522,897	524,514
(3) Percent error between cost in (2) and optimal cost of serving unaggregated demand (optimality error)	—	1.09%	0.32%	1.20%	1.51%
(4) Percent of error between values in (1) and (2) (cost-estimate error)	—	−1.13%	−1.42%	−1.79%	−7.66%

Table 12.4. Values for the ... Problem

	No Aggregation (500 Points)	Aggregation to 200 Points	Aggregation to 150 Points	Aggregation to 100 Points	Aggregation to 50 Points
First set of aggregation patterns					
(1) "Cost" and locations of optimal solution for given level of aggregation	$Z = 411{,}820$ (656, 415) (1344, 3061) (562, 1609) (2509, 620) (1501, 6156) (5347, 1285)	$Z = 409{,}082$ (635, 416) (1265, 3156) (588, 1627) (2408, 593) (1507, 6459) (4998, 1295)	$Z = 406{,}947$ (581, 417) (1244, 3016) (533, 1583) (2280, 569) (1567, 5953) (4706, 1338)	$Z = 401{,}002$ (638, 430) (1267, 3145) (597, 1692) (2525, 617) (1581, 6605) (4994, 1270)	$Z = 344{,}401$ (541, 589) (3883, 734) (1966, 492) (842, 2089) (6727, 1547) (1736, 5719)
(2) True cost of serving unaggregated demand from locations in (1)	411,820	412,636	413,449	413,477	429,293
(3) Percent error between cost in (2) and optimal cost of serving unaggregated demand (optimality error)	—	0.20%	0.40%	0.40%	4.24%
(4) Percent of error between values in (1) and (2) (cost-estimate error)	—	−0.86%	−1.57%	−3.02%	−19.77%
Second set of aggregation patterns					
(1) "Cost" and locations of optimal solution for given level of aggregation	$Z = 411{,}820$ (656, 415) (1501, 6156) (1344, 3061) (562, 1609) (2509, 620) (5347, 1285)	$Z = 404{,}118$ (539, 425) (1403, 6212) (1289, 3179) (601, 1657) (2188, 561) (4528, 1255)	$Z = 403{,}267$ (568, 442) (1680, 6826) (1251, 3219) (604, 1658) (2260, 539) (4529, 1374)	$Z = 400{,}181$ (674, 439) (1506, 5536) (2018, 2957) (2580, 549) (5569, 948)	$Z = 370{,}119$ (405, 575) (776, 2369) (1831, 4526) (1433, 780) (2747, 522) (5494, 1455)
(2) True cost of serving unaggregated demand from locations in (1)	411,820	413,122	415,438	419,607	429,933
(3) Percent error between cost in (2) and optimal cost of serving unaggregated demand (optimality error)	—	0.32%	0.88%	1.89%	4.34%
(4) Percent of error between values in (1) and (2) (cost-estimate error)	—	−2.18%	−2.93%	−4.63%	−13.91%

presents the results obtained for the second set of aggregation patterns. (In each table, all "costs" and coordinates are rounded to the nearest unit.)

The column headings of these tables are self-explanatory, but the row headings are not. The entries in the first row are the objective-function values (called "costs" and given as Z) and locations that resulted from solving the p-median problem for the level of aggregation listed at the top of each column. In Table 12.1, for example, the first entry in row 1 is the total distance value and location that resulted from solving the 1-median problem for the unaggregated demand; the second entry in row 1 is the total distance value and location that resulted from solving the 1-median problem using the first-generated pattern of 200 aggregated data points; and so on.

The entries in the second row are the true total travel distances that would result if the unaggregated demand were served from the locations given in row 1. In Table 12.1, for instance, the third entry in row 2 is the total travel distance required if the unaggregated demand were served from the location found by solving the 1-median problem for the 150 aggregated demand points. The difference between this value and the optimal distance value for serving the unaggregated demand is the error, caused by aggregation, in the objective-function value. This type of error will be called the optimality error. For each level of aggregation, this error is given in the third row as a percentage of the optimal value determined by using the unaggregated data.

Aggregation causes another type of error that the literature has not dealt with. This error is the difference between the entry in row 2 and the value entry in row 1 in each column. An example will show that analysis of this error is important.

Suppose that a manager uses 150 aggregated demand points to determine where a business should locate a single service center. Using Table 12.1, the manager finds that the center should be located at (1308, 1263) and that the total travel distance to serve all demand from this point will be 941,036. The manager must realize, however, that this is the value for serving the *aggregated* demand from this location. The true value for serving the *unaggregated* demand from this location is given in row 2, column 3 as 944,431. The difference between these values (3,395) represents how much more it will "cost" the business to serve demand than the manager thought it would. This type of error will be called the cost-estimate error. The fourth rows of Tables 12.1–12.4 show these errors as percentages of the values in row 2.

Both the cost-estimate and optimality errors will be discussed further, but first the locations given in Tables 12.1–12.4 will be examined briefly.

Figs. 12.2, 12.3, and 12.4 are graphs showing, for each level of aggregation, the service-center locations given in the first parts of Tables 12.2–12.4. For example, in Fig. 12.2, the two end points of the line segment marked with stars represent the locations of the two service centers found by solving the 2-

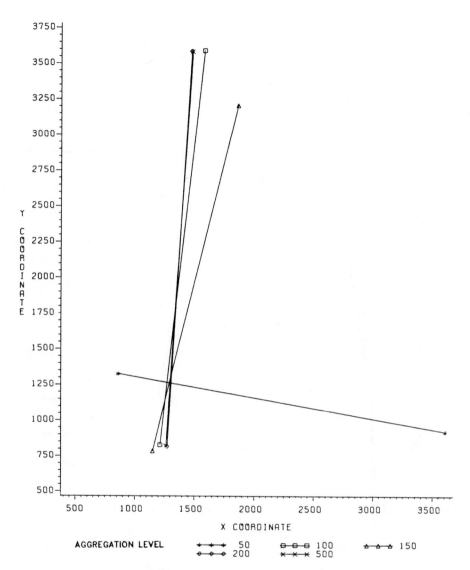

Figure 12.2. 2-median locations.

median problem for the first 50-point aggregation pattern; similarly, the end
points of the line segment marked with squares represent the locations of the
two service centers found by solving the 2-median problem for the first 100-
point aggregation pattern; and so on. Graphs for the second set of aggregation
patterns are not presented, but the coordinates given in the second parts of

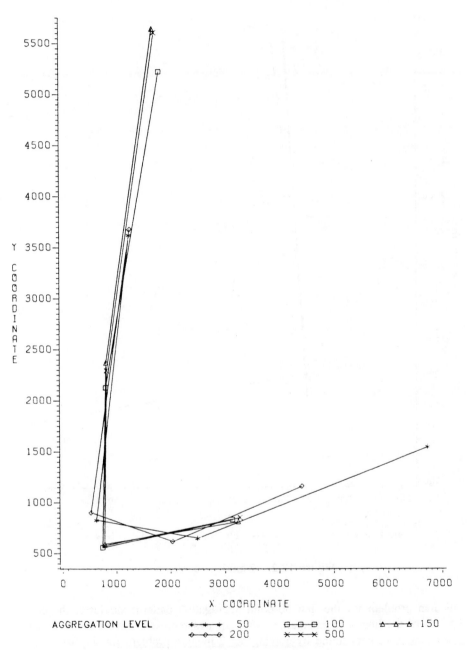

Figure 12.3.   4-median locations.

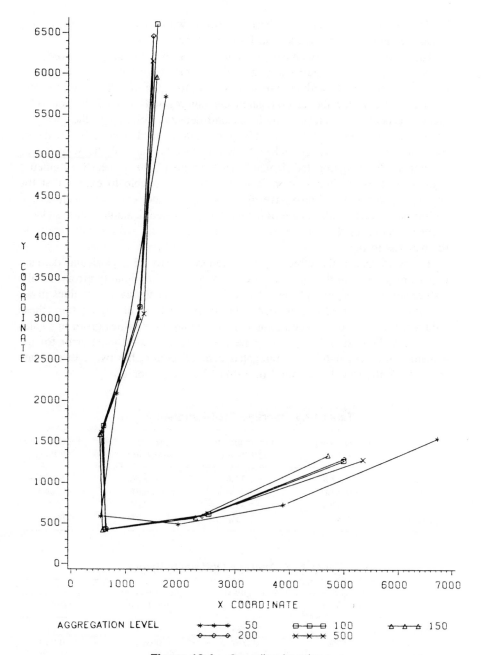

Figure 12.4.   6-median locations.

Tables 12.2–12.4 indicate that these locations, with some exceptions, display much the same patterns as those in Figs. 12.2–12.4.

The difference in service-center locations was not formally analyzed, but from Figs. 12.2–12.4 one basic conclusion seems valid; with only one exception—the 50-point, 6-median case—for a specific value of p the service-center location patterns for the unaggregated and aggregated demand do not differ greatly, especially in areas of high demand density. Pointing to the locations of the two centers in Fig. 12.2 for 50 aggregated demand points may challenge this conclusion. One must realize however, that because of the way it was generated, the unaggregated demand can be expected to be nearly symmetric with respect to the line $y = x$. It is therefore reasonable to expect that the aggregated demand will also have this property. Hence, one can conclude that the service-center location pattern for the 50 aggregated demand points is essentially the same as the other patterns in Fig. 12.2. The same results apply to the patterns in Fig. 12.3.

To try to reduce the effects that the idiosyncracies of a particular demand pattern may have on the analysis of cost-estimate and optimality errors, results concerning these errors will be discussed in terms of the average values given in Tables 12.5 and 12.6. For each value of p and level of aggregation, these values were found by averaging the corresponding error values given in Tables 12.1–12.4. For example, to determine the average cost-estimate error for the 4-median problem with 100 aggregated demand points, the two cost-estimate errors from the row 4, column 4 of Table 12.3 were averaged.

Table 12.5. Average Cost-Estimate Errors

	Aggregation to 200 Points	Aggregation to 150 Points	Aggregation to 100 Points	Aggregation to 50 Points
1 center	−0.29%	−0.32%	−0.58%	−0.82%
2 centers	−0.40%	−0.70%	−0.98%	−2.04%
4 centers	−0.88%	−1.35%	−1.76%	−8.70%
6 centers	−1.52%	−2.25%	−3.82%	−16.84%

Table 12.6. Average Optimality Errors

	Aggregation to 200 Points	Aggregation to 150 Points	Aggregation to 100 Points	Aggregation to 50 Points
1 center	0.00%	0.00%	0.00%	0.01%
2 centers	0.43%	0.20%	0.04%	0.96%
4 centers	1.06%	0.18%	0.75%	2.42%
6 centers	0.26%	0.64%	1.14%	4.29%

In most cases, the average cost-estimate errors given in Table 12.5 are relatively small. In 13 of the 16 cases, the cost-estimate errors are under -2.30%. Notice, however, that for the 50-point case, the average cost-estimate errors for the 4- and 6-median problems jump to -8.70% and -16.84%, respectively.

Fig. 12.5 is a series of graphs of the cost-estimate errors presented in Table 12.5. For a given level of aggregation, the cost-estimate errors for the values

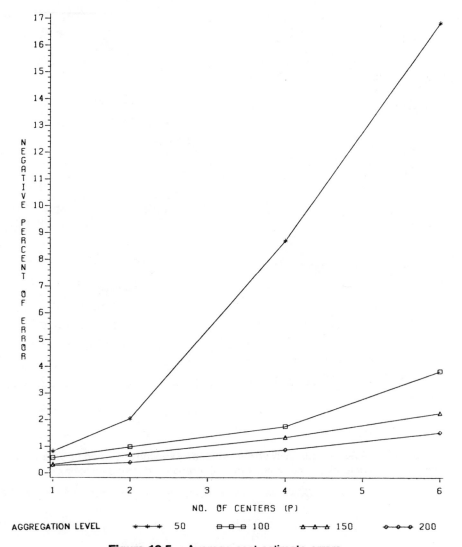

Figure 12.5. Average cost-estimate errors.

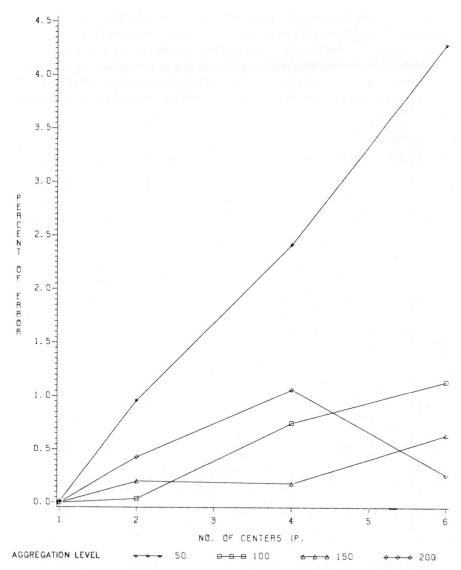

Figure 12.6.   Average optimality errors—caused by serving unaggregated demand from locations determined as optimal from aggregated demand locations.

of p examined in this study (p = 1, 2, 4, 6) were connected to form the four individual curves shown. Even a quick glance at these curves shows two significant trends. First, for a fixed aggregation level, the cost-estimate error increases smoothly as the number of centers increases, and the rate of this increase is higher for higher levels of aggregation. Second, for a fixed number of centers, the cost-estimate error increases as the level of aggregation increases and, in accordance with the previous finding, the rate of this increase is higher for higher values of p.

The average optimality errors given in Table 12.6 are, with two exceptions, even smaller in absolute value than the average cost-estimate errors. Moreover, the two largest average optimality errors are only 2.42% and 4.29%, while all other errors are under 1.2%. Again, the two largest errors were for the 50-point, 4-median problem and the 50-point, 6-median problem.

The optimality-error curves in Fig. 12.6 were drawn using the data in Table 12.6 and the same procedure as for the cost-estimate error curves. The one main conclusion that can be drawn from Fig. 12.6 is that the optimality-error curves do not behave very well. The 50- and 100-point curves are both strictly increasing, but their rates of increase do not follow any regular pattern. Moreover, both the 150- and 200-point curves fluctuate as the number of centers increases. Further, in the 4- and 6-center cases there is no connection between the level of aggregation and the size of the average optimality error.

CONCLUSIONS

Figs. 12.2–12.4 indicate, in some contrast to Goodchild's (1979) conclusion, that data aggregation may have little effect on the patterns of locations of service centers. Accounting for symmetry (something that should also be done for Goodchild's study) makes it clear that service-center location patterns may remain stable across all levels of aggregation.

Analyzing the magnitudes of cost-estimate errors indicates that larger errors occur when the p-median problem is solved, using highly aggregated data, for higher values of p. Such analysis also shows that as the level of aggregation and the value of p increase, the errors increase in a systematic manner.

Analyzing the magnitude of optimality errors indicates that these errors do not increase in a systematic manner. Rather, with one exception, for a given level of aggregation the optimality error is largest for the largest value of p examined in this study.

Finally, comparison of cost-estimate and optimality errors shows that for a fixed level of aggregation and value of p, the cost-estimate error is almost always larger than the corresponding optimality error. This result, coupled with the findings on the relative minimal variance of service-center locations, indicates that research on data aggregation in p-median problems should examine cost-estimate errors as well as the more traditional optimality and service-center location errors.

REFERENCES

Goodchild, M. F., 1979, The aggregation problem in location-allocation, *Geographical Analysis* **11**:240–255.

Gould, P., S. Nordbeck, and B. Rystedt, 1971, Data sensitivity and scale experiments in locating multiple facilities, in *Multiple Location Analysis*, G. Tornqvist, S. Nordbeck, B. Rystedt, and P. Gould, eds., Lund Studies in Geography Series C, General, Mathematical, and Regional Geography, The Royal University of Lund, Sweden. Lund, pp. 67–84.

Hillsman, E. L., 1981, *The p-Median Structure as a Unified Linear Model for Location-Allocation Analysis,* Oakridge National Laboratory, Tennessee.

Hillsman, E. L., 1979, A system for location-allocation analysis, Ph.D. diss, Department of Geography, University of Iowa, Iowa City.

Hillsman, E. L., and R. Rhoda, 1978, Errors in measuring distances from populations to service centres, *Annals of Regional Science* **12**:74–88.

Rushton, G., M. F. Goodchild, and L. M. Ostresh Jr., eds., 1973, *Computer Programs for Location-Allocation Problems,* Monograph 6, Department of Geography, University of Iowa, Iowa City.

13

SELECTING THE OBJECTIVE FUNCTION IN LOCATION-ALLOCATION ANALYSES

Gerard Rushton

University of Iowa

Whether to adopt the outcomes of a location-allocation analysis often hinges on our faith in the validity of the assumptions underlying the choice of the objective function. The more difficult it is to predict how providers and consumers will respond to any change in the locational arrangement of providers, the more difficult it is to select an appropriate objective function. The analyst's inability, in many cases, to select an objective function that captures the essential characteristics of the problem can lead to elegant but useless solutions (Grayson, 1973).

This chapter examines the assumptions often made about the expected outcomes of improved locations of a service. The chapter tries to identify the characteristics of locational problems for which forecasting these outcomes is most difficult, and it discusses approaches to solving the problem. The conclusion illustrates an application of a location-allocation model. In this case decision makers considered the solutions to the location problem as they had originally formulated it and, as a consequence, suggested an alternative formulation of the problem and objective function. The case was an attempt to develop a plan for the organization of primary health care resources in Iowa. The plan was to address the following objectives: the needs of the rural population, who desired adequate geographical accessibility to health services; the concerns of health care providers, who preferred to make maximal use of the existing system of health care resources, and who questioned whether working conditions would be adequate and economic conditions viable in isolated rural practice locations; and the concerns of the state's health planners, who wanted to reduce the number of areas that were "health manpower shortage areas."

PROVIDER BEHAVIOR IN LOCATION-ALLOCATION MODELS

In the simplest type of location problem, the performance of the service system of interest appears to depend solely on the responses of the providers. The consumer response is assumed to be independent of provider locational arrangement, and the variability in the provider's response arises solely from the provider's locational relationship to the demand to be served.

A good example of this case is fire department deployment analysis where the probability of a fire occurring is assumed to be independent of the locations of fire fighting resources. The damage from a fire is, though, assumed to be related to the response time of the fire fighting units. As Mirchandani and Reilly (chap. 12 herein) observe, however, "Studies relating fire-unit response time to fire damage have been limited to date." Selection of the objective function in these studies has been based on ad hoc assumption of the relationship between response time and fire severity, (proxy attributes of performance) (see Kolesar and Walker, 1974; Walker, Chaiken, and Ignall, 1979, pp. 79–87; Schreuder, 1981); on expert views gathered in a systematic way (see Keeney, 1973); or on insurance standards (Insurance Services Office, 1974). Only a few analyses have defined the objective function on the basis of studies that express fire damage as a function of characteristics of property and response time (see Ignall, Rider, and Urbach, 1978; Corman et al., 1976.

In addition to the question of how to incorporate response time in the objective function, there is the question of what kind of damage (people or property) should be protected. Schilling et al. (1980), for example, solved the locational deployment of fire fighting units in Baltimore as a multiobjective function. Property damage and personal casualties were weighted, so that decision makers would know the degree to which any given locational deployment would achieve spatial coverage of either property or people. By choosing the relative weights assigned to coverage of people and property in the objective function, decision makers could choose the importance they wished to assign to each.

On the matter of which locational arrangement should be selected, ReVelle, Cohon, and Shobrys (1981) describe a means of comparing alternative fire equipment deployments that are known to be good on the modeled criteria, according to their value on seven criteria of interest. These "value paths" allow alternative solutions to be compared on many objectives. They are useful in selecting the appropriate objective function because they enable decision makers both to compare alternative measures of the performance of different solutions and to reveal their preferences for unmodeled objectives. ReVelle,

Cohon, and Shobrys, however, do not complete this process by incorporating the decision makers revealed preferences in a revised objective function. Schilling, ReVelle, and Cohon (1983) take the analysis method an important step further when they show how alternative solutions—all good with respect to the defined criteria—can be displayed in a "decision space" that shows all the locations occurring in the alternative solutions. Decision makers can then focus on the generally few locations that differ in the solutions. At this stage they may introduce additional criteria into their evaluations to help make selections.

The apparently successful application of location-allocation models to the deployment of fire equipment and personnel contrasts sharply with the array of difficulties faced by researchers who have attempted to use such models to find the optimal locations of health resources. In a critical review of applications of operational research methods in health services planning, Rosenhead (1978) argues that only rarely have problems been accurately represented. He asserts that expansions of health facilities seldom involve the simultaneous location or relocation of many facilities; instead, change is introduced deliberately and incrementally while decision makers often carefully observe the responses of providers and consumers to new locations. The locational objectives may well change as the system develops, and such a pattern of decision making preserves flexibility for the future (see Goldberg, 1975). For example, the three kinds of health facilities in the U.S. that "have grown dramatically in the past 10 years" (Ermann and Gabel, 1985)—free-standing emergency centers, free-standing ambulatory surgery centers, and multihospital systems—have done so in a competitive locational environment where the provider's objective is to find locations that will give the greatest return. Such an environment contrasts with a system for which the goal is to optimize some systemwide measure of performance (see Wendell and McKelvey, 1981). Accordingly, planners of such health facilities do not appear to have benefited from location-allocation modeling of the traditional type characterized by simultaneous location or relocation and optimization of systemwide performance measures. When the provider considers the value of a location in relation to its potential to compete with existing or possible future competitors, traditional objective functions are inappropriate. In such cases, providers consider the key model element to be the interaction between their choice of location and the choices of other providers (see Eaton and Lipsey, 1979; Ghosh and Craig, 1984; Von Hohenbalken and West, 1984).

In selecting a new location, providers may also attempt to determine whether they can attract consumers from other providers by selecting a particular type of location rather than a location that is optimal with respect to the location of demand. Most of the approximately 2000 free-standing emergency centers (FECs) located in the United States between 1973 and 1985, for example, were placed near hospitals with busy emergency departments (Ermann and Gabel, 1985, p.

403), yet they did not affect hospital emergency department use (Ferber and Becker, 1983, p. 429). Apparently, the intention of most decision makers in locating free-standing emergency centers was to capture an existing market that was poorly served by the existing system. Instead, studies have confirmed that a different market segment (the young, mobile, and affluent) was attracted to these centers. These centers were bringing to the American health-care system a new delivery mode for primary health care. Their original target market, people who were using hospital emergency rooms for routine health care, were apparently not the people who were attracted to the new centers.

The locational strategy of providers in this example is not at all similar to that in a conventional location-allocation model that optimizes some geographical relationship of providers to consumers. Rather, provider behavior is closer to that found in competitive location theory, where location decisions depend on assumptions about the locational selection strategy of alternative providers (Hay, 1976; Eaton and Lipsey, 1975; Ghosh and Craig, 1984). Such circumstances make it difficult to formulate an appropriate objective function—even one couched as a competitive model seeking locations that are optimal in relation to the actual or possible locations of other providers. In this example most of the service providers evidently believed that other providers would be likely to alter their behaviors in response to the new service mode. Ermann and Gabel (1985, pp. 403–404) cite cases where competing hospitals significantly lowered their prices for comparable services or organized competing centers in response to the growth of FECs in their neighborhoods. Selecting the ''correct'' objective function therefore would involve an anticipation or prediction of the new behavior patterns that might emerge in the new system.

In this view, an activity's pattern of locations is altered to realize a desired outcome by changing some combination of spatial and other aspects of the activity's organization. The question to be solved thus becomes what organizational changes will cause both service providers and consumers in the system to behave so as to bring about the desired fundamental objectives.

CONSUMER BEHAVIOR IN LOCATION-ALLOCATION MODELS

In traditional location-allocation applications, consumers are assigned to locations according to some rule, such as nearness. A classic strategy for increasing the number of objective functions that location-allocation models can solve has been to incorporate an algorithm developed for the spatial-assignment problem (where locations are fixed) into a location-allocation model (where locations are variable). Different consumer spatial behavior patterns can then be incorporated in the objective function. Goodchild and Massam (1969), for example, incorporated the linear programming algorithm for the transportation problem in Cooper's (1964) alternating location-allocation algorithm to create the first optimal location algorithm that found optimal locations subject to capacity constraints. Cooper (1972) later rediscovered this algorithm. These spatial-assignment algorithms can also be incorporated in other heuristic location-allocation algorithms. This approach has been adopted by others, although as a general strategy for capturing complex consumer behavior in an optimal location model, it still offers much potential for future work. Consider, for example, the spatial-allocation rule for determining school attendance areas. In contrast to early studies, which used linear programming to assign students to their closest school subject to capacity constraints (Heckman and Taylor, 1969; Yeates, 1963), more recent studies have recognized that typically, several goals need to be satisfied and often conflict. Goal programming models have been applied successfully to the problem (Knutson et al., 1980; Lee and Moore, 1977; Saunders, 1981; Sutcliffe, Board, and Cheshire, 1984). These models allow decision makers to assign weights to the realization of each goal. At this time, however, no studies appear to have simultaneously solved the location or relocation problem using goal programming as a spatial-assignment tool in place of more traditional objective functions.

A number of researchers have recently noted that in many service systems, consumers freely choose providers and that their choices often differ from those postulated in traditional location-allocation models. Important as these developments are, several unsolved problems exist in this literature. When a spatial-interaction model (Coelho and Wilson, 1976; Hodgson, 1978; Leonardi, 1983; Leonardi and Tadei, 1984) is used in a location-allocation model, it is not clear what parameters should be used, since the parameters of the model that assigns consumers to locations depend on the spatial structure of alternatives in the system where the model was calibrated (Fotheringham and Webber, 1980). To try to deal with this problem, some have attempted to forecast these parameters as a function of local characteristics of the system (Baxter, 1985; Fotheringham, 1984; Ghosh, 1984).

The problem in location-allocation modeling is to find a model of spatial choice that, although based on empirical work in one context, is likely to accurately predict spatial choices in a new and different context. Future work is likely to parallel spatial-choice modeling's evolution toward more individual-level, behaviorally based models of choice (Eagle, 1984; Golledge and Rushton, 1984; Horowitz, 1983; Smith, 1983; Timmermans, 1984). In these procedures, individual-level discrete-choice models, multiattribute preference models, decision rules, or spatial-search algorithms replace the aggregate-data-based context-dependent spatial-choice predictions of the spatial-interaction or random utility models used in location-allocation models to date (Coelho and Wilson, 1976; Hodgson, 1978; Leonardi and Tadei, 1984). This new spatial-choice literature is concerned with such issues as the identity of the set of alternatives that an individual actually considers (choice-set generation) (Damm and Lerman, 1981; Forer and Kivell, 1981); whether spatial choices are the result of cognitive interpretations of the decision context (Pipkin, 1981); and whether the spatial-choice decision can be adequately modeled when it is divorced from other, related decisions (Burnett and Hanson, 1979). These concerns are particularly apposite to location-allocation modeling because they focus on the generality and robustness of spatial-choice models as applied to new spatial contexts rather than simply on such models' goodness of fit in empirical contexts. In selecting objective functions for location problems involving the free choice of providers, objective functions must accurately represent the spatial-choice rules consumers will use in judging alternative service locations.

PROBLEMS IN FORMULATING THE OBJECTIVE FUNCTION

Although decision makers have often been carefully consulted to determine criteria to be optimized, in only a few cases have formal analytical techniques been used to capture their preferences. The approach described by Keeney (1973) asks decision makers to evaluate abstractly defined service systems and to rate the expected levels of performance of the hypothetical systems that have defined scores on the several performance criteria. The rating values (or statements of indifference between systems that have different values on the criteria of interest) can be used to formulate a multiattribute utility function (Edwards, 1977; Keeney, 1973). A location-allocation model that incorporates such a utility function as its objective function will often give a solution that reveals unexpected outcomes when used in an analysis in a specific geographical area, and these outcomes may cause decision makers to revise their objective function. In a multiattribute utility function, when decision makers express their preferences for attributes of the solution they do not know the cost of meeting the properties of interest. Thus fire chiefs may claim to be indifferent between stated combinations of response times for various pieces of equipment (see Keeney, 1973, but they may discover, after the locational analysis, that the cost of improving one kind of response time is so much greater than the cost of improving another that they change their preference. The decision maker needs to know what a given amount of resources can achieve in terms of improvement in one performance measure as compared with another.

The preference function describes the decision maker's estimation of how different combinations of performance measures lead to equal outcomes. Thus, if an improvement in one performance measure is more expensive than an improvement in another, the more expensive solution should be replaced by the least expensive one in the final solution at the point where greatest utility would result.

Should this feedback of what is attainable given the geography of the area influence the decision maker's utility function? When attainable outcomes are outside anticipated levels, the decision maker's utility function becomes moot. Then utility functions need to be developed that are not directly based on past experience. The likelihood that this situation will occur in typical location-allocation applications is rather great. The simultaneous location and relocation of facilities in a system is likely to result in performance measures that are outside the decision maker's normal experience. If this were not the case, it is unlikely that the kinds of major changes in system efficiency usually expected would occur following the implementation of a location-allocation problem solution. Thus decision makers asked to express their preferences for attributes of a future service delivery system are likely to think in terms of past, familiar systems and to have considerable difficulty in comparing performance measures of a hypothetical unfamiliar system. Often, though, novel service delivery systems need to be developed with properties unlike traditional systems. Fischhoff and Goitein (1984) discuss the use of formal analytical techniques to accurately capture the judgments of decision makers for complex, novel situations. They conclude that these complex expert judgments should not simply be measured but need to be found through a reactive process. This process requires questions posed by the analyst that "shape (not just reveal) the structure of their clients' decision" (p. 511). Fischoff and Goitein also describe the typical biases that decision makers bring to the judgment of the expected performance level of hypothetical situations:

If the precise task and situation have never occurred before in exactly the present form, the estimator must judge which past experiences are similar enough to the present instance to be included in the data base. The estimator is likely to exclude data regarding saliently atypical situations (e.g., the long time it took to perform a similar task during a February snowstorm) but not less saliently atypical ones (e.g., the short time it took to perform the task when an unusually small number of problems was encountered). The judgment of which situations and tasks are similar enough to be relevant thus can bias the construction of the archive of past experience (1984, p. 508).

In changing the objective function, the decision maker must first consider the people responsible for the performance criterion that analyses have shown to be most easily improved. The system needs to induce behavioral changes in these people so that their new preference function will still be consistent with the system's fundamental objectives. In the fire fighting deployment case, for example, a location-allocation analysis might show that it was possible to markedly improve the response time of the first engine companies but not so easy to improve that of the first ladders. This information might then induce changes in organization that would lead to more effective service being performed by first-engine responders. When this change could clearly be achieved, the utility function of ladders and engines would need to be revised. In this case an implication of the model solution—that the price of improving one of the performance criterion was very high and the price of improving a second criterion was much less—would be apparent only after the solution was known. Yet this information is important in the formulation of the objective function and thus is needed before the solution can be found. An iterative strategy of formulating the problem, finding a solution, reformulating the problem, and so on, may be more appropriate for solving the fundamental objectives. Sutcliffe, Board, and Cheshire (1984) report that the choice of weights by decision makers in solving a problem of allocating children to secondary schools "was also conditioned by aspects of the results produced in successive runs. . . . in particular the trade-off these results suggested as between conflicting goals. Thus one reason for the low weight on sex mix is the comparative ease with which that goal could be nearly met" (p. 727).

Such a process likely happens far more frequently than is reported in the literature, where the classic research paradigm of problem definition, formulation, and solution reigns supreme (Rinnooy Kan, 1984; Rushton, 1984). We must acknowledge that the mutually adjusting behavior patterns of providers and consumers in many human-services delivery problems can, at best, be incompletely represented in a prescriptive location-allocation model. Further, we must acknowledge that some of the problems of implementing a given prescriptive model may become evident only after model solutions have been examined. The problems revealed in solving the prescriptive model may require novel solutions not evident from past experience and may represent "complicated design tasks where expert knowledge is only implicitly available and not readily extractable in mathematical form" (Rinnooy Kan, 1984, p. 67).

AN ILLUSTRATION: IMPROVING ACCESS TO PRIMARY HEALTH CARE

The Problem

The 1960s and 1970s saw increasing concern in rural America about declining access to primary health care as the numbers of doctors practicing in rural areas declined. Although linked to an overall shortage of physicians in the United States, the problem was compounded by the fact that the majority of new medical graduates were born in urban areas, trained in medical specialty fields, and usually, had served their residencies in large urban hospitals. Practicing solo general medicine in isolated rural communities did not attract them. One solution to this problem was to train the graduate to practice in a group of at least three doctors who would serve a group of villages from a regional service center.

In discussing the problem with health planners and others in Iowa in the period 1974–1976, two key criteria emerged as important in identifying the sites that might be suitable for these group practices (Rushton et al., 1976). Geographical accessibility of the places was clearly important, but so were the health resources already present to support group practices. Extensive discussions with health planners revealed five factors as existing resources that if present, would facilitate the development of a regional primary care center. Each was weighted in a multiattribute utility function elicited from a group of health planners, and each was scaled to values between zero and one (Rushton et al., 1976, pp. 16–24). The five measures of health resources were

1. Current number of primary care physicians under age 60.
2. Hospital's occupancy rate (0 if site has no hospital) and related hospital information.
3. Statistical forecast of a place's potential for maintaining or increasing its number of primary care physicians (see Henderson et al., 1977).
4. Population of the place.
5. Number of people per physician in the local health service area.

The most accessible places in Iowa were computed as a p-median problem using an interchange, heuristic location-allocation algorithm (Teitz and Bart, 1968). Demand areas were defined as the 1970 populations of townships or small towns, and their locational coordinates were measured. The 2700 data units so defined comprised the set of places to be allocated. Distances were computed using the Manhattan distance metric, which generally corresponds well with the orientation and structure of the Iowa road system. The 210 possible service sites included all places with more than 1000 people and smaller places in areas far from any urban center. The 68 places identified as most accessible (Fig. 13.1a) were different from the places judged to have the greatest potential based on their health resources (Fig. 13.1b).

The challenge of development was to find multicriteria plans that included places both accessible to the rural population and possessed of well-developed resources that would help in the community's development or continuation as a regional or primary care center.

Developing the Multicriteria Regional Plans

We found that decision makers neither could nor wanted to formulate systemwide performance measures to describe a general trade-off function showing the degree to which they would be willing to sacrifice part of the accessibility criterion in return for a plan based on places with better health resources. In other words, we found that the decision makers could not identify, a priori, an objective

Criterion - minimize distance to population (p-median)

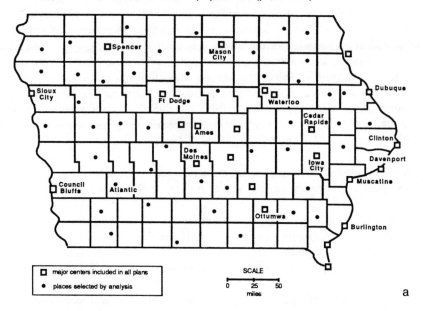

a

Criterion - maximize medical resource-values of places

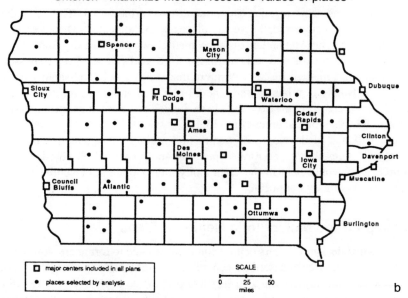

b

Figure 13.1.   Possible regional primary health care centers in Iowa: single-criteria solutions.

function that related to the global properties of the solution space. We found that they were concerned about the properties of the global solution space, but that they were even more concerned with the trade-off of the two attributes in the local solution space. In fact, we found that the decision makers could decide that the global solution was appropriate only by examining the kinds of changes in the proposed location plan (the decision space) at the local level.

As in many formulations of a location-allocation problem, essential features of the problem could be translated as integral steps in a solution algorithm. This approach is the one we took. We argued that the total performance of a plan is a derived measure from a process of decision making. This process is what produces the systemwide characteristics, although there may be a mechanism whereby this derived measure is fed back to decision makers in the hope that the decision making process will be modified.

One can argue about the degree to which such local location decision strategies might optimize systemwide characteristics. For some location-allocation models, there is a literature on this point. Most of the early heuristic algorithms for the p-median problem were cast in local decision-making terms, although their authors appeared to be interested only in the degree to which the process led to the optimization of a global objective function. For these researchers, that the alternating algorithm (Cooper, 1964; Maranzana, 1964) frequently terminated in a local minimum for the objective function, for example, was a source of concern and a reason to develop better algorithms to solve the p-median problem, rather than a description of a local decision-making process in which providers relocated to be more accessible to their local market areas and consumers selected the centers that most conveniently could serve them. From our point of view, however, if the process of local location decision making within the alternating algorithm represents reality, it is interesting to know that the result of such decision making over a prolonged period would lead to the system outcome of a local minimum solution to the p-median problem.

Description of Trade-Off Algorithm

The algorithm designed to find the trade-offs between gains and losses in accessibility to local health resources is explicitly spatial (Hillsman, 1979, 1980). It substitutes for places within a solution all candidate places not in the solution and makes a replacement when the criterion value of an objective function allows it. It terminates when no further pairwise substitution will improve the objective function. Unlike conventional heuristic location-allocation algorithms, however, this algorithm has no global objective function. Instead, it involves the sequential solution of a series of local optimization problems. Although the algorithm is unorthodox in this respect, its design reflects several reasonable assumptions and objectives.

First, we wanted to simulate a kind of behavior that seems common among decision makers who are presented with an optimal spatial pattern. This behavior includes: questioning the pattern; testing the pattern by making small changes to it; accepting changes that improve the pattern; and evaluating each of these changes on its own, local merits, independent of other changes and without explicit concern for the change's impact on measures of the quality of the entire pattern. When the decision makers have access to an information system to measure the local and global effects of these changes, this behavior combines optimization and unstructured sensitivity analysis.

Thus, we approach the problem as a local decision-making one. We allow decision makers to consider the global solution space that results from applying their local preference functions, and we allow them to decide to modify their local preferences accordingly. We want to stress that in any such modification of local preferences to achieve a new solution in the global space, the decision maker is essentially exercising a third preference that involves modifying local preferences in response to the returns in the global solution space. We speculate that in some cases the decision maker evaluates the global returns that follow from a change in the local preference structure and continues to change the local preference structure so long as the returns justify it. In other cases, the decision maker may have no interest in the global solution space.

We assumed that decision makers begin by ignoring the accessibility factor and consider locating health centers at the locations with the highest resource characteristics (Fig. 13.1b). Their overriding concern initially is to meet the obvious accessibility needs without sacrificing too much of the plan's high-resource–site characteristics. Next, we assumed that decision makers consider each potential site that is not in the solution set and compute how much (if any) accessibility the entire system would gain if each nonincluded site were substituted for any site in the solution set. That is, all pairwise substitutions are considered as a function of accessibility gain and place resource loss, (Fig. 13.2). The requirement that accessibility must improve prevents the algorithm from trading the two attributes in both directions. This requirement, applied to a finite number of potential center locations, ensures the algorithm's termination.

The function chosen to evaluate and rank potential changes to the pattern of centers is an ellipse (Hillsman, 1980, p. 32). The health-resource values of places involved in changes are measured along the vertical axis, and the relative accessibility-change values are measured along the horizontal axis. The accessibility gain is standardized as the ratio of the increase in distance that people would have to travel if a given place is removed from the solution and the decrease in distance that occurs when a place is added to the solution. Subtracting this ratio from one yields a measure of relative accessibility change ranging from zero to one, with small relative changes receiving smaller values than larger ones.

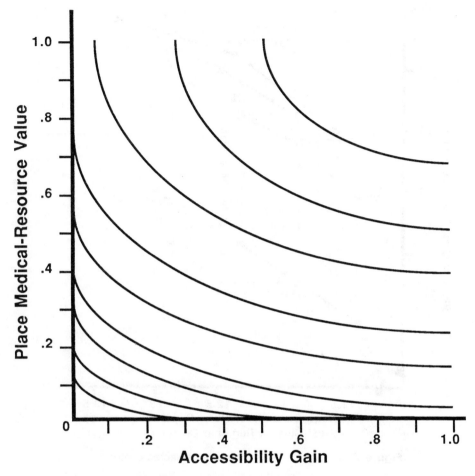

Figure 13.2.   Trade-off preference function.

The values of the two single-criterion solutions (Fig. 13.1) in the global solution space are shown as the extreme points in Fig. 13.3. The solid squares in that figure show the criteria values for other solutions computed by the algorithm. The clear "elbow" on this graph shows it is possible in this case to have solutions that are close to optimal with respect to both criteria.

One of the multicriteria solutions, derived in this case from the trade-off function in Fig. 13.2, is shown in Fig. 13.4. The 23 largest places in Iowa are, in this case, arbitrarily constrained to be included in the compromise plan. Another 26 places are identical in both maps, and the remaining 19 places are different.

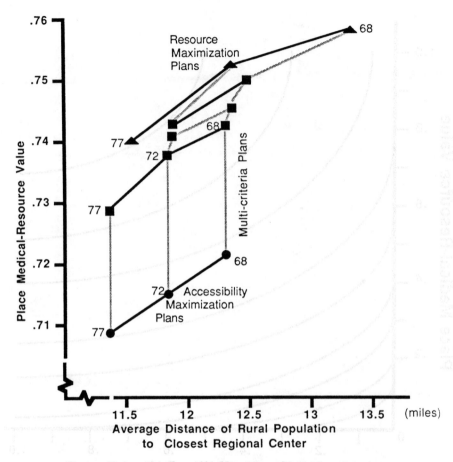

Figure 13.3.   Solution space for three selected center plans.

Solving these problems for plans that ranged from 55 centers to 102 revealed that in plans that involved more than 68 places, some centers did not have service populations of adequate size (estimated at 17,000 people) to support a group-practice site. For this reason the 68-center plan was, for a time, thought to be the one best suited for implementation (Rushton et al., 1976, pp. 50–53).

In this illustration, location-allocation modeling is a decision-making process in which the local decision-making function may be modified based on the location pattern's place in the global solution space. Whereas traditional location-allocation models viewed the decision steps within an algorithm as merely steps to optimize the global objective function, here the roles of these two important features are reversed. The decision steps are the focus, and the place

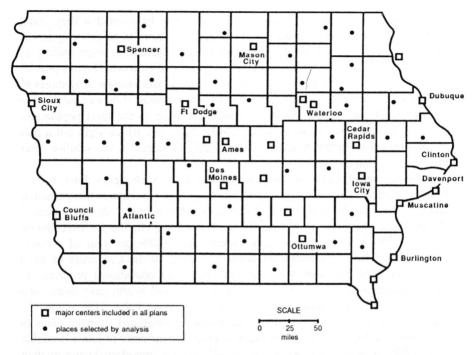

Figure 13.4.   Possible regional health care centers in Iowa: multicriteria solution.

of the solution in the global solution space is a secondary consideration that may be used to modify the preferences exercised in the decision steps. Also, the location-allocation methodology models a process, rather than optimizing an objective function in a global solution space.

Several implications follow. What is the relationship between processes and ends generated? Do dissimilar processes lead to essentially similar ends? Which processes lead to severe local minimum problems with respect to given objective functions in the global solution space? Given a certain desirable end, what process is most likely to reach that end? What ends are existing processes likely to reach? Can regulations be devised that will lead local decision making to solutions with desirable characteristics in the global solution space? Is it possible to model processes when the local preference function changes within a study area (possibly as a consequence of socioeconomic geographic variations)?

Modification of the Objective Function

The 1976 study advocated the development of the 68 places, but two problems arose when the discussion turned, at the state-health-planning level, to whether and how such a plan should be implemented. Many communities not recognized as regional primary health care centers in the plan believed that they could support a physician and did not look forward to journeys to neighboring towns for their primary health needs. Many of these communities were willing to provide facilities to physicians who would practice in their town. Doctors providing more specialized health services in the larger towns foresaw advantages to their practices and more admissions to their local hospitals if they could provide primary health services in locations more accessible to the rural population. These two aspects of consumer and provider behavior, which were not a part of the original objective function, more clearly belonged in the solution after these results were studied.

Two mechanisms were available for implementing such a plan. First, Iowa's Office of Health Planning recommended to the federal government that particular areas of the state be designated "medical health manpower shortage areas." Physicians who choose to practice in such areas are eligible for cancellation of up to 50% of their federal educational loans—10% per year of service. Second, the Office of Community Based Programs of the University of Iowa College of Medicine provided consulting help and educational programs to Iowa communities on the subject of the regionalized health care model and the role of group-practices in it. A recent study showed that 51% of the 123 physicians who established practices in federally designated shortage areas in Iowa between 1980 and 1984 indicated that the "opportunity to join a desirable partnership or group practice in the area" was an important factor in their decision to locate there (Peters, 1985, p. 10). Thus, evidence exists that the trend in provider decision making supports the gradual change in the state's rural health care system from isolated solo providers to organized group-practice arrangements.

These group practices are not always in single locations, however. In 23 cases discovered to date in Iowa (personal communication, Roger Tracy, Director, Office of Community Based Programs), the physician, although located alone in a community that dominates his or her practice, is nevertheless part of a larger group-practice arrangement that includes patient cross-referrals and from which physicians will cover for colleagues who are away on vacation or receiving further medical education. This evolution in how primary health care providers locate and organize is being further studied as an example of how behavior patterns of providers and consumers evolve to change the original problem formulation of an optimal location problem.

CONCLUSION

The objective function of a location-allocation algorithm often incompletely represents the fundamental objective of the problem to be solved. Early decisions by the analyst that result in the specific problem formulation are critical to the success of any project. For service location problems, the formulation is often defined as a structural relationship between providers and consumers, when in fact the fundamental objective is to design a service-delivery system in which providers and consumers will act together to optimize some outcome. Location-allocation modeling must therefore identify the behavioral patterns that exist at present and must state the fundamental goals of locational change as expected future behaviors of providers and consumers. It is not easy to discover exactly what these future behavior patterns should be and to identify the mechanisms available for inducing them given current system characteristics. The analyst could adopt the role advocated by Fischhoff and Goitein (1984), which explicitly designs interaction between the analyst and the decision maker to assist decision makers in clarifying their views and uncovering possible contradictions.

This diagnosis of the problems that location-allocation modelers must solve is consistent with the view that modeling in this area is evolving. The early work focused on finding patterns of locations that optimized a given structural spatial relationship. Next came a focus on optimizing the process of service delivery. The final focus will be on the realization of desirable outcomes.

We hope that this chapter contributes to the search for "a broader spectrum of meaningful objective functions" in location-allocation research (Leonardi and Tadei, 1984, p. 429). Such objective functions will typically describe how providers and consumers will respond to changes in locational arrangements. The attraction terms in the consumer spatial-choice component of the models will often be endogenously determined by the models. Difficulties in formulation will continue to be related to the nonstationary nature of parameters in spatial-choice models.

REFERENCES

Baxter, M. J., 1985, Misspecification in spatial interaction models: further results, *Environment and Planning, A* **17**:673–678.

Burnett, P. and S. Hanson, 1979, Rationale for an alternative mathematical approach to movement as complex human behavior, *Transportation Research Record* **723**:11–24.

Coelho, J. D. and A. G. Wilson, 1976, The optimum location and size of shopping centres, *Regional Studies* **10**:413–421.

Cooper, L., 1964, Heuristic methods for location-allocation problems, *SIAM Review* **6**:37–53.

Cooper, L., 1972, The transportation-location problem, *Operations Research* **20**:94–108.

Corman, H., E. Ignall, K. Rider, and S. Stevenson, 1976, Fire casualties and their relation to fire company response distance and demographic factors, *Fire Technology* **12(3)**,193–203.

Damm, D. and S. R. Lerman, 1981, A theory of activity scheduling behaviour, *Environment and Planning, A* **13**:703–718.

Eagle, T. C., 1984, Parameter stability in disaggregate retail choice models: experimental evidence, *Journal of Retailing* **60**:101–123.

Eaton, B. C. and R. G. Lipsey, 1975, The principle of minimum differentiation reconsidered: some new developments in the theory of spatial competition, *Review of Economic Studies* **42**:27–49.

Eaton, B. C. and R. G. Lipsey, 1979, The theory of market preemption: the persistence of excess capacity and monopoly in growing spatial markets, *Economica* **46**:149–158.

Edwards, W., 1977, How to use multiattribute utility measurement for social decision making, *IEEE Transactions on Systems, Man, and Cybernetics* **SMC-7**:326–340.

Ermann, D. and J. Gabel, 1985, The changing face of American health care, *Medical Care* **23**:401–420.

Ferber, M. and L. Becker, 1983, The impact of freestanding emergency centers on hospital emergency department use, *Annals Emergency Medicine* **12**:429–433.

Fischhoff, B. and B. Goitein, 1984, The informal use of formal models, *Academy of Management Review* **9**:505–512.

Forer, P. C. and H. Kivell, 1981, Space-time budgets, public transport, and spatial choice, *Environment and Planning, A* **13**:497–509.

Fotheringham, A. S., 1984, Spatial flows and spatial patterns, *Environment and Planning, A* **16**:529–543.

Fotheringham, A. S. and M. J. Webber, 1980, Spatial structure and the parameters of spatial interaction models, *Geographical Analysis* **12**:33–46.

Ghosh, A., 1984, Parameter nonstationarity in retail choice models, *Journal of Business Research* **12**:425–436.

Ghosh, A. and C. S. Craig, 1984, A location allocation model for facility planning in a competitive environment, *Geographical Analysis* **16**:39–51.

Goldberg, M. A., 1975, On the inefficiency of being efficient, *Environment and Planning, A* **7**:921–939.

Golledge, R. G. and G. Rushton, 1984, A review of analytic behavioural research in geography, in *Geography and the Urban Environment*, vol. 6, *Progress in Research and Applications*, D. T. Herbert and R. J. Johnston, eds., Wiley, New York, pp. 1–43.

Goodchild, M. F. and B. H. Massam, 1969, Some iterative least-cost models of spatial administrative systems in Southern Ontario, *Geografiska Annaler* **52B**:86–94.

Grayson, C. J., Jr., 1973, Management science and business practice, *Harvard Business Review* **51**(4):41–48.

Hay, D. A., 1976, Sequential entry and entry-deterring strategies in spatial competition, *Oxford Economic Papers* **29**:240–257.

Heckman, L. B. and H. M. Taylor, 1969, School rezoning to achieve racial balance: a linear programming approach, *Socio-Economic Planning Sciences* **3**:127–133.

Henderson, W. G., G. J. Meneley, J. A. Kohler, and G. Rushton, 1977, A statistical model of changes in locations of primary-care dentists in a rural state, *Journal of Public Health Dentistry* **37**:189–199.

Hillsman, E. L., 1979, A system for location-allocation analysis, Ph.D. diss., Department of Geography, University of Iowa, Iowa City.

Hillsman, E. L., 1980, *Heuristic Solutions to Location-Allocation Problems*, Monograph #7, Department of Geography, University of Iowa, Iowa City.

Hodgson, M. J., 1978, Toward more realistic allocation in location-allocation models: an interaction approach, *Environment and Planning, A* **10**:1273–1285.

Horowitz, J. L., 1983, Evaluation of discrete-choice random-utility models as practical tools of transportation systems analysis, *Transportation Research Board Special Report* **201**:127–136.

Ignall, E., K. Rider, and R. Urbach, 1978, Fire severity and response distance: initial findings, R-2013-NYC, Rand Corporation, Santa Monica.

Insurance Services Office, 1974, *Grading Schedule for Municipal Fire Protection*, Insurance Services Office, New York.

Keeney, R. L., 1973, A utility function for the response times of engines and ladders to fires, *Urban Analysis* **1**:209–222.

Knutson, D. L., L. M. Marquis, D. N. Ricchiute, and G. J. Saunders, 1980, A goal programming model for achieving racial balance in public schools, *Socio-Economic Planning Sciences* **14**:109–116.

Kolesar, P. and W. Walker, 1974, An algorithm for the dynamic relocation of fire companies, *Operations Research* **22**:249–274.

Lee, S. M. and L. J. Moore, 1977, Multi-criteria school busing models, *Management Science* **23**:703–715.

Leonardi, G., 1983, The use of random-utility theory in building location-allocation models, in *Locational Analysis of Public Facilities*, J.-F. Thisse and H. G. Zoller, eds., North-Holland, Amsterdam, pp. 357–383.

Leonardi, G. and R. Tadei, 1984, Random utility demand models and service location, *Regional Science and Urban Economics* **14**:399–431.

Maranzana, F. E., 1964, On the location of supply points to minimize transport costs, *Operational Research Quarterly* **15**:261–270.

Peters, B. A., 1985, What factors influence physicians to locate in rural "health manpower shortage areas"? unpublished paper, Department of Geography, University of Iowa, Iowa City.

Pipkin, J. S., 1981, The concept of choice and cognitive explanations of spatial behavior, *Economic Geography* **57**:315–331.

ReVelle, C., J. Cohon, and D. Shobrys, 1981, Multiple objective facility location, *Sistemi Urbani* **3**:319–343.

Rinnooy Kan, A. M. G., 1984, Operations research: prospects and limitations of optimal decision making, in *The Quest for Optimality*, J. H. P. Paelinck and P. H. Vossen, eds., Gower, Brookfield, Vermont, pp. 63–69.

Rosenhead, J., 1978, Operational research in health services planning, *European Journal of Operational Research* **2**:75–85.

Rushton, G., 1984, Use of location-allocation models for improving the geographical accessibility of rural services in developing countries, *International Regional Science Review* **9**:217–240.

Rushton, G., K. J. Dueker, E. L. Hillsman, J. A. Kohler, and G. J. Meneley, 1976, A statewide plan for regional primary medical and dental care centers in Iowa, Tech. Rep. 7, Health Services Research Center, University of Iowa, Iowa City.

Saunders, G., 1981, An application of goal programming to the desegregation busing problem, *Socio-Economic Planning Sciences* **15**:291–294.

Schilling, D., C. ReVelle, J. Cohon, and D. Elzinga, 1980, Some models for fire protection locational decisions, *European Journal of Operational Research* **5**:1–7.

Schilling, D., C. ReVelle, and J. Cohon, 1983, An approach to the display and analysis of multiobjective problems, *Socio-Economic Planning Sciences* **17**:57–63.

Schreuder, J. A. M., 1981, Application of a location model to fire stations in Rotterdam, *European Journal of Operational Research* **6**:212–219.

Smith, T. R., 1983, Computational process models of individual decision-making, in *Cities and Regions as Nonlinear Decision Systems,* R. W. Crosby, ed., Westview Press. Boulder, Colorado, pp. 175–210.

Sutcliffe, C., J. Board, and P. Cheshire, 1984, Goal programming and allocating children to secondary schools in Reading, *Journal of the Operational Research Society* **35**:719–730.

Teitz, M. B. and P. Bart, 1968, Heuristic methods for estimating the generalized vertex median of a weighted graph, *Operations Research* **16**:955–961.

Timmermans, H., 1984. Decompositional multiattribute preference models in spatial choice analysis: a review of some recent developments, *Progress in Human Geography* **8**:189–221.

Von Hohenbalken, B. and D. S. West, 1984, Predation among supermarkets: an algorithmic locational analysis, *Journal of Urban Economics* **15**:244–257.

Walker, W. E., J. M. Chaiken, and E. J. Ignall, 1979, *Fire Department Deployment Analysis,* North-Holland, New York.

Wendell, R. E. and R. D. McKelvey, 1981, New perspectives in competitive location theory, *European Journal of Operational Research* **6**:174–182.

Yeates, M., 1963, Hinterland delimitation: a distance minimizing approach, *Professional Geographer* **15**:7–10.

ABOUT THE CONTRIBUTORS

John R. Beaumont is head of Consultancy at Pinpoint Analysis Limited and a research fellow at Birkbeck College, University of London. Previously, he taught at Keele University and was a management consultant at Coopers & Lybrand. He has published extensively on various aspects of locational analysis and is the coauthor of three books on spatial analysis.

Thomas L. Bell is professor of geography at the University of Tennessee, Knoxville. He obtained his doctorate from the University of Iowa. His research interests include location theory, urban and economic geography, and the application of location-allocation modeling. He has published in a variety of journals and is the coauthor of an economic geography text.

Paul Casillas is a graduate student in the Department of Mathematics at Oregon State University. He recently received the master's degree in geography at the University of Iowa.

Richard L. Church is professor of geography and chairman of the department at the University of California at Santa Barbara. He obtained the doctorate in environmental engineering and geography from the Johns Hopkins University. He has also taught at the University of Tennessee, Knoxville. His research interests include systems modeling applied to a variety of environmental, economic, urban, and health-related topics. He has published extensively in engineering, transportation, operations research, and geographical journals.

Mark S. Daskin is associate professor of civil engineering and transportation at Northwestern University. Previously, he taught at the University of Texas at Austin. His research interests lie in developing vehicle routing algorithms and in facility location modeling. His papers have appeared in *Transportation Research* and *Transportation Journal* as well as in other journals.

David J. Eaton is professor of public affairs and geography at the University of Texas at Austin. Previously he served on the staffs of the President's Council on Environmental Quality and the National Science Foundation. His research interests include location analysis, water resource development, energy technologies, and policy applications of mathematical programming.

Avijit Ghosh is associate professor of marketing at New York University's Graduate School of Business Administration. He received the Ph.D. in geography from the University of Iowa. His research focuses on the application of location-allocation modeling in designing retail outlet networks and on the interface between location-allocation modeling and location theory. He is editor of the *Journal of Retailing*.

Michael F. Goodchild is professor of geography at the University of Western Ontario. He specializes in the application of mathematics, statistics and computing to spatial problems, particularly to location-allocation, site selection, and spatial-interaction models. His interests also extend to automated cartography, geographic information systems and quantitative aspects of physical geography.

Victor Ginsburgh is professor of economics at the Université Libre de Bruxelles, Belgium. His papers have appeared in various economic journals including *Econometrica, European Economic Review,* and *International Economic Review.* He is coauthor with J. Waelbroeck of *Activity Analysis and General Equilibrium Modeling* (Amsterdam, North-Holland).

Sidheswar Jena is systems analyst at the IPITATA Sponge Iron Company, India. He previously worked at the Indian Institute of Management, Bangalore. His research interests are in developing management information systems for regional planning.

Michael Kuby, a university fellow at Boston University, is working towards the Ph.D. in geography. Recently he worked as a geographer/operations research analyst for the Institute of Water Resources of the Army Corps of Engineers. He is currently working on hierarchical location-allocation models and on models of coal transportation.

Keumsook Lee is working toward the Ph.D. in geography at Boston University. She holds the B.A. and M.A. in geography from Sungshin Women's University, Korea. She is presently completing her dissertation entitled "Intertemporal Multicommodity Models for Planning Transportation Infrastructure Developments."

Sara L. McLafferty is assistant professor of geography at Columbia University. She received the Ph.D. from the University of Iowa. Her research interests are in location of public services, with special reference to health facilities.

She is also interested in the application of location-allocation models in locational analysis and location theory.

Pitu Mirchandani is associate professor in the Electrical, Computer, and Systems Engineering Department and in the Operations Research and Statistics Program at Rensselaer Polytechnic Institute. He obtained the Ph.D. from the Massachusetts Institute of Technology. His research includes the application of operations research and systems engineering to physical distribution, operations management, transportation and manufacturing. He is a coauthor of *Location on Networks* (MIT Press, Cambridge Mass., 1979).

Valerin T. Noronha recently completed the doctorate in geography at the University of Western Ontario. The dissertation proposed a mathematical unification of the concepts of spatial interaction and functional regions and has application in our understanding of spatial behavior. He has recently been involved in developing location-allocation software for microcomputers.

Morton E. O'Kelly is assistant professor of geography at Ohio State University, Columbus. His research interests are in optimization for locational and transportation analysis. A central concern in his published papers is the analysis of the structure of discretionary travel.

Jeffrey Osleeb is associate professor of geography at Boston University. Previously, he taught at Ohio State University, Columbus, and was a researcher at the U. S. Department of Energy. His papers have appeared in geography, operations research, and planning journals. He is the coauthor of a book on the Soviet steel industry and coeditor with Samuel Ratick of *Locational Decisions: Methodology and Applications* (J. C. Baltzer, Basel, forthcoming).

Pierre Pestieau is professor of economics of the Université de Liege, Belgium. His papers have appeared in various journals, including *Econometrica, Journal of Economic Theory,* and *International Economic Review*. He is coeditor with M. Marchand and H. Tulkens of *The Performance of Public Enterprise* (North-Holland, Amsterdam, 1984). His main research interests are in public economics and income distribution.

Samuel Ratick is associate professor of geography and associate director of the Center for Energy and Environmental Studies at Boston University. Previously, he was an operations research analyst for the Environmental Protection Agency and the Legislative Assistant for Energy and Environment to Senator Moynihan. He has published in operations research, geography, and environmental journals and is coeditor with Jeffrey Osleeb of *Locational Decisions: Methodology and Applications* (J. C. Baltzer, Basel, forthcoming).

John M. Reilly is the director of Planning and Development at the Capital District Transport Authority in Albany, N. Y., and instructor in the Department of Geography at the State University of New York at Albany. He obtained the doctorate in Urban and Environmental Studies at Rensselaer Polytechnic Institute. His professional interests are in the application of systems theory, computers, and communication technology to urban services.

Gerard Rushton is professor of geography at the University of Iowa, Iowa City. His research area is location theory, behavioral geography, and the geographical organization of human service systems in developing countries. He is the author of a number of articles on these topics and the coeditor with Reginald Golledge of *Spatial Choice and Spatial Behavior* (Ohio State University Press, Columbus, 1976). He is the author of *Optimal Location of Facilities* (COMpress, Wentworth: N.H., 1979).

Vinod K. Tewari is professor and coordinator of the Centre for Human Settlements and Environmental Studies, Indian Institute of Management, Bangalore. He is interested in the management and policy issues related to human settlement systems. He has published widely on these topics and is the coauthor of two books.

Jacques-Francois Thisse is professor of regional science at the Université Catholique de Louvain, Louvain-la-Neuve, Belgium. His papers have appeared in economics, operations research, and regional science journals. He is coeditor with H. Zoller of *Locational Analysis of Public Facilities* (North-Holland, Amsterdam, 1983). His current research interests are in spatial economics, location-allocation models, and new industrial economics.

INDEX